Mechanobiology of the Endothelium

Mechanobiology of the Endothelium

Editor

Helim Aranda-Espinoza

Fischell Department of Bioengineering
University of Maryland
College Park, MD
USA

CRC Press
Taylor & Francis Group
Boca Raton London New York

CRC Press is an imprint of the
Taylor & Francis Group, an **informa** business

A SCIENCE PUBLISHERS BOOK

CRC Press
Taylor & Francis Group
6000 Broken Sound Parkway NW, Suite 300
Boca Raton, FL 33487-2742

First issued in paperback 2019

ISBN-13: 978-1-4822-0724-8 (hbk)
ISBN-13: 978-0-367-37778-6 (pbk)

Library of Congress Cataloging-in-Publication Data

Mechanobiology of the endothelium / editor, Helim Aranda-Espinoza.
   p. ; cm.
   Includes bibliographical references and index.
   ISBN 978-1-4822-0724-8 (hardcover : alk. paper)
   I. Aranda-Espinoza, Helim, editor.
   [DNLM: 1. Endothelium. 2. Biomechanical Phenomena. 3. Biophysical Phenomena.
QS 532.5.E7]

   QP88.45
   611'.0187--dc23   2014028248

Visit the Taylor & Francis Web site at
http://www.taylorandfrancis.com

and the CRC Press Web site at
http://www.crcpress.com

Preface

The endothelium is responsible for regulating the exchange of material from the blood to tissues and organs. It is exposed to shear stress from the blood stream as well as pressure from tissues and smooth muscle cells. It is also exposed to different mechano-chemical environments, such as blood and extracellular matrix. It is quite an incredible task for this cell monolayer to maintain homeostasis. Therefore, endothelial cells are remarkable mechanotranducers. They need to convert chemical signals into mechanical outputs on a daily basis, as well as the reverse: translate the mechanical forces they feel into biological and chemical signals to adapt or respond accordingly. The most evident example of such a response is the remodeling of the endothelial cell cytoskeleton in response to shear stresses from the blood flow. Because of these types of responses, the endothelium has been the subject of intense research from biological, biochemistry, biophysical, and mathematical perspectives. However, the combination of these fields, with the emergence of mechanobiology as a scientific discipline, has opened the possibility of studying the endothelium with a broader perspective.

Mechanobiology is a relatively new scientific discipline. It merges mechanics and biology to study systems where forces elicit a biological response or a biological function is subjected to specific forces. When the response or the forces in the biological system change, a pathological condition can ensue, as in atherosclerosis. Therefore, studying the mechanobiology of these biological systems offers the opportunity to obtain a fundamental understanding and the possibility to translate this understanding into disease diagnosis and/or treatments.

The endothelium is an excellent example of where the study of mechanobiology is already being put to use. For instance, aspects of mechanobiology have been used in the design of heart valves and studying the endothelium of the valves. Mechanobiology has also been applied to the study of angiogenesis and the forces that affect the mechanobiology of the endothelium. Many facets of mechanobiology of the endothelium are reviewed in this book, including the examples cited above. Other highlights of the book include: studies toward understanding the scientific basis of the endothelial cells as mechanosensors, using the endothelium as a means to deliver drugs to specific tissues, studies on the role of endothelium biomechanics in cancer metastasis and atherosclerosis, and research on understanding cell transmigration. Because of space constraints, some specific areas have been left out, but I hope that this book still manages to cover important topics for readers that are starting to delve into the study of endothelium mechanobiology and for experienced researchers that need an overview of the topics covered here.

The team of authors consists of experts in their respective fields and the work devoted to each chapter is greatly appreciated. The editor has learned a lot on each subject and it is expected that the reader will learn as much. Each chapter is independent of one another and therefore they can be read individually and not necessarily in sequence. Please enjoy!

August 2014   Helim Aranda-Espinoza

Contents

1

The Endothelium as a Mechanosensor

Rachel E. Neubrander[1] and Brian P. Helmke[1,2,3,]*

The endothelium is the single layer of cells that lines blood and lymph vessels and the heart. Because of its location at these interfaces between tissue and blood or lymph, the endothelium serves several important physiological functions that maintain tissue homeostasis. Most importantly, the endothelium provides a barrier that regulates transport of fluid, macromolecules, proteins, and cells into and out of tissues. Second, the endothelium is a key player in autoregulation of tissue perfusion in response to changes in blood pressure, potentiating blood flow rate through vessels by controlling their diameters. In normal circumstances, the endothelium also prevents blood clotting.

Both transport barrier and autoregulation functions involve complex interactions between biochemical signaling in endothelial cells and the local biophysical environment in the blood vessel wall. For example, the blood-brain barrier (BBB) that separates blood from cerebrospinal fluid consists of endothelial cells with tight junctions that limit transport to small macromolecules. Tight junctions, which consist of a complex

[1]Department of Biomedical Engineering, University of Virginia, Charlottesville, VA 22908.
[2]Robert M. Berne Cardiovascular Research Center, University of Virginia, Charlottesville, VA 22908.
[3]NanoSTAR Institute, University of Virginia, Charlottesville, VA 22908.
*Corresponding author: helmke@virginia.edu

of transmembrane receptors and intracellular signaling molecules, are stabilized by connection to an actively contracting cytoskeleton that maintains consistent tension among adjacent endothelial cells. Without cell-generated tension, tight junctions are destabilized, and the endothelial barrier permeability increases.

The link between physical forces and chemical signaling is perhaps even clearer in endothelium-dependent autoregulation. When an increase in blood pressure upstream of a tissue causes an increase in blood flow, the endothelium responds by releasing nitric oxide. In the small arteries, nitric oxide diffuses from the endothelial cells to the underlying vascular smooth muscle cells and causes a reduction in their steady contractile force, or tone. Smooth muscle relaxation corresponds to vasodilation, an increased vessel diameter that returns the flow rate to normal at the new higher level of blood pressure.

In both of these examples, endothelial physiology involves physical forces associated with blood flow or with maintaining a physical barrier. In autoregulation, the endothelium serves as a *mechanosensor*: it senses changes in blood flow and responds by chemically stimulating vasoactivity. In barrier function, endothelial cells generate physical forces that determine transport properties across the cell monolayer. How does the endothelium sense physical stimuli and transduce them into physiological functions? What happens when the endothelium senses incorrectly or when local variations in physical forces cause the endothelium to function differently? This chapter explores these questions at the cellular and subcellular levels and lays the groundwork for understanding physiological and pathological functions of endothelial mechanobiology that are described in later chapters.

Mechanical Forces Acting on Vascular Endothelium

The blood vessel wall experiences mechanical forces due to blood flow and pressure as well as forces in the surrounding tissues (Fig. 1). Blood pressure forces act perpendicular to the endothelial surface in the outward direction from the blood vessel centerline. These distending forces are balanced by reaction forces in the surrounding tissue and/or by active contraction in the case of muscle tissue. The vector sum of these forces due to *transmural pressure* is the net normal (perpendicular) force acting in the radial direction on the endothelium. The transmural pressure determines the steady state blood vessel diameter or whether the vessel even remains open to blood flow. At the cellular level, the net normal force may act to compress the height of endothelial cells.

The inflation of blood vessels by transmural pressure also causes normal forces in the circumferential and longitudinal directions that are associated with stretch of the elastic blood vessel wall. The most well-characterized

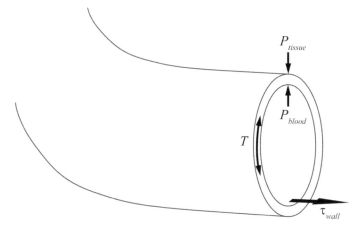

Figure 1. Mechanical stresses acting on the endothelium and blood vessel wall include blood pressure P_{blood}, pressure P_{tissue} exerted on the vessel wall by the surrounding tissue, hoop stress T, and wall shear stress τ_{wall}.

wall stretch is the circumferential component, which is associated with a *hoop stress T* that can be roughly estimated from the Law of Laplace: $T = \dfrac{\Delta P \cdot r}{2t}$, where ΔP is the transmural pressure, r is the radius of the blood vessel, and t is the wall thickness. Both hoop stress and longitudinal stress have been associated with long-term artery wall adaptation and growth (Nichol et al. 2005, Humphrey et al. 2009, Gruionu et al. 2012). As with transmural pressure, the wall strain varies temporally during the cardiac cycle. In straight segments of artery, the wall strain can be approximated as a uniaxial waveform in the circumferential direction. In artery regions near bifurcations or high curvature, the wall strain must be represented in two dimensions as a biaxial stretch profile. In these regions, the endothelium experiences more complex strain profiles that vary significantly from region to region in the artery.

The third category of force acting on the endothelium is shear force that results from flowing blood interacting with the endothelial surface. In most physiological cases, blood is a viscous fluid whose flow patterns in blood vessels can be represented as *laminar flow*, which means that the flow profile looks like layers of fluid sliding past each other (Fig. 2A). The *shear rate* $\dot{\gamma}$ is the rate of change of fluid speed parallel to the vessel wall with respect to distance between fluid layers. In large blood vessels (diameter > 150 µm), blood can be represented as a *Newtonian fluid* with constant viscosity μ. In this case, the shear stress τ (force per unit area) exerted by one layer of fluid on its neighbor is proportional to the shear rate between the layers, as given by Newton's law of viscosity: $\tau = \mu\dot{\gamma}$. Assuming steady, unidirectional laminar flow in a blood vessel with circular cross-section,

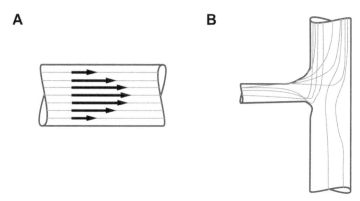

Figure 2. Fluid streamlines representing (A) unidirectional laminar flow through a straight, cylindrical tube, with arrows showing relative velocity of the layers, and (B) the disturbed flow pattern near a tube bifurcation.

the flowing blood exerts a wall shear stress τ_{wall} on the endothelium that is approximated by $\tau_{wall} = \dfrac{4\mu Q}{\pi r^3}$, where Q is the volumetric flow rate through the blood vessel and r is the vessel radius.

The walls of arteries are subjected to pulsatile forces associated with heart rhythm. The blood pressure waveform is the most well-known of these pulsatile force profiles. Perhaps more importantly for endothelial physiology, the wall shear stress imposed by blood on the endothelium varies both spatially and temporally. Spatial variations in wall shear stress are associated with curved path lines of blood flow in regions near bifurcations (Fig. 2B) or high curvature (Wootton and Ku 1999). *Disturbed flow* is a term that encompasses the many possible spatial variations of wall shear stress that result from multidirectional velocity profiles, vortices, etc. in these regions. Temporal variations in wall shear stress are associated with cardiac cycle. Unidirectional shear stress waveforms, defined as *pulsatile flow*, occur in relatively straight regions of arteries. In regions near bifurcations or higher wall curvature, the direction of shear stress may reverse during diastole as small traveling vortices form, defined to be an *oscillatory flow* profile. The combination of oscillatory and disturbed flow profiles results in a heterogeneous shear stress distribution focused on localized regions of the endothelium.

Focal Nature of Endothelial Mechanosensing

Spatial heterogeneity in mechanical forces acting on the endothelium corresponds with physiological endothelial function or with focused locations of endothelial dysfunction in disease. For example, vascular dysfunction and atherosclerotic lesions tend to originate in sections of large arteries near bifurcations or regions of high curvature, which are associated with complex or disturbed flow profiles (Caro et al. 1969, Glagov et al. 1988), and the local profile of hemodynamic forces in these regions promotes inflammatory signaling cascades associated with atherogenesis. In contrast, cellular mechanoadaptation to higher magnitude, pulsatile wall shear stress in other arterial segments guides atheroprotective gene expression profiles and signaling mechanisms (Gimbrone et al. 1997, Davies et al. 2002). Thus, focal mechanosensing at a subcellular length scale must be related to regional functional adaptation at the tissue level (Davies et al. 2003).

Focal mechanosensing occurs in response to changes in wall shear stress profile on a range of time scales. Within seconds after acute onset of shear stress, inward-rectifying potassium channels are activated, leading to rapid hyperpolarization of the plasma membrane (Olesen et al. 1988). Recovery and depolarization depends on activation of plasma membrane chloride channels (Barakat et al. 1999). Calcium ion influx through transient receptor potential vanilloid type 4 (TRPV4) (Köhler et al. 2006) and volume-regulated anion current (Romanenko et al. 2002) are also activated by shear stress in a manner that is potentiated by both membrane cholesterol content and interaction with cortical actin cytoskeleton (Byfield et al. 2004). Both tyrosine kinase receptors (Shay-Salit et al. 2002, Tzima et al. 2005) and G protein-coupled receptors (Chachisvilis et al. 2006) are also activated within seconds after onset of shear stress. Both the cell surface glycocalyx (Florian et al. 2003, Thi et al. 2004) and the biophysical properties of the membrane itself (Haidekker et al. 2000, Butler et al. 2001, Byfield et al. 2004) are capable of modulating rapid responses of ion channels, cell surface receptors, downstream signaling, and cytoskeleton remodeling.

Rapid activation of cell surface receptors and plasma membrane ion channels leads to regulation of a number of intracellular signaling pathways on a time scale of minutes to hours. Some key nodes in crosstalk among these signaling networks include mitogen-activated protein kinases (MAPKs) such as extracellular signal-regulated kinase (ERK) and c-Jun N-terminal kinase (JNK), cell growth and cell cycle kinases such as phosphoinositide3-kinase (PI3K) and Akt, and the Rho and Ras families of small G proteins. These networks specify the activity profile of a set of transcription factors that includes nuclear factor κB (NF-κB), activator protein–1 (AP-1), β-catenin, and peroxisome proliferator-activated receptor γ (PPAR-γ) (Gimbrone et al. 1997, Chien et al. 1998). Since these transcription factors encode mostly genes such as leukocyte adhesion receptors that are associated with endothelial inflammation, endothelial adaptation to the local hemodynamic force profile depends on whether activation of transcription factors is transient, such as during exercise, or is sustained for longer periods, as is the case in disturbed flow.

Endothelial cell genotype and phenotype adapt to fluid shear stress profile on a time scale of 24–72 h in well-controlled *in vitro* flow experiments. Sustained disturbed flow profiles are achieved temporally by oscillating the flow direction or spatially by including in the flow chamber a backward-facing step, which creates a standing vortex. Endothelial cells maintained in this environment exhibit sustained expression of genes and proteins associated with leukocyte adhesion (Nagel et al. 1994, Gerszten et al. 1998), oxidative stress (Wei et al. 1999, Dai et al. 2007), and vascular permeability (Phelps and DePaola 2000, Orr et al. 2007). In addition, endothelial cells downregulate genes and proteins associated with preventing thrombosis (Chiu and Chien 2011).

Strikingly, flow disturbances regulate these events on the length scale of a few cells. For example, near the edge of an atherosclerotic lesion in vivo, cell shapes are elongated and aligned with the main flow direction but polygonal and without preferred orientation just 3–4 cells from the adjacent disturbed flow region (Davies et al. 1976). In a flow chamber with a backward-facing step to create a standing vortex, downregulation of connexin-43 expression and disassembly of gap junctions limits cell-cell communication within a few cell diameters of the flow stagnation point. Gap junctions are unaffected downstream of the stagnation point (DePaola et al. 1999). These observations suggest that individual cells sense local forces and adapt their communication and signaling profiles in a manner that limits coordinated function in the endothelial monolayer. In support of this idea, modeling of the shear stress distribution as a function of surface topography demonstrates that significant heterogeneity exists both from cell to cell and within the surface of individual cells (Barbee et al. 1994, 1995). This subcellular distribution of force across the endothelial surface provides the boundary condition for elucidating molecular mechanisms of mechanotransmission from the extracellular milieu to intracellular sites of mechanochemical signal activation.

Mechanotransmission through the Cytoskeleton

The endothelial cell cytoskeleton is integral to the process of adaptation to mechanical forces. This integrated network of microfilaments, intermediate filaments, and microtubules provides shape and mechanical structure to the cell, and active remodeling of cytoskeletal networks is required for adaptation of cell shape in response to mechanical forces. The cytoskeleton also plays a role in the transmission of forces throughout the cell, facilitating mechanotransduction at intracellular sites that are not directly exposed to an externally applied force.

Dynamic cytoskeletal remodeling associated with shape adaptation occurs in response to both shear stress and stretch. Unidirectional shear stress induces increased turnover of actin monomers as well as structural reorganization of microfilaments, including actin ruffling, polarized edge extension, and alignment of stress fibers (Osborn et al. 2006, Choi and Helmke 2008). When cyclic stretch oriented perpendicular to the flow direction is added to mimic arterial wall dynamics, similar patterns of actin reorganization occur (Zhao et al. 1995). Shear stress also induces polarized stabilization of microtubules and reorientation of the microtubule organizing center (MTOC), which regulates planar cell polarity (McCue et al. 2006). These dynamic remodeling processes facilitate shape change and are required for cell alignment to unidirectional mechanical forces, since blocking cytoskeletal dynamics prevents alignment. For example, treating monolayers of endothelial cells with cytochalasin to depolymerize the actin cytoskeleton, nocodozole to destabilize microtubules, or taxol to stabilize microtubules prevents alignment in response to shear stress (Malek and Izumo 1996). Taxol also inhibits shear stress-induced lamellipodial protrusion and mechanotaxis in single endothelial cells (Hu et al. 2002). Although cells exposed to disturbed flow patterns maintain a polygonal rather than an aligned shape, cytoskeletal remodeling still occurs. Instead of aligned actin stress fibers, *in vivo* these cells have fewer stress fibers and more peripheral bands of actin (Kim et al. 1989).

Cytoskeletal networks are critical not only for controlling cell shape remodeling but also for transmitting mechanical forces to the mechanotransduction sites that initiate the signaling that is required for cells to respond and adapt to shear stress and stretch. Although mechanosensitive proteins, which are activated by mechanical forces to cause biochemical signaling, exist throughout the cell, two primary sites of mechanotransduction are the cell-matrix and cell-cell contacts. The proteins in focal adhesions and adherens junctions, especially adhesion proteins such as integrins and platelet–endothelial cell adhesion molecule 1 (PECAM-1), are responsible for activating many mechanosensitive pathways that are involved in both structural remodeling and inflammation. In order for mechanotransduction to occur at these adhesive sites, externally applied force must be transmitted to the mechanosensory proteins. In the case of substrate stretch, force can be directly transferred from the substrate to bound integrins within focal adhesions, but force must be transmitted indirectly to adherens junctions. When shear stress is applied, forces must be transmitted from the luminal surface of the cell to the focal adhesions along the basal surface and to adherens junctions. The actin cytoskeleton serves as the primary connector of all these structures and provides a pathway for transmission of forces throughout the cell. If the actin cytoskeleton is disrupted, force transmission no longer occurs and mechanosensing is inhibited. For example, local application of force at a subcellular length scale rapidly activates Src not only at that site but also at other remote sites within the cell (Na et al. 2008). If cytochalasin or latrunculin is used to disrupt the actin cytoskeleton, this long-range activation of Src no longer occurs. Under shear stress, disruption of the actin cytoskeleton with cytochalasin prevents the activation of extracellular signal-regulated kinases (ERK-1/2), which act as important regulators of inflammatory signaling (Takahashi and Berk 1996). Shear forces are transmitted not only throughout the cell but through the cell to the extracellular matrix by a mechanism that can also be inhibited by the addition of latrunculin (Mott and Helmke 2007).

When external forces are applied, they cause deformation throughout the endothelial cell cytoskeleton. Local forces can be transmitted throughout the cell to distant sites of mechanotransduction. The magnitude and distribution of transmitted forces depends on the organization and mechanical properties of the cytoskeleton. Because the spatial distribution of cytoskeletal filaments network is not uniform, the deformations transmitted through the cell depend on the subcellular organization and mechanical properties of the cytoskeleton. For example, deformation of the intermediate filament network in response to shear stress onset has been measured (Helmke et al. 2000). A change in the distribution of mechanical strain that is calculated from these deformations exhibits localized high intensity strains, even though the applied shear stress is spatially uniform (Helmke et al. 2003). This strain focusing effect may serve to amplify small forces and channel them to sites of mechanotransduction. Similar effects on the actin cytoskeleton have been observed in response to both fluid shear stress and substrate stretch (Mott and Helmke 2007, Huang et al. 2010). Shear forces can also be transmitted through the cell to the underlying matrix. For example, deformation of fibronectin matrix fibrils occurs under shear stress and is inhibited by the disruption of the actin cytoskeleton. Transmission of forces to the extracellular matrix may also play a role in mechanotransduction, since changes in fibronectin fibril strain can regulate binding of soluble fibronectin, which could in turn modulate cell signaling (Klotzsch et al. 2009, Kubow et al. 2009). Because the organization and structure of the cytoskeleton is so critical for force transmission, alterations to the pre-existing cytoskeletal organization and material properties will influence the distribution of applied forces within the cell and the efficiency of force transmission.

Cytoskeletal remodeling is mainly controlled through mechanosensitive signaling pathways, but cytoskeletal dynamics can be regulated directly by applied forces. Direct regulation of microtubule stability by changes in strain regulates activity of Rho and RacGTPases, which are important in regulating actin remodeling (Putnam et al. 2003). Increased force on microtubules increases their stability, suggesting that a gradient in applied force could influence cell polarity through the polarized stabilization of microtubules in the direction of the force. It has also been suggested that large changes in strain can also increase turnover of actin stress fibers, which has been modeled mathematically by a mechanism that incorporates competition between applied stretch and stress fiber relaxation (Hsu et al. 2010, Kaunas et al. 2010).

Mechanosensing at Adherens Junctions

The adherens junction is a critical site for shear stress–induced mechanotransduction in endothelial cells. Although it remains to be shown whether junction proteins experience external forces directly or indirectly due to transmission through the cytoskeleton or plasma membrane, molecular mechanisms that initiate mechanochemical signaling have recently been solved.

The primary mechanosensory complex at adherens junctions consists of platelet–endothelial cell adhesion molecule 1 (PECAM-1), vascular endothelial cell cadherin (VE-cadherin), and vascular endothelial growth factor receptor 2 (VEGFR-2) (Tzima et al. 2005). This mechanosensory complex is necessary and sufficient to confer the ability to activate inflammatory signaling pathways and to align cell shape in response to shear stress. In this complex, PECAM-1 acts as the primary mechanosensor, as demonstrated by both the application of shear stress and direct mechanical pulling on PECAM-1 with anti-PECAM–coated beads (Osawa et al. 2002). When detergent-extracted cytoskeletons are stretched, PECAM-1 is phosphorylated by the Src family kinase Fyn (Chiu et al. 2008). When force is applied, PECAM-1 is rapidly phosphorylated and activates Src kinases, which then phosphorylate VEGFR-2 (Tzima et al. 2005). This process requires the presence of VE-cadherin but does not require VE-cadherin ligation. The activated VEGFR-2 then phosphorylates and activates phosphatidylinositol-3-OH kinase (PI3K), which activates integrins to promote ligand binding. If this signaling pathway is blocked at any of these points, integrin activation and ligand binding do not occur. As a result, downstream effects such as NF-κB activation and alignment of cell shapes are prevented. The downstream signaling due to activation of this mechanosensory complex is implicated in the development of atherosclerotic lesions in the vasculature. PECAM-null/ApoE-null mice develop smaller atherosclerotic lesions than ApoE-null mice, which are often used as a mouse model for atherosclerosis (Harry et al. 2008). Endothelial cell cadherins may also act as primary mechanosensors in addition to simply functioning as adaptors for signals from PECAM-1. For example, force applied via E-cadherin–coated beads induces local cytoskeletal reinforcement through the recruitment of vinculin (LeDuc et al. 2010). Similar responses involving VE-cadherin occur, suggesting that this behavior might be relevant in endothelial cells (Huveneers et al. 2012). Although the role of VE-cadherin in the shear stress–dependent mechanosensory complex is ligation-independent, VE-cadherin ligation is necessary for stretch-induced proliferation of confluent endothelial cells, suggesting that VE-cadherin may have a more direct mechanosensory function (Liu et al. 2007).

Mechanotransduction through adhesion proteins at adherens junctions is critical for stimulating structural adaptation to shear stress. Mechanical signals transduced at the adherens junction activate signaling through integrins at focal adhesions, which are mechanosensors themselves.

Mechanosensing at Focal Adhesions

Mechanosensory proteins located in focal adhesions are responsible for the activation of many signaling pathways in response to externally applied forces. Evidence for the mechanosensitivity of focal adhesions structures includes remodeling to grow in the direction of an applied force (Galbraith et al. 2002). Shear stress promotes turnover and reorganization of focal adhesions to align their shapes in parallel with cell alignment in the direction of flow. In both subconfluent and confluent layers of endothelial cells, exposure to several hours of shear stress causes a decrease in the number of focal adhesions per cell, but the individual area per focal adhesion increases so that total adhesion area per cell remains approximately constant (Davies et al. 1994, Li et al. 2002). In subconfluent cell layers, shear stress also induces polarized formation of new focal adhesions near the downstream edge of the cell (Li et al. 2002).

As the link between the intracellular cytoskeleton and the extracellular matrix, integrins are ideally positioned to act as mechanosensors within focal adhesions. Integrin-dependent signaling has been implicated in responses to a variety of mechanical forces. Integrin-mediated mechanotransduction occurs through conformational changes in both the intracellular and extracellular domains of the protein. Integrins exist on the cell surface in either an inactive, low-affinity conformation or an active, high-affinity conformation. This conformational switch may undergo "inside-out activation" when interactions in the cytoplasmic portion of the integrin cause changes in the extracellular domains. For example, talin binding to the cytoplasmic tail of the integrin beta subunit causes the conformational switch from the inactive to active state (Tadokoro et al. 2003). Ligand binding to these activated integrins triggers additional downstream signaling, which is called "outside-in activation". In response to shear stress, both PI3K-dependent "inside-out" activation of integrins and subsequent ligand binding and "outside-in" signaling are required for structural remodeling and regulation of inflammatory pathways. If integrin activation is blocked, or if integrins are unable to bind their ligands, "outside-in" signaling is prevented, and cell shape alignment and ERK/

JNK activation do not occur (Tzima et al. 2001, Tzima et al. 2002). Similarly, in endothelial cells exposed to substrate stretch, blocking integrin binding prevents cell shape alignment and MAPK activation (Hirayama and Sumpio 2007). Although new integrin ligation is an essential in shear stress and stretch induced signaling, ligand binding is not the only way for mechanical forces to stimulate "outside-in" integrin signaling. Application of force to ligand-bound integrins can also activate mechanosignaling. For example, when force is applied to bound $\alpha 5 \beta 1$ integrins, FAK phosphorylation is increased (Friedland et al. 2009).

Integrin activation through affinity modulation is also intimately linked to integrin clustering, or avidity modulation, which can also contribute to ligand binding. Integrin activation and subsequent ligand binding promotes cluster formation (Cluzel et al. 2005). These clusters of bound integrins nucleate the formation of nascent focal contacts by recruiting both signaling and scaffolding proteins such as FAK, Src, paxillin, and vinculin. Focal contacts mature into focal adhesions as more proteins are recruited and force is applied through links to the actin cytoskeleton. Under shear stress and stretch, mechanotransduction occurs both through established focal adhesions and the formation of new focal complexes based on newly activated integrins. In addition, direct application of force to integrins promotes clustering. In magnetic bead pulling experiments, applying force caused the recruitment of $\alpha 5 \beta 1$ integrins to a fibronectin-coated bead. Interestingly, more integrins were recruited to beads coated with multivalent fibronectin fragments than to those coated with monomeric fibronectin. This suggests that ligand presentation also plays a role in controlling integrin clustering and, therefore, formation of focal adhesions (Roca-Cusachs et al. 2009). Integrin clustering is essential to shear stress–induced mechanotransduction, since integrin clusters form the basis of focal adhesions and recruit other proteins to stabilize the adhesions and activate downstream signaling.

Talin is a focal adhesion protein that facilitates the conformational activation of integrins, stabilizes activated, ligand-bound integrin clusters that nucleate focal contacts, and links integrin clusters to the actin cytoskeleton, both directly and through recruitment of vinculin and paxillin (Jiang et al. 2003, Cluzel et al. 2005, Zhang et al. 2008). Talin also plays a mechanosensory role within established focal adhesions. When force is applied via bead pulling, talin mediates force-dependent cytoskeletal reinforcement of focal adhesions (Giannone et al. 2003). Both myosin-dependent cell contractility and externally applied forces can induce a conformational change in talin that exposes a vinculin binding site (del Rio et al. 2009). This mechanism contributes to force-induced reinforcement of focal adhesions, since vinculin recruitment reinforces the link between integrins and the actin cytoskeleton that is initiated by talin (Galbraith et al.

2002). Potentiation of conformational activation by talin may be integrin-specific. Under shear stress, activation of α2β1integrins by collagen binding suppresses αVβ3 integrin activation through a talin-mediated mechanism (Orr et al. 2006). Since talin directly interacts with integrins and can be conformationally activated by force, it plays an important role in force-induced focal adhesion remodeling.

Another component of focal adhesions that is critical for reinforcement of focal adhesions and intracellular structural remodeling is vinculin. Vinculin interacts with talin to link focal adhesions to the actin cytoskeleton, thereby facilitating transmission of forces through the cytoskeleton as well as structural reinforcement of adhesions. Vinculin has a head domain and tail domain, and in the cytosol exists in an autoinhibited state in which the head and tail domain interact with each other (Johnson and Craig 1995). Upon activation by the phospholipid phosphatidylinositol-4,5-bisphosphate (PIP_2), this inhibitory interaction is disrupted, and vinculin is recruited to focal adhesions through interaction of the head domain with talin (Gilmore and Burridge 1996, Bass et al. 1999). Vinculin activation has also been proposed to occur through simultaneous binding of talin to the vinculin head domain and actin to the vinculin tail domain (Chen et al. 2006). Force-induced exposure of cryptic vinculin binding sites in talin also regulates vinculin recruitment in response to applied forces (Riveline et al. 2001, del Rio et al. 2009). Since vinculin is so critical in the linkage between focal adhesions and the cytoskeleton, it is an important regulator of force transmission and cell contractility. When force is applied, vinculin is responsible for transferring mechanical forces that promote increased contractility to the cytoskeleton (Mierke et al. 2008). Reducing vinculin expression inhibits cell spreading and reduces cell stiffness (Goldmann et al. 1998). Whether due to cell contractility or externally applied forces, vinculin is necessary for force transmission between focal adhesions and the cytoskeleton.

Src family kinases (SFKs) are activated immediately downstream of integrin ligand binding and clustering. Important members of this family of kinases include Src, Yes, and Fyn, and these kinases phosphorylate a variety of targets. SFKs are activated by interaction with the clustered cytoplasmic domains of the beta subunit of integrins. Different types of SFKs interact with different beta subunits. For example, in a Chinese hamster ovary (CHO) model system, Src and Fyn selectively interact with β3 subunits, and Yes interacts with β1, β2, and β3 subunits (Arias-Salgado et al. 2003). However, this specificity may be cell-type dependent, since shear stress-induced Src activation requires β1 (rather than β3) integrin ligation (Yang et al. 2011). Src activity is also closely linked with focal adhesion kinase (FAK) activity.Src and FAK form a complex in which Src promotes FAK activity and, in turn, FAK promotes Src activity (Calalb et al. 1995, Schaller et al. 1999). Src is activated within 1 minute of the onset of unidirectional laminar shear stress. However, both increases and decreases in Src activity after 10 minutes of shear stress have been reported, suggesting that Src-mediated mechanosignaling is modulated by factors other than shear stress (Jalali et al. 1998, Radel and Rizzo 2005). For example, Src is also activated by locally pulling on integrins, as shown by magnetic twisting cytometry using fibronectin-coated beads (Wang et al. 2005). Src activation is required for activating pathways responsible for alignment to shear stress, such as those involving p130Cas and p190RhoGAP, and for activation of JNK and ERK pathways that control inflammatory signaling (Jalali et al. 1998, Okuda et al. 1999, Yang et al. 2011).

FAK is central to many adhesion-dependent signaling pathways and is a major regulator of both motility and proliferation. FAK is recruited to sites of integrin clustering by talin or paxillin. FAK is autophosphorylated at these sites, creating a site where Src can bind and phosphorylate other residues to further activate FAK. When interacting with α5β1 integrins, FAK is phosphorylated only when tension is applied or clustered integrins are tethered to a surface. It is unknown whether this mechanism is α5β1-specific or is relevant to other integrins. In addition to this integrin-dependent activation mechanism, FAK can be activated by growth factor receptors through an integrin-independent mechanism involving ezrin (Poullet et al. 2001). Both shear stress and substrate stretch cause increased FAK phosphorylation and kinase activity (Ishida et al. 1996, Yano et al. 1996, Wu et al. 2007), although the time courses of activation vary. In migrating cells, FAK modulates stability and turnover of focal adhesions, thereby regulating both motility and proliferation. Similarly, under shear stress and substrate stretch, FAK activity is required for alignment with the direction of force (Ishida et al. 1996, Yano et al. 1996, Naruse et al. 1998, Wu et al. 2007). FAK activation regulates both Rho and RacGTPases. In confluent endothelial cell monolayers, phosphorylated FAK recruits p190RhoGAP by interaction with p120RasGAP to transiently inactivate Rho at the onset of shear stress. In single endothelial cells under shear stress, FAK is recruited to new adhesion sites on the downstream edge of the cell, where it facilitates downstream mechanotaxis and the activation of Rho to prevent apoptosis (Li et al. 2002, Wu et al. 2007). FAK also contributes to shear stress-induced inflammatory signaling by interacting with Grb2 and Sos to activate MAPK signaling networks that include JNK and ERK (Li et al. 1997).

The focal adhesion protein p130Cas, which is phosphorylated by SFKs, is critical in establishing planar cell polarity in response to mechanical stimuli. Alignment of cell shape relative to the direction of applied force is a defining result of endothelial mechanotransduction, and p130Cas is suggested to be a mechanosensitive component of focal adhesions that is required for this alignment process. Under shear stress, p130Cas participates in the activation of Rac, which is required for lamellipodial protrusion, mechanotaxis and alignment. In single endothelial cells and wounded monolayers exposed to shear stress, polarized p130Cas phosphorylation activates Rac at the downstream edge of the cell. Localized Rac activation establishes a new direction of planar cell polarity that is defined by the direction of flow and facilitates downstream mechanotaxis. This polarity mechanism is controlled by paxillin dephosphorylation on the upstream side of the cell, which limits paxillin and p130Cas phosphorylation and subsequent Rac activation to the downstream edge (Zaidel-Bar et al. 2005). Substrate stretch causes molecular stretching of p130Cas proteins anchored in focal adhesions, which increases the ability of SFKs to phosphorylate them (Sawada et al. 2006). This SFK-dependent phosphorylation of p130Cas is required for cells to reorient perpendicular to the stretch direction (Niediek et al. 2012). In addition, p130Cas can also bind FAK to facilitate SFK-dependent FAK activation (Kostic and Sheetz 2006). This mechanism may enhance FAK-dependent signaling under shear stress and stretch.

Other proteins associated with focal adhesions play a modulatory role in mechanosensing by interacting with integrins to regulate their signaling. For example, caveolin, which mainly serves as the coat protein of cell surface caveolae, interacts with β1 integrins to regulate Rho activity and intracellular structural remodeling in response to shear stress. At the onset of shear stress, β1 integrin activation causes the SFK-dependent phosphorylation of caveolin. Caveolin phosphorylation increases its association with β1 integrin and promotes recruitment of the kinase Csk. Recruitment of Csk causes the inactivation of SFKs by phosphorylating a negative regulatory site, and Csk binding is required for shear stress-induced myosin phosphorylation and formation of aligned stress fibers (Radel and Rizzo 2005). The transient receptor potential vanilloid-4 (TRPV4) calcium channel has also been linked to β1 integrin–mechanotransduction. Force applied to β1 integrins through magnetic beads induces the influx of calcium through TRPV4 channels nearly instantaneously. Stimulation of calcium influx is associated with cytoskeletal deformation and involved the transmembrane amino acid transporter heavy chain CD98 (Matthews et al. 2010). Calcium influx then stimulates activation of PI3K, resulting in integrin activation and binding. This mechanism is important in mediating stretch-induced remodeling, since knocking down TRPV4 expression blocks stretch-induced integrin activation and cell alignment (Thodeti et al. 2009). TRPV4 channel activation in response to force applied to integrins could widely impact signaling within the cell since calcium is a second messenger in a variety of signaling pathways.

Focal adhesions integrate a variety of mechanical and biochemical signals to facilitate adaptation of endothelial cell structure, as well as inflammation and survival, in response to shear stress and substrate stretch.

Integrins are the primary mediators of mechanotransduction, but several other mechanosensitive proteins, including talin, p130Cas and SFKs, play a role in sensing mechanical signals. Other components of focal adhesions, including vinculin, FAK, and paxillin, may not be directly mechanosensitive but are absolutely required for the biochemical signaling cascades and structural remodeling initiated by mechanotransducers. Finally, integrin-dependent mechanotransduction can be modulated by interactions with other focal adhesion-associated proteins, such as caveolin, syndecans, and TRPV4 channels.

Modulation of Endothelial Mechanosensing by the Extracellular Matrix

The extracellular matrix (ECM) is an important modulator of endothelial mechanotransduction. Since many shear stress and stretch-induced responses require integrin-dependent signaling, the ECM regulates these responses by interacting directly with integrins. Both *in vivo* and *in vitro*, ECM composition has been shown to regulate both structural remodeling and inflammatory signaling in response to shear stress.

In atherosclerosis, the ECM in lesions is different from that found in healthy regions of the vasculature. In even very early lesions, fibronectin is deposited in the ECM. In contrast, the ECM in healthy regions is composed of mainly collagen (Feaver et al. 2010). Fibronectin and collagen bind different sets of integrins and therefore stimulate different signaling pathways. When unidirectional shear stress is applied *in vitro*, endothelial cells interacting with fibronectin activate JNK signaling faster and align more quickly than endothelial cells interacting with collagen, suggesting that ECM composition regulates both structural remodeling and inflammatory signaling (Hahn et al. 2011). Furthermore, cross-talk exists between integrins that bind to different types of ECM. When collagen-binding integrins are activated, they suppress the activation of fibronectin binding integrins, and vice-versa. This integrin cross-talk results in a non-linear response to ECM density when endothelial cells interact with mixtures of collagen and fibronectin (Orr et al. 2006). These studies reveal the importance of ECM composition in modulating endothelial responses to shear stress. Even within ECM of a single composition, there can be varying signals from different integrins. For example, $\alpha 5\beta 1$ and $\alpha V\beta 3$ integrins both bind fibronectin; however, these two types of integrins respond differently to applied forces. In magnetic bead pulling assays, $\alpha 5\beta 1$ is responsible for adhesion strength but $\alpha V\beta 3$ activates adhesion reinforcement (Roca-Cusachs et al. 2009). In addition, $\alpha 4\beta 1$ integrin also binds to fibronectin, but it plays a specialized role in directing planar cell polarity. These studies suggest that not only the ECM composition but also any factors regulating the profile of integrins ligated will modulate responses to mechanical forces.

Independently of shear stress and substrate stretch, the physical properties of the ECM can also directly stimulate mechanotransduction pathways through integrins. Both the stiffness of the ECM and the density of integrin ligands in ECM regulate structural remodeling. Studies using ECM protein–coated polyacrylamide gel substrates of varying stiffness show that increasing substrate stiffness or ligand density promotes assembly and stiffening of cytoskeletal networks, cell spreading, contractility, and proliferation (Wang and Ingber 1994, Reinhart-King et al. 2003, Yeung et al. 2005, Pompe et al. 2009). As demonstrated by nanopatterning to control the distribution of fibronectin, the presentation and local density rather than the average density of integrin ligands controls proliferation and spreading (Slater and Frey 2008). A gradient in ECM density also promotes endothelial motility towards regions of higher ECM density, a behavior termed "haptotaxis". In the presence of shear stress, the mechanosignaling that induces downstream mechanotaxis can overcome the ECM signals that induce haptotaxis and cause endothelial cells to migrate against a gradient in ECM density (Hsu et al. 2005). Many of the same mechanosensory pathways involved in adapting to shear stress or stretch are stimulated by changes in substrate stiffness and ligand density, suggesting that microenvironmental cues may modulate integrin-dependent signaling required for adaptation to shear stress and stretch.

Summary

How does the endothelium sense and transduce mechanical stimuli into physiological (and pathological) functions? At the subcellular length scale, mechanosensors include junction-associated mechanosensory complexes and integrins located at focal adhesions. These mechanosensors are uniquely positioned to interact directly with a large number of signaling proteins and structures that modulate stress and strain transmission between the inside and outside of the cell. Intracellular mechanotransmission through the cytoskeleton decentralizes mechanosignaling, promoting integration of several signaling networks that lead to control of endothelial phenotype. These signaling networks regulate functions such as gene expression and cholesterol transport. Outside the cell, integrins sense and interact with the ECM to modulate endothelial signaling to immune cells and assembly of new blood vessels. Thus, the endothelium as a mechanosensor plays a critical role in vascular homeostasis of tissues from heart valves to tumors.

References

Arias-Salgado, E.G., S. Lizano, S. Sarkar, J.S. Brugge, M.H. Ginsberg and S.J. Shattil. 2003. Src kinase activation by direct interaction with the integrin β cytoplasmic domain. *Proc. Natl. Acad. Sci. USA* 100: 13298–13302.

Barakat, A.I., E.V. Leaver, P.A. Pappone and P.F. Davies. 1999. A flow-activated chloride-selective membrane current in vascular endothelial cells. *Circ. Res.* 85: 820–828.

Barbee, K.A., P.F. Davies and R. Lal. 1994. Shear stress-induced reorganization of the surface topography of living endothelial cells imaged by atomic force microscopy. *Circ. Res.* 74: 163–171.

Barbee, K.A., P.F. Davies and R. Lal. 1995. Subcellular distribution of shear stress at the surface of flow aligned and non-aligned endothelial monolayers. *Am. J. Physiol.* 268: H1765–H1772.

Bass, R.B., M.D. Coleman and J.J. Falke. 1999. Signaling domain of the aspartate receptor is a helical hairpin with a localized kinase docking surface: cysteine and disulfide scanning studies. *Biochemistry* 38: 9317–9327.

Butler, P.J., G. Norwich, S. Weinbaum and S. Chien. 2001. Shear stress induces a time- and position-dependent increase in endothelial cell membrane fluidity. *Am. J. Physiol.* 280: C962–969.

Byfield, F.J., H. Aranda-Espinoza, V.G. Romanenko, G.H. Rothblat and I. Levitan. 2004. Cholesterol depletion increases membrane stiffness of aortic endothelial cells. *Biophys. J.* 87: 3336–3343.

Calalb, M.B., T.R. Polte and S.K. Hanks. 1995. Tyrosine phosphorylation of focal adhesion kinase at sites in the catalytic domain regulates kinase activity: a role for Src family kinases. *Mol. Cell. Biol.* 15: 954–963.

Caro, C.G., J.M. Fitz-Gerald and R.C. Schroter. 1969. Arterial wall shear and distribution of early atheroma in man. *Nature* 223: 1159–1161.

Chachisvilis, M., Y.-L. Zhang and J.A. Frangos. 2006. G protein-coupled receptors sense fluid shear stress in endothelial cells. *Proc. Natl. Acad. Sci. USA* 103: 15463–15468.

Chen, H., D.M. Choudhury and S.W. Craig. 2006. Coincidence of actin filaments and talin is required to activate vinculin. *J. Biol. Chem.* 281: 40389–40398.

Chien, S., S. Li and Y.J. Shyy. 1998. Effects of mechanical forces on signal transduction and gene expression in endothelial cells. *Hypertension* 31: 162–169.

Chiu, J.-J. and S. Chien. 2011. Effects of disturbed flow on vascular endothelium: pathophysiological basis and clinical perspectives. *Physiol. Rev.* 91: 327–387.

Chiu, Y.-J., E. McBeath and K. Fujiwara. 2008. Mechanotransduction in an extracted cell model: Fyn drives stretch- and flow-elicited PECAM-1 phosphorylation. *J. Cell Biol.* 182: 753–763.

Choi, C.K. and B.P. Helmke. 2008. Short-term shear stress induces rapid actin dynamics in living endothelial cells. *Mol. Cell. Biomech.* 5: 247–258.

Cluzel, C., F. Saltel, J. Lussi, F. Paulhe, B.A. Imhof and B. Wehrle-Haller. 2005. The mechanisms and dynamics of αvβ3 integrin clustering in living cells. *J. Cell Biol.* 171: 383–392.

Dai, G., S. Vaughn, Y. Zhang, E.T. Wang, G. Garcia-Cardena and M.A. Gimbrone, Jr. 2007. Biomechanical forces in atherosclerosis-resistant vascular regions regulate endothelial redox balance via phosphoinositol 3-kinase/Akt-dependent activation of Nrf2. *Circ. Res.* 101: 723–733.

Davies, P.F., D.C. Polacek, C. Shi and B.P. Helmke. 2002. The convergence of haemodynamics, genomics, and endothelial structure in studies of the focal origin of atherosclerosis. *Biorheology* 39: 299–306.

Davies, P.F., M.A. Reidy, T.B. Goode and D.E. Bowyer. 1976. Scanning electron microscopy in the evaluation of endothelial integrity of the fatty lesion in atherosclerosis. *Atherosclerosis* 25: 125–130.

Davies, P.F., A. Robotewskyj and M.L. Griem. 1994. Quantitative studies of endothelial cell adhesion: directional remodeling of focal adhesion sites in response to flow forces. *J. Clin. Invest.* 93: 2031–2038.

Davies, P.F., J. Zilberberg and B.P. Helmke. 2003. Spatial microstimuli in endothelial mechanosignaling. *Circ. Res.* 92: 359–370.

del Rio, A., R. Perez-Jimenez, R. Liu, P. Roca-Cusachs, J.M. Fernandez and M.P. Sheetz. 2009. Stretching single talin rod molecules activates vinculin binding. *Science* 323: 638–641.

DePaola, N., P.F. Davies, W.F. Pritchard, Jr., L. Florez, N. Harbeck and D.C. Polacek. 1999. Spatial and temporal regulation of gap junction connexin43 in vascular endothelial cells exposed to controlled disturbed flows *in vitro*. *Proc. Natl. Acad. Sci. USA* 96: 3154–3159.

Feaver, R.E., B.D. Gelfand, C. Wang, M.A. Schwartz and B.R. Blackman. 2010. Atheroprone hemodynamics regulate fibronectin deposition to create positive feedback that sustains endothelial inflammation. *Circ. Res.* 106: 1703–1711.

Florian, J.A., J.R. Kosky, K. Ainslie, Z. Pang, R.O. Dull and J.M. Tarbell. 2003. Heparan sulfate proteoglycan is a mechanosensor on endothelial cells. *Circ. Res.* 93: 14.

Friedland, J.C., M.H. Lee and D. Boettiger. 2009. Mechanically activated integrin switch controls α5β1 function. *Science* 323: 642–644.

Galbraith, C.G., K.M. Yamada and M.P. Sheetz. 2002. The relationship between force and focal complex development. *J. Cell Biol.* 159: 695–705.

Gerszten, R.E., Y.C. Lim, H.T. Ding, K. Snapp, G. Kansas, D.A. Dichek, C. Cabanas, F. Sanchez-Madrid, M.A. Gimbrone, Jr., A. Rosenzweig and F.W. Luscinskas. 1998. Adhesion of monocytes to vascular cell adhesion molecule-1-transduced human endothelial cells: implications for atherogenesis. *Circ. Res.* 82: 871–878.

Giannone, G., G. Jiang, D.H. Sutton, D.R. Critchley and M.P. Sheetz. 2003. Talin1 is critical for force-dependent reinforcement of initial integrin-cytoskeleton bonds but not tyrosine kinase activation. *J. Cell Biol.* 163: 409–419.

Gilmore, A.P. and K. Burridge. 1996. Regulation of vinculin binding to talin and actin by phosphatidyl-inositol-4-5-bisphosphate. *Nature* 381: 531–535.

Gimbrone, M.A., N. Resnick, T. Nagel, L.M. Khachigian, T. Collins and J.N. Topper. 1997. Hemodynamics, endothelial gene expression, and atherogenesis. *Ann. NY Acad. Sci.* 811: 1–10.

Glagov, S., C.K. Zarins, D.P. Giddens and D.N. Ku. 1988. Hemodynamics and atherosclerosis. Insights and perspectives gained from studies of human arteries. *Arch. Pathol. Lab. Med.* 112: 1018–1031.

Goldmann, W.H., R. Galneder, M. Ludwig, W. Xu, E.D. Adamson, N. Wang and R.M. Ezzell. 1998. Differences in elasticity of vinculin-deficient F9 cells measured by magnetometry and atomic force microscopy. *Exp. Cell Res.* 239: 235–242.

Gruionu, G., J.B. Hoying, A.R. Pries and T.W. Secomb. 2012. Structural remodeling of the mouse gracilis artery: coordinated changes in diameter and medial area maintain circumferential stress. *Microcirculation: no-no.*

Hahn, C., C. Wang, A.W. Orr, B.G. Coon and M.A. Schwartz. 2011. JNK2 promotes endothelial cell alignment under flow. *PLoS ONE* 6: e24338.

Haidekker, M.A., N. L'Heureux and J.A. Frangos. 2000. Fluid shear stress increases membrane fluidity in endothelial cells: a study with DCVJ fluorescence. *Am. J. Physiol.* 278: H1401–H1406.

Harry, B.L., J.M. Sanders, R.E. Feaver, M. Lansey, T.L. Deem, A. Zarbock, A.C. Bruce, A.W. Pryor, B.D. Gelfand, B.R. Blackman, M.A. Schwartz and K. Ley. 2008. Endothelial cell PECAM-1 promotes atherosclerotic lesions in areas of disturbed flow in ApoE-deficient mice. *Arterioscler. Thromb. Vasc. Biol.* 28: 2003–2008.

Helmke, B.P., R.D. Goldman and P.F. Davies. 2000. Rapid displacement of vimentin intermediate filaments in living endothelial cells exposed to flow. *Circ. Res.* 86: 745–752.

Helmke, B.P., A.B. Rosen and P.F. Davies. 2003. Mapping mechanical strain of an endogenous cytoskeletal network in living endothelial cells. *Biophys. J.* 84: 2691–2699.

Hirayama, Y. and B.E. Sumpio. 2007. Role of ligand-specific integrins in endothelial cell alignment and elongation induced by cyclic strain. *Endothelium* 14: 275–283.

Hsu, H.-J., C.-F. Lee, A. Locke, S.Q. Vanderzyl and R. Kaunas. 2010. Stretch-induced stress fiber remodeling and the activations of JNK and ERK depend on mechanical strain rate, but not FAK. *PLoS ONE* 5: e12470.

Hsu, S., R. Thakar, D. Liepmann and S. Li. 2005. Effects of shear stress on endothelial cell haptotaxis on micropatterned surfaces. *Biochem. Biophys. Res. Comm.* 337: 401–409.

Hu, Y.L., S. Li, H. Miao, T.C. Tsou, M.A. del Pozo and S. Chien. 2002. Roles of microtubule dynamics and small GTPase Rac in endothelial cell migration and lamellipodium formation under flow. *J. Vasc. Res.* 39: 465–476.

Huang, L., P.S. Mathieu and B.P. Helmke. 2010. A stretching device for high-resolution live-cell imaging. *Ann. Biomed. Eng.* 38: 1728–1740.

Humphrey, J.D., J.F. Eberth, W.W. Dye and R.L. Gleason. 2009. Fundamental role of axial stress in compensatory adaptations by arteries. *J. Biomech.* 42: 1–8.

Huveneers, S., J. Oldenburg, E. Spanjaard, G. van der Krogt, I. Grigoriev, A. Akhmanova, H. Rehmann and J. de Rooij. 2012. Vinculin associates with endothelial VE-cadherin junctions to control force-dependent remodeling. *J. Cell Biol.* 196: 641–652.

Ishida, T., T.E. Peterson, N.L. Kovach and B.C. Berk. 1996. MAP kinase activation by flow in endothelial cells. Role of β1 integrins and tyrosine kinases. *Circ. Res.* 79: 310–316.

Jalali, S., Y.-S. Li, M. Sotoudeh, S. Yuan, S. Li, S. Chien and J.Y.-J. Shyy. 1998. Shear stress activates p60src-Ras-MAPK signaling pathways in vascular endothelial cells. *Arterioscler. Thromb. Vasc. Biol.* 18: 227–234.

Jiang, G., G. Giannone, D.R. Critchley, E. Fukumoto and M.P. Sheetz. 2003. Two-piconewton slip bond between fibronectin and the cytoskeleton depends on talin. *Nature* 424: 334–337.

Johnson, R.P. and S.W. Craig. 1995. F-actin binding site masked by the intramolecular association of vinculin head and tail domains. *Nature* 373: 261–264.

Kaunas, R., Z. Huang and J. Hahn. 2010. A kinematic model coupling stress fiber dynamics with JNK activation in response to matrix stretching. *J. Theor. Biol.* 264: 593–603.

Kim, D.W., B.L. Langille, M.K. Wong and A.I. Gotlieb. 1989. Patterns of endothelial microfilament distribution in the rabbit aorta *in situ*. *Circ. Res.* 64: 21–31.

Klotzsch, E., M.L. Smith, K.E. Kubow, S. Muntwyler, W.C. Little, F. Beyeler, D. Gourdon, B.J. Nelson and V. Vogel. 2009. Fibronectin forms the most extensible biological fibers displaying switchable force-exposed cryptic binding sites. *Proc. Natl. Acad. Sci. USA* 106: 18267–18272.

Köhler, R., W.-T. Heyken, P. Heinau, R. Schubert, H. Si, M. Kacik, C. Busch, I. Grgic, T. Maier and J. Hoyer. 2006. Evidence for a functional role of endothelial transient receptor potential V4 in shear stress-induced vasodilatation. *Arterioscler. Thromb. Vasc. Biol.* 26: 1495–1502.

Kostic, A. and M.P. Sheetz. 2006. Fibronectin rigidity response through Fyn and p130Cas recruitment to the leading edge. *Mol. Biol. Cell* 17: 2684–2695.

Kubow, K.E., E. Klotzsch, M.L. Smith, D. Gourdon, W.C. Little and V. Vogel. 2009. Crosslinking of cell-derived 3D scaffolds up-regulates the stretching and unfolding of new extracellular matrix assembled by reseeded cells. *Integrative Biology* 1: 635–648.

LeDuc, Q., Q. Shi, I. Blonk, A. Sonnenberg, N. Wang, D. Leckband and J. de Rooij. 2010. Vinculin potentiates E-cadherin mechanosensing and is recruited to actin-anchored sites within adherens junctions in a myosin II–dependent manner. *J. Cell Biol.* 189: 1107–1115.

Li, S., P.J. Butler, Y. Wang, Y. Hu, D.C. Han, S. Usami, J.-L. Guan and S. Chien. 2002. The role of the dynamics of focal adhesion kinase in the mechanotaxis of endothelial cells. *Proc. Natl. Acad. Sci. USA* 99: 3546–3551.

Li, S., M. Kim, Y.L. Hu, S. Jalali, D.D. Schlaepfer, T. Hunter, S. Chien and J.Y.-J. Shyy. 1997. Fluid shear stress activation of focal adhesion kinase. Linking to mitogen-activated protein kinases. *J. Biol. Chem.* 272: 30455–30462.

Liu, W.F., C.M. Nelson, J.L. Tan and C.S. Chen. 2007. Cadherins, RhoA, and Rac1 are differentially required for stretch-mediated proliferation in endothelial versus smooth muscle cells. *Circ. Res.* 101: e44–e52.

Malek, A.M. and S. Izumo. 1996. Mechanism of endothelial cell shape change and cytoskeletal remodeling in response to fluid shear stress. *J. Cell Sci.* 109: 713–726.

Matthews, B.D., C.K. Thodeti, J.D. Tytell, A. Mammoto, D.R. Overby and D.E. Ingber. 2010. Ultra-rapid activation of TRPV4 ion channels by mechanical forces applied to cell surface β1 integrins. *Integrative Biology* 2: 435–442.

McCue, S., D. Dajnowiec, F. Xu, M. Zhang, M.R. Jackson and B.L. Langille. 2006. Shear stress regulates forward and reverse planar cell polarity of vascular endothelium *in vivo* and *in vitro*. *Circ. Res.* 98: 939–946.

Mierke, C.T., P. Kollmannsberger, D.P. Zitterbart, J. Smith, B. Fabry and W.H. Goldmann. 2008. Mechano-coupling and regulation of contractility by the vinculin tail domain. *Biophys. J.* 94: 661–670.

Mott, R.E. and B.P. Helmke. 2007. Mapping the dynamics of shear stress-induced structural changes in endothelial cells. *Am. J. Physiol.* 293: C1616–C1626.

Na, S., O. Collin, F. Chowdhury, B. Tay, M. Ouyang, Y. Wang and N. Wang. 2008. Rapid signal transduction in living cells is a unique feature of mechanotransduction. *Proc. Natl. Acad. Sci. USA* 105: 6626–6631.

Nagel, T., N. Resnick, W.J. Atkinson, C.F. Dewey, Jr. and M.A. Gimbrone, Jr. 1994. Shear stress selectively upregulates intercellular adhesion molecule-1 expression in cultured human vascular endothelial cells. *J. Clin. Invest.* 94: 885–891.

Naruse, K., T. Yamada, X.R. Sai, M. Hamaguchi and M. Sokabe. 1998. pp125FAK is required for stretch dependent morphological response of endothelial cells. *Oncogene* 17: 455–463.

Nichol, J.W., M. Petko, R.J. Myung, J.W. Gaynor and K.J. Gooch. 2005. Hemodynamic conditions alter axial and circumferential remodeling of arteries engineered *ex vivo*. *Ann. Biomed. Eng.* 33: 721–732.

Niediek, V., S. Born, N. Hampe, N. Kirchgeßner, R. Merkel and B. Hoffmann. 2012. Cyclic stretch induces reorientation of cells in a Src family kinase- and p130Cas-dependent manner. *Eur. J. Cell Biol.* 91: 118–128.

Okuda, M., M. Takahashi, J. Suero, C.E. Murry, O. Traub, H. Kawakatsu and B.C. Berk. 1999. Shear stress stimulation of p130cas tyrosine phosphorylation requires calcium-dependent c-Src activation. *J. Biol. Chem.* 274: 26803–26809.

Olesen, S.-P., D.E. Clapham and P.F. Davies. 1988. Hemodynamic shear stress activates a K⁺ current in vascular endothelial cells. *Nature* 331: 168–170.

Orr, A.W., M.H. Ginsberg, S.J. Shattil, H. Deckmyn and M.A. Schwartz. 2006. Matrix-specific suppression of integrin activation in shear stress signaling. *Mol. Biol. Cell* 17: 4686–4697.

Orr, A.W., R. Stockton, M.B. Simmers, J.M. Sanders, I.J. Sarembock, B.R. Blackman and M.A. Schwartz. 2007. Matrix-specific p21-activated kinase activation regulates vascular permeability in atherogenesis. *J. Cell Biol.* 176: 719–727.

Osawa, M., M. Masuda, K.-i. Kusano and K. Fujiwara. 2002. Evidence for a role of platelet endothelial cell adhesion molecule-1 in endothelial cell mechanosignal transduction: is it a mechanoresponsive molecule? *J. Cell Biol.* 158: 773–785.

Osborn, E.A., A. Rabodzey, C.F. Dewey and J.H. Hartwig. 2006. Endothelial actin cytoskeleton remodeling during mechanostimulation with fluid shear stress. *Am. J. Physiol.* 290: C444–C452.

Phelps, J.E. and N. DePaola. 2000. Spatial variations in endothelial barrier function in disturbed flows *in vitro*. *Am. J. Physiol.* 278: H469–476.

Pompe, T., S. Glorius, T. Bischoff, I. Uhlmann, M. Kaufmann, S. Brenner and C. Werner. 2009. Dissecting the impact of matrix anchorage and elasticity in cell adhesion. *Biophys. J.* 97: 2154–2163.

Poullet, P., A. Gautreau, G. Kadaré, J.-A. Girault, D. Louvard and M. Arpin. 2001. Ezrin interacts with focal adhesion kinase and induces its activation independently of cell-matrix adhesion. *J. Biol. Chem.* 276: 37686–37691.

Putnam, A.J., J.J. Cunningham, B.B.L. Pillemer and D.J. Mooney. 2003. External mechanical strain regulates membrane targeting of Rho GTPases by controlling microtubule assembly. *Am. J. Physiol.* 284: C627–C639.

Radel, C. and V. Rizzo. 2005. Integrin mechanotransduction stimulates caveolin-1 phosphorylation and recruitment of Csk to mediate actin reorganization. *Am. J. Physiol.* 288: H936–H945.

Reinhart-King, C.A., M. Dembo and D.A. Hammer. 2003. Endothelial cell traction forces on RGD-derivatized polyacrylamide substrata. *Langmuir* 19: 1573–1579.

Riveline, D., E. Zamir, N.Q. Balaban, U.S. Schwarz, T. Ishizaki, S. Narumiya, Z. Kam, B. Geiger and A.D. Bershadsky. 2001. Focal contacts as mechanosensors: externally applied local mechanical force induces growth of focal contacts by an mDia1-dependent and ROCK-independent mechanism. *J. Cell Biol.* 153: 1175–1186.

Roca-Cusachs, P., N.C. Gauthier, A. del Rio and M.P. Sheetz. 2009. Clustering of α5β1 integrins determines adhesion strength whereas αvβ3 and talin enable mechanotransduction. *Proc. Natl. Acad. Sci. USA* 106: 16245–16250.

Romanenko, V.G., P.F. Davies and I. Levitan. 2002. Dual effect of fluid shear stress on volume-regulated anion current in bovine aortic endothelial cells. *Am. J. Physiol.* 282: C708–718.

Sawada, Y., M. Tamada, B.J. Dubin-Thaler, O. Cherniavskaya, R. Sakai, S. Tanaka and M.P. Sheetz. 2006. Force Sensing by mechanical extension of the Src family kinase substrate p130Cas. *Cell* 127: 1015–1026.

Schaller, M.D., J.D. Hildebrand and J.T. Parsons. 1999. Complex formation with focal adhesion kinase: a mechanism to regulate activity and subcellular localization of Src kinases. *Mol. Biol. Cell* 10: 3489–3505.

Shay-Salit, A., M. Shushy, E. Wolfovitz, H. Yahav, F. Breviario, E. Dejana and N. Resnick. 2002. VEGF receptor 2 and the adherens junction as a mechanical transducer in vascular endothelial cells. *Proc. Natl. Acad. Sci. USA* 99: 9462–9467.

Slater, J.H. and W. Frey. 2008. Nanopatterning of fibronectin and the influence of integrin clustering on endothelial cell spreading and proliferation. *J. Biomed. Mater. Res.* A 87A: 176–195.

Tadokoro, S., S.J. Shattil, K. Eto, V. Tai, R.C. Liddington, J.M. de Pereda, M.H. Ginsberg and D.A. Calderwood. 2003. Talin binding to integrin ß tails: a final common step in integrin activation. *Science* 302: 103–106.

Takahashi, M. and B.C. Berk. 1996. Mitogen-activated protein kinase (ERK1/2) activation by shear stress and adhesion in endothelial cells. Essential role for a herbimycin-sensitive kinase. *J. Clin. Invest.* 98: 2623–2631.

Thi, M.M., J.M. Tarbell, S. Weinbaum and D.C. Spray. 2004. The role of the glycocalyx in reorganization of the actin cytoskeleton under fluid shear stress: A "bumper-car" model. *Proc. Natl. Acad. Sci. USA* 101: 16483–16488.

Thodeti, C.K., B. Matthews, A. Ravi, A. Mammoto, K. Ghosh, A.L. Bracha and D.E. Ingber. 2009. TRPV4 channels mediate cyclic strain–induced endothelial cell reorientation through integrin-to-integrin signaling. *Circ. Res.* 104: 1123–1130.

Tzima, E., M.A. del Pozo, W.B. Kiosses, S.A. Mohamed, S. Li, S. Chien and M.A. Schwartz. 2002. Activation of Rac1 by shear stress in endothelial cells mediates both cytoskeletal reorganization and effects on gene expression. *EMBO J.* 21: 6791–6800.

Tzima, E., M.A. del Pozo, S.J. Shattil, S. Chien and M.A. Schwartz. 2001. Activation of integrins in endothelial cells by fluid shear stress mediates Rho-dependent cytoskeletal alignment. *EMBO J.* 20: 4639–4647.

Tzima, E., M. Irani-Tehrani, W.B. Kiosses, E. Dejana, D.A. Schultz, B. Engelhardt, G. Cao, H. DeLisser and M.A. Schwartz. 2005. A mechanosensory complex that mediates the endothelial cell response to fluid shear stress. *Nature* 437: 426–431.

Wang, N. and D.E. Ingber. 1994. Control of cytoskeletal mechanics by extracellular matrix, cell shape, and mechanical tension. *Biophys. J.* 66: 2181–2189.

Wang, Y., E.L. Botvinick, Y. Zhao, M.W. Berns, S. Usami, R.Y. Tsien and S. Chien. 2005. Visualizing the mechanical activation of Src. *Nature* 434: 1040–1045.

Wei, Z., K. Costa, A.B. Al-Mehdi, C. Dodia, V. Muzykantov and A.B. Fisher. 1999. Simulated ischemia in flow-adapted endothelial cells leads to generation of reactive oxygen species and cell signaling. *Circ. Res.* 85: 682–689.

Wootton, D.M. and D.N. Ku. 1999. Fluid mechanics of vascular systems, diseases, and thrombosis. *Annu. Rev. Biomed. Eng.* 1: 299–329.

Wu, C.-C., Y.-S. Li, J.H. Haga, R. Kaunas, J.-J. Chiu, F.-C. Su, S. Usami and S. Chien. 2007. Directional shear flow and Rho activation prevent the endothelial cell apoptosis induced by micropatterned anisotropic geometry. *Proc. Natl. Acad. Sci. USA* 104: 1254–1259.

Yang, B., C. Radel, D. Hughes, S. Kelemen and V. Rizzo. 2011. p190 RhoGTPase-activating protein links the β1 integrin/caveolin-1 mechanosignaling complex to RhoA and actin remodeling. *Arterioscler. Thromb. Vasc. Biol.* 31: 376–383.

Yano, Y., J. Geibel and B.E. Sumpio. 1996. Tyrosine phosphorylation of pp125FAK and paxillin in aortic endothelial cells induced by mechanical strain. *Am. J. Physiol.* 271: C635–C649.

Yeung, T., P.C. Georges, L.A. Flanagan, B. Marg, M. Ortiz, M. Funaki, N. Zahir, W. Ming, V. Weaver and P.A. Janmey. 2005. Effects of substrate stiffness on cell morphology, cytoskeletal structure, and adhesion. *Cell Motil. Cytoskel.* 60: 24–34.

Zaidel-Bar, R., Z. Kam and B. Geiger. 2005. Polarized downregulation of the paxillin-p130CAS-Rac1 pathway induced by shear flow. *J. Cell Sci.* 118: 3997–4007.

Zhang, X., G. Jiang, Y. Cai, S.J. Monkley, D.R. Critchley and M.P. Sheetz. 2008. Talin depletion reveals independence of initial cell spreading from integrin activation and traction. *Nat. Cell Biol.* 10: 1062–1068.

Zhao, S., A. Suciu, T. Ziegler, J.E. Moore, E. Bürki, J.-J. Meister and H.R. Brunner. 1995. Synergistic effects of fluid shear stress and cyclic circumferential stretch on vascular endothelial cell morphology and cytoskeleton. *Arterioscler. Thromb. Vasc. Biol.* 15: 1781–1786.

Mechanobiology of the Endothelial Nucleus

Elizabeth A. Booth-Gauthier,[1] Stephen Spagnol[1] and Kris Noel Dahl[2,]*

Introduction

Mechanical forces are an integral part of endothelial cell physiology and development, and there are distinct functional roles for shear stress. Endothelial cells are involved in many aspects of vascular physiology including barrier functions that separate the circulating materials from the underlying tissue and select what components will enter and exit the blood stream, vasoconstriction and vasodilation to maintain blood pressure, angiogenesis, inflammation, and atherosclerotic response. Cellular function requires maintaining homeostasis as well as being able to respond to mechanical and chemical stimuli in short and long times. Under normal physiological conditions endothelial cells experience a wide range of shear stress depending on their location within the circulatory system. Generally, laminar shear stress results in an atheroprotective response by the

[1]Department of Chemical Engineering, Carnegie Mellon University, Pittsburgh, PA.
[2]Department of Biomedical Engineering, Carnegie Mellon University, Pittsburgh, PA.
*Corresponding author: krisdahl@cmu.edu

endothelium. Conversely, disturbed shear stress including low shear stress, oscillatory flow, and flow reversals results in an atherosclerotic response.

Endothelial cells respond to shear stress with changes in gene expression as well as alterations to cell morphology and structures within the cell. In this work we will present the nuclear response of endothelial cells to shear stress related to nuclear structure, nuclear mechanics and gene expression. Nuclei reorganize in cells under shear stress and cells change gene expression. Related work has shown evidence correlating gene expression with position of genes within the nucleus and even gene translocation with differential expression. Thus changes in gene expression and nuclear mechanics associated with response to shear stress may be inherently related.

Nuclear Structures

While much attention has been dedicated to structure and mechanics of the extracellular matrix and the cytoskeleton, there are also mechanically relevant structures within the nucleus. The nucleus is the largest organelle of the cell, and the nucleus is also significantly stiffer than the rest of the cell (Dahl et al. 2008). Studies of endothelial cells under compression modeled as solids show that the nucleus is 5 times stiffer than the cytoplasm (Caille et al. 2002). Also, the nuclei of endothelial cells are mechanically responsive and stiffen with exposure to shear stress (Deguchi et al. 2005).

The nucleoskeleton

The nucleoskeleton is the mechanical structural component of the nucleus, primarily at the inner nuclear membrane. The nucleoskeleton is involved in DNA replication, cell division, and chromatin organization (Mattout-Drubezki et al. 2003, Verstraeten et al. 2007). The stiffness of the nucleoskeleton derives primarily from the nuclear lamina, which is a filamentous meshwork of intermediate filaments composed of lamin proteins. Lamins are type V intermediate filament proteins with a rod-shaped domain flanked by globular domains on the N and C termini. The C-termini of lamins are much larger than cytoplasmid intermediate filaments and contain the nuclear localization sequence and an Ig-fold structure (Herrmann et al. 2003, Herrmann et al. 2003). The tail domain is also the site of protein-protein and protein-DNA binding.

Lamins assemble into coiled-coil dimers that assemble both linearly head-to-tail and laterally into staggered rope-like structures and form the mostly-disorganized meshwork of the nuclear lamina. In standard intermediate filament assembly, dimer assembly form staggered lateral associations to form an an intermediate filament (Buehler et al. 2008, Goldman et al. 2008).

However, *in vitro* assembly of the lamina nucleoskeleton is not possible since lamins form unique ordered structures such as paracrystalline arrays (Moir et al. 1991, Heitlinger et al. 1992). Due to the head to tail assembly method of the coiled-coils lamin filaments lack polarity, and the large tail domains which extend from the central structures allows for multiple binding sites along the lamina. The lamina is organized only in amphibian oocytes as a two dimensional, orthogonal network (Aebi et al. 1986, Akey 1989).

There are two main types of lamin proteins. A-type lamins, primarily lamin A and lamin C, splice variants of the same gene *LMNA*. B-type lamins, primarily lamin B1 and lamin B2, are encoded from different genes. Lamin A is primary contributor to the mechanical stability of the nucleus (Gruenbaum et al. 2000). Loss of lamin A causes nuclear weakness and rupture (Lammerding et al. 2004). Conversely, association of lamin A in model nuclear systems induces sheets of thicker filaments on the nuclear lamina, significantly increasing the rigidity of the nucleus (Dahl et al. 2006). There are hundreds of mutations in *LMNA*, which lead to a host of disorders. We will discuss endothelial-relevant mutations and syndromes below. Loss of B-type lamins is usually lethal for organisms and for cells (Harborth et al. 2001). However, B-type lamins are thought to regulate gene expression, and model systems with loss of these lamins do not show altered nuclear mechanics (Lammerding et al. 2006).

A-type lamins and B-type lamins form primarily independent networks that assemble independently and appear to have different nuclear functions. Lamin filaments are homopolymers of either lamin A or lamin B (Delbarre et al. 2006). A- and B-type lamins also have different binding partners at the inner nuclear membrane. In nuclear envelope reassembly after mitosis, B-type lamins are incorporated into the nucleoskeleton earlier than A-type lamins (Moir et al. 2000). A-type lamin filaments are 15 nm in diameter and form thick sheets and bundles whereas B-type lamins are 7 nm in diameter assemble into more regular, dispersed structures (Goldberg et al. 2008).

Genome and chromatin

Chromatin, a complex of mainly DNA and histone proteins, is the major component of the nuclear interior. DNA winds around histones to pack the nearly 1 m of DNA into the nucleus while still leaving regions accessible for transcription. The DNA-histone complex, called the nucleosome, is a 100 kDa, 10 nm histone octamer (made of histone proteins H3, H4, H2A and H2B) complexed with a DNA molecule which folds two-times around (Richmond et al. 1984). Condensed nucleosome structures form chromatin. Expanded chromatin reconstituted from histones and DNA has the appearance of 10 nm thick beads on a string, but the 30 nm thick condensed chromatin fiber is the mechanically relevant structure of the genome (Ostashevsky 2002). Recent studies suggest that there may be larger chromatin domains which exist as domains above chromatin fibers but below chromosomes (Hu et al. 2009).

Chromatin is separated functionally and structurally into euchromatin and heterochromatin. Euchromatin is a more open form which is more readily translatable and is transcribed first (Gilbert et al. 2004). Heterochromatin, by contrast is typically condensed by virtue of expression of proteins such as Heterochromatin Protein 1 (HP1) and DNA and histones modified by acetylation and methylation (for example, trimethylation of histone H3 lysine 9 is commonly known to be associated with heterochromatin) to shift between euchromatin and heterochromatin (Georgatos et al. 2009). In general, chromatin found in heterochromatin possess a higher density of nucleosomes per DNA than in euchromatin which is closer to the 30-nm fiber seen *in vitro* (Bassett et al. 2009). Heterochromatin is stabilized by lamin proteins at the nucleoskeleton, highlighting how interconnected structures are in the nucleus (Gruenbaum et al. 2005). In many cells, a large fraction of heterochromatin lies at the nuclear envelope. Lamins can interact with chromatin directly via 30–40 nonspecific bp DNA segments (Moir et al. 2000, Stierle et al. 2003) and indirectly (Gant et al. 1997) or via lamin binding proteins (Segura-Totten et al. 2002). LEM-domain proteins bind DNA through BAF (barrier to autointegration function) and LBR (lamin B-receptor) binds lamin B, DNA, histones and HP1 (Ye et al. 1997, Worman 2005). Chromatin is further organized into chromosomes (~100–200 Mbp), which are important functional nuclear units both for mitosis and for organization of the interphase nucleus (Misteli 2007).

Integration of nucleoskeleton and cytoskeleton

The nucleus is not structurally or mechanically isolated from cytoskeletal structures; the linker of nucleoskeleton and cytoskeleton complex (LINC) provides a direct physical route for mechanotransduction (Fig. 1). A complex of SUN-domain and nesprin proteins interconnects actin, microtubule and intermediate filament cytoskeletal structures with the nucleoskeleton of the cell. Within the nucleus, SUN1 and SUN2 proteins interact with the nucleoskeleton including lamins, lamin binding proteins, nuclear pore complexes, and other proteins (Zastrow et al. 2004, Razafsky et al. 2009). Large, multi-domain nesprin proteins are found on the outer nuclear membrane and connect to actin via the N-terminal actin binding of nesprins 1 and 2. Nesprin-3 contains a site that binds to plectin, which associates to intermediate filaments (Crisp et al. 2006, Ketema et al. 2007). Nesprin 4, present in specialized cells, interacts with light chain kinesin and is suggested to be involved in microtubule-dependent movement (Roux et al. 2009).

Extracellular forces such as shear stress can be initially sensed by the cell at either cell-cell connections or through integrins and other adhesion molecules. Integrins are physically coupled to the cytoskeleton and may transduce forces directly into the nucleus. Cytoskeletal elements interact with the nuclear lamina through nesprins and the SUN complex, which span the nuclear membrane (Dahl et al. 2010). The LINC complex provides direct physical connection from the cytoskeleton into the chromatin at the nuclear interior and is hypothesized to play a role in gene expression.

Mechanics of Nuclear Structures

The different structures of the nucleus have different mechanical features that contribute to cellular mechanics and impact reorganization of cells experiencing extracellular mechanical stress.

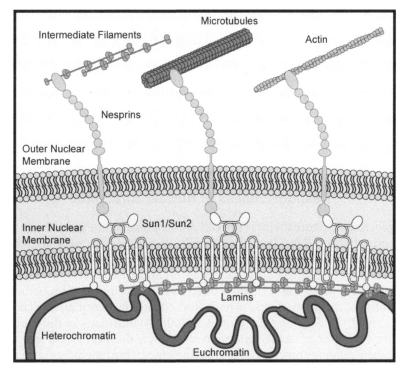

Figure 1. Schematic of nucleoskeleton-cytoskeleton interconnections. Cytoskeletal filament systems connect to the outer nuclear membrane via nesprin proteins. The direct connection is then maintained through SUN 1/2 protein complexes, which bind to inner nuclear membrane nesprins, transmembrane proteins and lamins. Lamins then bind directly and indirectly to chromatin in the nuclear interior.

Color image of this figure appears in the color plate section at the end of the book.

Mechanics of the nucleoskeleton

Intermediate filament proteins have a persistence length between 100–1000 nm (Hohenadl et al. 1999). The nucleoskeleton found in model systems with limited chromatin has an elasticity of 25 mN/m, and the nucleoskeleton deforms as a two-dimensional elastic material (Dahl et al. 2004). Since expression of A-type lamins varies among cell types and levels of A-type impact nuclear mechanics, there is likely a large variation in nucleoskeletal stiffness in human cells under various conditions. However, the largest source of mechanical variation in human cell types comes from the contribution of the genome mechanics.

Mechanics of the genome

The persistence length of DNA is ~50 nm, similar to other biological polymers for the same molecular density (Hegner et al. 2002). As discussed earlier, the primary mechanical element in the genome is condensed chromatin, which has a functional persistence length of ~30 nm (Cui et al. 2000). At forces under 20 pN, chromatin fibers can be extended reversibly with a stretch modulus of 150 pN (Bennink et al. 2001). Chromatin and chromosomes are impacted by cofactors including histones, non-histone tethering elements (Poirier et al. 2000), condensing complexes (Marko et al. 2003) and salt (Poirier et al. 2002) alters experimental conditions. Elastic moduli of 250 Pa (Poirier et al. 2000) to 5,000 Pa (Houchmandzadeh et al. 1997) have been reported for chromosomes.

Responsiveness of Nuclear Structures

Organized structures within the cell are responsive to extracellular force including shear flow. Endothelial cells show active remodeling of the cytoskeleton, cell-cell adhesions and cell-matrix interactions. There are also numerous changes in intracellular organization in cells under shear stress. The mechanical activation of genes is called mechanotransduction. In endothelial cells, researchers have studied the chemical pathways in cells activated by shear stress (Lehoux et al. 2006). While many of the chemical pathways leading to gene regulation are known, the mechanical receptors, and their relative contributions, are still being determined. Some include: the glycocalyx, stretch activated ion channels, cytoskeletal reorganization and changes in adhesion molecules.

Functional changes of cells with shear stress

Generally, laminar shear stress results in an atheroprotective response by the endothelium and disturbed shear stress results in an atherosclerotic response. Endothelial cells often experience pulsatile flow but an average forward shear stress on the order of 10–30 dyn/cm^2 will show an anti-atherosclerotic response (Davies 1995). In order to maintain proper function of the circulatory system, each cell needs to show appropriate mechanotransductive response to shear stress. Under laminar shear stress endothelial cells show an atheroprotective response by altering gene expression, inhibiting growth of endothelial cells and underlying smooth muscle cells, discouraging adhesion and migration of leukocytes through the endothelium, and decreasing accumulation of low density lipoproteins (LDL) (Traub et al. 1998). Endothelial cells under consistent laminar shear stress undergo 'adaptation' to shear stress. Adaptation is the down regulation of some genes under shear stress and the expression of a new set of genes in response to the new shear stress condition (Resnick et al. 2003). Adaptation is accompanied by the alignment of the endothelium in the direction of flow and the corresponding changes in the cytoskeleton and cell polarity.

Structural changes of cells with shear stress

Endothelial cells are mono-nuclear cells that exhibit contact inhibition and thus form a single monolayer between the luminal space of the circulatory system and the underlying smooth muscle cells. Upon application of laminar shear stress endothelial cells reorient from a polygonal cell monolayer to anextended cell shape parallel to the direction of the shear stress. Alignment takes about 12–24 hours at 12–20 dyn/cm^2 laminar shear stress. Interestingly, the cells respond within minutes by decreasing their movement and increasingtheir actin turnover rate threefold. Even after hours of shear stress, motilityhas increased but the actin remolding is still highly dynamic, suggesting acontinual shape optimization in response to shear stress. Shear stress also induces reorganization of the microtubules and intermediate filaments. The microtubule organizing center (MTOC) will orientto the downstream side of the nucleus relative to the direction of flow (Tzima et al. 2003). The consistent orientation of the MTOC downstream of the nucleus demonstratesthat the cell's polarity can be regulated by mechanical force (Chien 2006).The regulation of polarity for a group of cells is important for endothelium-wide remodel ingincluding processes like wound healing and angiogenesis *in vivo*.

Nuclear structural changes with shear stress

The nucleus is also a mechanoresponsive structure within the cell (Fig. 2). In response to shear stress, endothelial cells attenuate to reduce the force on their nuclei. If the cells experience shear stress for an extended period of time, 24 hours, they upregulate the expression of lamins, thickening the lamina and stiffening the cells nuclei (Philip et al. 2008). In particular, lamin A has been shown to be upregulated and relocated closer to the nuclear periphery (Philip and Dahl 2008). Isolated endothelial cell nuclei exposed to 24 hours of 20 dyn/cm² shear stress showed an increased Young's modulus from 0.4kPa to 0.6kPa (Deguchi et al. 2005). All of the above responses by the nucleus are mechanically induced suggesting that the nucleus is also a force sensing organelle.

The nucleus is the largest and stiffest organelle in most mammalian cells, including the endothelial cell, and can greatly influence cell mechanics (Caille et al. 2002). Simulations of the cell response to shear stress suggest that a simple model of minimizing the shape of the nucleus to reduce drag is sufficient to model the morphological changes experienced by cells (Hazel et al. 2000). This suggests that the nucleus does, in fact, feel and respond to forces over long times. Deguchi and colleagues showed that shear stress aligns nuclei from endothelial cells, and the shape is retained even after nuclei are removed from cells (Deguchi et al. 2005). These global arrangements are correlated with a nearly 50% increase in nuclear stiffness

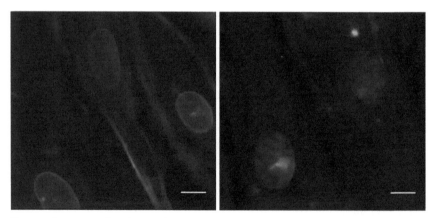

Figure 2. Response of nucleus to shear stress stimulation. Control HUVECs (left) show actin stress fibers (green; phalloidin) with a centralized nucleus with DNA (blue; DAPI) and lamin A (red; immunolabeling for lamin A/C). HUVECs treated with 10 dyn/cm² for 12 hours showed nuclear reorganization in the direction of flow along with actin filament structures.

Color image of this figure appears in the color plate section at the end of the book.

(Deguchi et al. 2005). Studies from our lab have shown that shear stress upregulates structural proteins of the nucleus and redistributes them into a more load-bearing configuration at the nuclear periphery (Philip et al. 2008).

Thus, the nucleus is capable of, reorganizing in response to the stress (Deguchi et al. 2005) and changing internuclear protein organization (Philip and Dahl 2008). Based on the knowledge that nuclear positioning is related to gene expression (Dundr et al. 2001), it is possible that this reorganization could impact the expression profile of the cell in a probabilistic way.

Changes in gene expression with shear stress

In nearly all cell types, force on the cell is able to alter gene expression by chemical factors (such as NF-kB, mitogen-activated protein (MAP) kinase, etc.) activated through chemical pathways induced by strains on the cell membrane or cell cytoskeleton through transcriptional activators and repressors (Papadaki et al. 1997). Most mechanotransduction pathways have mechanosensors at the plasma membrane which are activated and produce a chemical pathway to induce a response in the cytoplasm or change in gene expression. Chemical signaling in cells typically occurs quickly (within minutes-hours) and is attenuated by homeostasis pathways over time, otherwise constant stimulation of genes could lead to oncogenesis (Tzima et al. 2002, Petzold et al. 2009). So as cells transition from unsheared to sheared their responses are transient and yet the effects are permanent, and the "character" of the cell changes in expression patterns (Brooks et al. 2002).

From DNA microarrays, chemical pathways have been mapped in endothelial cells exposed to shear stress. Gene expression changes associated with laminar flow, not disturbed flow (Brooks et al. 2002), include genes associated with inflammation; HUVEC exposed to steady laminar flow tend to show reduced levels of most inflammation markers (Brooks et al. 2004). Cell cycle and proliferation genes are also increased or decreased.

Angiogenesis and the Nuclear Response

Angiogenesis is the formation of new blood vessels from the preexisting vasculature that occurs during developmental and normal physiology in response to hypoxia and nutrient deficiency in peripheral tissues (Cebe-Suarez et al. 2006). It is implicated in a variety of disease pathologies including cancer metastasis (Carmeliet et al. 2011), macular degeneration (Ng et al. 2005, Qazi et al. 2009) and cardiovascular disease (Webber et al.

2011). Angiogenesis is initiated by numerous biochemical signaling factors including a family of vascular endothelial growth factors (VEGF), acidic and basic fibroblast growth factors (aFGF, bFGF), platelet-derived growth factor (PDGF), angiopoietins (Ang), interleukin-8 (IL-8), hepatocyte growth factor (HGF) and transforming growth factors alpha and beta (TGF-α, TGF-β) (Huang et al. 2004, Ng and Adamis 2005). VEGF stimulation is the most heavily studied pro-angiogenic pathway and it is critical for the development of normal vasculature. In fact, the loss of a single VEGF allele results in embryonic lethality (Ferrara et al. 1997).

Angiogenesis occurs through two physiologically distinct mechanisms: sprouting and intussusceptions (Djonov et al. 2000). Sprouting involves endothelial cell invasion and degradation of the basement membrane as the endothelium extends towards the hypoxic or nutrient deficient tissue via cell migration and proliferation. Subsequently, the formation of adherens junctions through vascular endothelial cadherins (VE-cadherins) stabilizes the newly formed vessels (Hillen et al. 2007). Intussusceptive angiogenesis is initiated by a transluminal bridge of cell-cell contacts that split a vessel in two (Burri et al. 2004) through a mechanism that is hypothesized to be flow-dependent (Hillen and Griffioen 2007).

Thus, inherent to angiogenesis is a prevalent mechanical response from cell-ECM and cell-cell signaling interactions through integrins and VE-cadherins, respectively, that alter the cytoskeleton. Recent evidence indicates that varying the balance of these two mechanical interactions can appreciably alter cell stiffness and morphology (Stroka et al. 2011). These mechanical effects correspond to different functional states of endothelial cells, especially with respect to migration and proliferation (Nelson et al. 2004). This effect is modulated through cytoskeleton-imposed changes in nuclear shape (Sims et al. 1992, Maniotis et al. 1997, Versaevel et al. 2012) that are suggested to play a role in the altered gene expression (Thomas et al. 2002, Lanctot et al. 2007) in parallel with biochemical signaling mechanisms (Lamalice et al. 2007).

Morphological response

The quiescent and angiogenic states of endothelial cells involve unique cytoskeletal mechanics that loosely correlate with cell morphology (Nelson et al. 2004, Reinhart-King et al. 2005, Stroka and Aranda-Espinoza 2011). During angiogenesis, cell migration leads to the cell elongation through p38-regulated actin rearrangement (Rousseau et al. 1997) and polarization (Lamalice et al. 2007) as endothelial cells break cell-cell contacts in favor of integrin-based cell-ECM contacts through focal adhesions (Carmeliet and Jain 2011). Specifically, VEGF stimulation leads to VE-cadherin phosphorylation and sequestration (Vestweber et al. 2009). This terminates the cell-cell contacts that are the basis of contact inhibited growth and proliferation (Nelson et al. 2003, Liu et al. 2010), and frees the cells to take up the cell-ECM connections that then control cytoskeletal mechanics. The cell-ECM contacts and the underlying biochemical signaling cascades they activate are necessary but not sufficient for the proper pro-angiogenic response (Nelson and Chen 2003, Gong et al. 2004, Stoletov et al. 2004, Shiu et al. 2005). Instead, mechanical activation leading to cytoskeletal dynamics appears to be a requisite part of the mechanism (Nelson and Chen 2003, Gong et al. 2004, Stoletov et al. 2004, Liu et al. 2010).

Endothelial cell elongation during the migratory phase of angiogenesis has been mimicked using fibronectin-coated micropatterned substrates to control cell shape. This has illuminated the role of cell shape and surface area on nuclear remodeling through a mechanism heavily influenced by actomyosin contraction and dynamics, but mediated by intermediate filaments (Rousseau et al. 1997, Reinhart-King et al. 2005, Shiu et al. 2005, Grevesse et al. 2012). Increased elongation results in lateral nuclear deformation by actin stress fibers, which occurs concomitantly with increased chromatin condensation and nuclear stiffening (Versaevel et al. 2012). These specific effects have also been linked to changes in gene expression and proliferative state in other cell types (Thomas et al. 2002). This is also consistent with other work demonstrating reorientation of nucleoli along the axis of applied strain using microbeads for micromanipulation of integrins in endothelial cells (Maniotis et al. 1997). Such studies highlight direct mechanical linkages between external mechanical events and subnuclear rearrangement that may underpin specific genetic responses in conjunction with the established biochemical gene activation.

These results are not unique to the pro-angiogenic state of endothelial cells. In the quiescent state the suppression of cell growth and proliferation by VE-cadherin cell-cell connections in a Rho/Rock-dependent manner has been well-established (Nelson and Chen 2003, Nelson et al. 2004, Shiu et al. 2005, Vestweber et al. 2009, Carmeliet and Jain 2011). More recently, the VE-cadherin has been shown to control the formation of tight junctions through upregulation of claudin-5 by limiting β-catenin nuclear translocation (Taddei et al. 2008). Furthermore, mechanical force transmitted through VE-cadherins also regulates the size of adherens junctions themselves (Liu et al. 2010). Thus, endothelial cells in the quiescent state also respond to mechanical cues that effect their nuclear mechanics and gene expression.

Gene response

VEGF-stimulated transitions in gene expression have been the predominate focus of pro-angiogenic genetic assays. An early study using cDNA microarrays identified 139 genes that were upregulated more than two-fold by 24-hour VEGF stimulation of HUVECs, of which 53 were upregulated after the first two hours of stimulation (Abe et al. 2001). The time dependence of these results exposes a possible correlation between the phases of angiogenesis and specific gene expression. This corresponded to a more recent study noting functional classes of genes upregulated by VEGF stimulation in HUVECs (Wary et al. 2003). These included time-specific expression of genes associated with endothelial cell motility (i.e., matrix metalloproteinases 1 and 2, myosin VIIA and α_2-integrin), transcriptional regulation (early growth factors 1–3) and DNA repair and replication (mini-chromosome maintenance proteins, CD3-epsilon-associated protein). Other work exposed tissue-specific gene expression in response to VEGF stimulation for myometrial microvascular endothelial cells (MMECs) (Weston et al. 2002). While tissue-specific function obviates unique VEGF-stimulated gene expression, the mechanisms regulating tissue-specific expression of the genome remain unclear.

Interestingly, lamin A is upregulated with the VEGF response (Yang et al. 2002), which may suggest interconnections of structure, mechanics and gene regulation within chemically stimulated nuclei.

The biochemical signaling mechanisms behind VEGF-stimulated gene expression are well documented in the literature (Leung et al. 1989, Ferrara and Davis-Smyth 1997, Risau 1997, Esser et al. 1998, Bussolati et al. 2001, Kliche et al. 2001, Matsumoto et al. 2001, Vasir et al. 2001, Yang et al. 2002, Cebe-Suarez et al. 2006, Matsumoto et al. 2006, Kiec-Wilk et al. 2010, Cao et al. 2011, Carmeliet and Jain 2011) as it has been a major focus of research for decades. While biochemical signaling is required for the proper genetic response, recent evidence indicates cell mechanics also play a role. In particular, the pro-angiogenic biochemical signaling involves cytoskeletal rearrangement through Rho/ROCK-dependent effects on actin that effect nuclear shape, as discussed previously. More direct evidence implicating the mechanical response in gene expression was divulged in the discovery of the effect of Rho-inhibitor p190RhoGAP on VEGF receptor 2 (VEGFR2) expression through control of transcription factors in a mechanism dependent on ECM elasticity (Mammoto et al. 2009). Subsequent work has shown that the invasive and proliferative stages of angiogenesis can be differentially triggered by varying the level of cell-substrate adhesion (Shen et al. 2011), with specific gene expression associated with each stage. This result points to another obvious mechanical role in the pro-angiogenic response. In aggregate, these findings highlight the interplay of mechanical and biochemical cues involved in cytoskeletal rearrangement, nuclear remodeling and gene expression that is a recurring theme throughout each aspect of the pro-angiogenic response (Fig. 3).

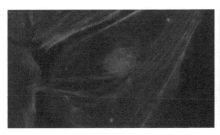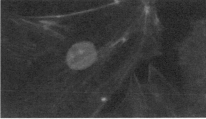

Figure 3. Response of nucleus to VEGF stimulation. Control HUVECs (left) show actin stress fibers (green; phalloidin) with a centralized nucleus with DNA (blue; DAPI) and low levels of lamin A (red; immunolabeling for lamin A/C). HUVECs treated with 50 ng/mL of VEGF for 24 hours showed increase lamin labeling suggesting reorganization of nuclear structure.

Color image of this figure appears in the color plate section at the end of the book.

Dysfunctions of Endothelial Response

This chapter has previously focused on endothelial morphological and genetic response to high wall shear stress (SS) (>12dyn/cm^2). Low levels of SS (<12dyn/cm^2) are also studied and have specific morphological and genetic endothelial cell response (Chatzizisis et al. 2007). Regions of low SS colocalize with atherosclerotic plaques suggesting that low SS is pathogenic in plaque formation and plaques initially arises from a dysfunctional endothelium (Libby 2002).

Pathological force

The vascular architecture is complicated and includes a variety of flow patterns, thus far we have only considered high levels of laminar SS (Resnick et al. 2003). Patterns characteristic of low shear stress include disturbed laminar flow where the flow initially separates, forms a zone of recirculation, then reattaches and turbulent shear stress characterized by flow velocity at a single position continually varying even though net overall flow is steady (Chatzizisis et al. 2007). Pulsatility can further compound the formation of low shear stress environments within the circulatory system (Vasava et al. 2012). Zones of recirculation and turbulent flow are often found near areas of curvature or bifurcations and can increase or decrease cell responses to SS (Zarins et al. 1983, Ziegler et al. 1998). Interestingly, endothelial cells are sensitive to the nature of the shear stress they experience, either turbulent or recirculatory, as evidence by the composition and morphology of atherosclerotic plaques (Enrico et al. 2009, Poepping et al. 2010). The low SS that the plaques form in could also help to mechanically stabilize the plaque so that it can continues to develop (Rouleau et al. 2010).

At high levels of SS endothelial cells upregulate the gene for endothelial nitric oxide synthase (eNOS), which increases the bioavailibilty of nitric oxide (NO), an anti-inflammatory, anti-apoptotic, anti-mitogenic, and anti-thrombic molecule (Harrison et al. 2006). In areas of low SS there is decreased availability of NO. In addition to the cells under low SS, cells adjacent to areas of recirculatory flow also show decreased expression of eNOS and free NO relative to cells adjacent to areas of fully developed flow potentially from intercellular communication (Plank et al. 2007). Under low SS, endothelial cells upregulate transcription factors associated with the uptake, synthesis (Liu et al. 2002) and permeability (Himburg et al. 2004) of the endothelium to low density lipoproteins, 'bad cholesterol' (Traub and Berk 1998). Zones of recirculation and turbulence increase the residence time of circulating LDL, facilitating infiltration into the endothelium (Khakpour et al. 2008). Low SS plays a role in signaling for the adhesion and infiltration of the endothelium by circulating leukocytes (Libby 2006). NF-κB is activated through transcription factors in the cytoplasm and translocated to the nucleus under low SS (Orr et al. 2005). NF-κB is involved in the upregulation of adhesion molecules; vascular cell adhesion molecule 1 (VCAM-1) and intercellular adhesion molecule 1(ICAM-1) (Chappell et al. 1998); upregulation of monocyte chemoattractant protein (MCP-1) (Chien 2003); and pro-inflammatory cytokines tumor necrosis factor α (TNF-α) and interleukins (De Caterina et al. 1995, Griffin et al. 2012, McCormick et al. 2001, Brooks et al. 2002). VCAM-1 and ICAM-1 mediate the rolling and adhesion of circulating leukocytes onto the endothelial surface, MCP-1 promotes transmigration of the leukocytes through the endothelium where they undergo structural and functional changes maintaining the inflammatory process and contributing to plaque formation (Libby 2002). Endothelial cells under low shear stress promote atherosclerotic plaque formation by a cascade of events starting with down regulation of eNOS, increased LDL accumulation, activation of NF-κB and subsequent upregulation of cell adhesion molecules, chemoattractants, and pro-inflammatory cytokines furthering plaque formation through leukocyte recruitment.

Morphological response

Under high levels of SS endothelial cells develop elongated spindle like morphology parallel to the direction of shear (Davies 1995). This enlongation decreases the apical cell height and the SS gradient across the endothelial cells (Barbee et al. 1994). However, under no or very low SS conditions the

SS gradient is low to begin with and cells maintain a polygonal shape similar to cells without shear stress (Rouleau et al. 2010). Low SS is conducive to mitotic (White et al. 2001) and apoptotic (Tricot et al. 2000) activity, contrary to what is found in endothelial cells under high SS. These activities are conducive to forming cells with a polygonal shape and may be responsible for widening the junctions between the cells (Chen et al. 1995). Increased junction space between the cells could result in increased LDL (Chien 2003) and leukocyte permeability further aggravating plaque formation. Under, low levels of SS the endothelial cells do not experience high apical stress, decrease their height, or reorient in the direction of SS. Instead they appear polygonal, with comparatively larger junctions between cells conducive to endothelium transmigration.

Gene response

From DNA microarrays, chemical pathways have been mapped in endothelial cells exposed to shear stress. Gene expression changes associated with disturbed flow, not laminar flow (Brooks et al. 2002), include genes associated with inflammation; HUVEC exposed to steady laminar flow tend to show reduced levels of most inflammation markers (Brooks et al. 2004). Cell cycle and proliferation genes are also increased or decreased. About 3% of endothelial genes are SS responsive under laminar SS and turbulent SS there is a decrease in the number of genes activated, about 1.1% (Ohura et al. 2003). Turbulent SS also specifically activates genes that are involved in the remodeling of the vasculature (Ohura et al. 2003).

Nuclear Specific Diseases and Endothelial Dysfunction

There are over 180 mutations in lamin A that lead to ~13 different diseases collectively termed laminopathies. Roughly, laminopathies caused by the expression of A-type lamin mutated forms can be divided into four categories: i) diseases of the striated muscle; ii) peripheral neuropathies; iii) lypodistrophy syndromes; iv) accelerated aging disorders (Worman et al. 2007). HGPS belongs to the last category of laminopathies, where patients age prematurely while experiencing degradation of load-bearing tissues and disease states related to the mechanophysiology of organs such as atherosclerosis and loss of bone mass (Mounkes et al. 2003).

HGPS and aging

One unique mutation in *LMNA* (C608T in exon 11) activates a cryptic splice donor site, resulting in a truncated lamin A protein, Δ50 lamin A or progerin, that lacks 50 internal amino acids near its C-terminus (De Sandre-Giovannoli et al. 2003, Eriksson et al. 2003). Due to incorrect post-translational processing (Korf 2008) progerin is more strongly localized to the nuclear membrane, and there are structural changes in the nucleus such as loss of interior chromatin condensation, changes in heterochromatin organization, nuclear envelope blebbing and increased thickness of the nuclear lamina (Goldman et al. 2004). Mechanically, this over-accumulation of progerin at the nuclear envelope decreases the lamina network's ability to deform (Dahl et al. 2006), and cells expressing progerin are less able to adapt to shear stress (Philip and Dahl 2008). Endothelial cells and smooth muscle cells are particularly sensitive in HGPS, and histological sections of patients show high levels of progerin expression (ref McClintock).

Processing of pre-lamin A into its mature form is well understood. Lamin A is first translated as a precursor molecule, prelamin A that possesses a C-terminal motif, CaaX, which acts as a receptor site for protein farnesylation and methylation. The cysteine residue is farnesylated and the three final amino acids are proteolytically cleaved. The protein is then methylated at the C-terminal cysteine residue. The partially processed protein is translocated to the inner nuclear membrane At the inner nuclear membrane, where the c-terminal 15 amino acids are cleaved by ZMPSTE24 (zinc metallopeptidaseSte24 homologue) protease to generate mature lamin A.

Over 80% of mutations giving rise to HGPS are caused by mutations in exon 11 of theLMNA gene. The DNA mutation activates a cryptic splice site, resulting in accumulation of a mutant form of lamin A, termed progerin. Progerin lacks a 50 amino acid section in the C-terminal region required for cleavage by ZMPSTE24. The loss of the 50 amino acids in the tail domain results in a loss of down stream processing and the maintained farnesyl group, which anchors the protein to the nuclear membrane during processing. The maintenance of this farnesyl group is hypothesized to be responsible for the accumulationof progerin at the nuclear membrane. Additionally, progerin has beenshown to accumulate in HGPS patients' nuclei as well as in the nuclei ofotherwise healthy adults as they age. As progerin accumulates the nucleus progressively shows signs of being structurally compromised including the formation of blebs, invaginations, thickening of the nuclear lamina, andclustering of nuclear pores. Six Farnesyl transferase inhibitors, which effectively release progerin from the nuclear membrane, improves the shape of the nuclei and their responsiveness to force. Structurally the accumulation of progerin is likely to causes in-homogeneities in the lamina structure and the formation of microdomains that are stiffer and more densely packed with lamins within the nuclear lamina.

Progerin alters the effectiveness of the LINC complex at a several locations. First, progerin decreases the effectiveness of the nuclear lamina in binding chomatin. Progerin also is less able to bind SUN protiens located at the inner nuclear membrane breaking communication with the LINC complexdirectly in two positions. This is further felt at the nuclear periphery where there is a decreased elasticity of the cytoskeletal structures. Further aggravating the HGPS disease state are defects in the DNA repair pathway resulting in increased double stranded breaks. An increase in reactiveoxygen species (ROS) found in HGPS patients is thought to be a majorcontributor to the formation of double stranded breaks. Inclusion of progerin results in alterations in nuclear and cellular structure that could lead to cell misregulation under different stimuli, chemical, mechanical, or otherwise. HGPS has been studied as a model aging system, however, with all models there are drawbacks and limitations. The HGPS model succeeds in several ways. First, it incorporates the nuclear mechanical defects from sporadic production and accumulation of progerin over an average adult lifetime. Progeria patients also show increased ROS, DNA damage, telomere defects, and cell senescence that are consistent with the normal aging process. However, they do not have all identified molecular pathways some of whichappear specific to aging. The study of progeria does provide an opportunity to understand a subset of aging mechanisms, many of which are nucleus specific, i.e., nuclear stiffening, telomere shortening, and increased DNA damage.

Other nuclear mutations

There are many other mutations in lamin A that cause systemic defects including in endothelial cells. Most of these mutations impact cellular senescence and a reduced ability of stressed cells to repair including the metabolically active endothelial cells. There are other disorders that impact genome function. Mutations in the protein WRN cause the accelerated aging disorder Werner syndrome. WRN impacts helicase and exonuclease activity leading to increased DNA damage (Kudlow et al. 2007). Ultimately cells collect gene defects leading to senescence or cancer. In vascular endothelium this impacts the cell's ability to respond to chemical cues including VEGF (Lutomska et al. 2008), discussed above.

Conclusions

In this chapter we have highlighted the role of the nucleus in endothelial cell mechanobiology. The nucleus acts as both a load-bearing organelle in endothelial cells, which resists and responses to shear stress. Also, the nucleus is a central hub for gene expression changes in response to chemical and mechanical forces. We suggest that this structural and functional interplay may allow a better understanding of how cells are able to respond to changes in extracellular mechanical changes on long time scales. These studies may allow us to better appreciate sustained response to angiogenic factors, response to pathological shear stress and the role of nuclear mutations on endothelial and vascular health.

References

Abe, M. and Y. Sato. 2001. cDNA microarray analysis of the gene expression profile of VEGF-activated human umbilical vein endothelial cells. *Angiogenesis* 4(4): 289–98.

Aebi, U., J. Cohn, L. Buhle and L. Gerace. 1986. The nuclear lamina is a meshwork of intermediate-type filaments. *Nature* 323(6088): 560–4.

Akey, C.W. 1989. Interactions and structure of the nuclear pore complex revealed by cryo-electron microscopy. *J. Cell Biol.* 109(3): 955–70.

Barbee, K.A., P.F. Davies and R. Lal. 1994. Shear stress-induced reorganization of the surface topography of living endothelial cells imaged by atomic force microscopy. *Circ. Res.* 74(1): 163–71.

Bassett, A., S. Cooper, C. Wu and A. Travers. 2009. The folding and unfolding of eukaryotic chromatin. *Curr. Opin. Genet. Dev.* 19(2): 159–65.

Bennink, M.L., S.H. Leuba, G.H. Leno, J. Zlatanova, B.G. de Grooth and J. Greve. 2001. Unfolding individual nucleosomes by stretching single chromatin fibers with optical tweezers. *Nat. Struct. Mol. Biol.* 8(7): 606–10.

Brooks, A.R., P.I. Lelkes and G.M. Rubanyi. 2002. Gene expression profiling of human aortic endothelial cells exposed to disturbed flow and steady laminar flow. *Physiol. Genomics* 9(1): 27–41.

Brooks, A.R., P.I. Lelkes and G.M. Rubanyi. 2004. Gene expression profiling of vascular endothelial cells exposed to fluid mechanical forces: relevance for focal susceptibility to atherosclerosis. *Endothelium* 11(1): 45–57.

Buehler, M.J. and T. Ackbarow. 2008. Nanomechanical strength mechanisms of hierarchical biological materials and tissues. *Comput. Meth. Biomech. Biomed. Eng.* 11(6): 595–607.

Burri, P.H., R. Hlushchuk and V. Djonov. 2004. Intussusceptive angiogenesis: its emergence, its characteristics, and its significance. *Dev. Dyn.* 231(3): 474–88.

Bussolati, B., C. Dunk, M. Grohman, C.D. Kontos, J. Mason and A. Ahmed. 2001. Vascular endothelial growth factor receptor-1 modulates vascular endothelial growth factor-mediated angiogenesis via nitric oxide. *Am. J. Pathol.* 159(3): 993–1008.

Caille, N., O. Thoumine, Y. Tardy and J.J. Meister. 2002. Contribution of the nucleus to the mechanical properties of endothelial cells. *J. Biomech.* 35(2): 177–87.

Cao, Y.H., J. Arbiser, R.J. D'Amato, P.A. D'Amore, D.E. Ingber, R. Kerbel, M. Klagsbrun, S. Lim, M.A. Moses, B. Zetter, H. Dvorak and R. Langer. 2011. Forty-Year Journey of Angiogenesis Translational Research. *Sci. T.M.* 3(114).

Carmeliet, P. and R.K. Jain. 2011. Molecular mechanisms and clinical applications of angiogenesis. *Nature* 473(7347): 298–307.

Cebe-Suarez, S., A. Zehnder-Fjallman and K. Ballmer-Hofer. 2006. The role of VEGF receptors in angiogenesis; complex partnerships. *Cell Mol. Life Sci.* 63(5): 601–15.

Chappell, D.C., S.E. Varner, R.M. Nerem, R.M. Medford and R.W. Alexander. 1998. Oscillatory shear stress stimulates adhesion molecule expression in cultured human endothelium. *Circ. Res.* 82(5): 532–9.

Chatzizis, Y.S., A.U. Coskun, M. Jonas, E.R. Edelman, C.L. Feldman and P.H. Stone. 2007. Role of endothelial shear stress in the natural history of coronary atherosclerosis and vascular remodeling: molecular, cellular, and vascular behavior. *J. Am. Coll. Cardiol.* 49(25): 2379–93.

Chen, Y.L., K.M. Jan, H.S. Lin and S. Chien. 1995. Ultrastructural studies on macromolecular permeability in relation to endothelial cell turnover. *Atherosclerosis* 118(1): 89–104.

Chien, S. 2003. Molecular and mechanical bases of focal lipid accumulation in arterial wall. *Prog. Biophys. Mol. Biol.* 83(2): 131–51.

Chien, S. 2006. Mechanical and chemical regulation of endothelial cell polarity. *Circ. Res.* 98(7): 863–5.

Crisp, M., Q. Liu, K. Roux, J.B. Rattner, C. Shanahan, B. Burke, P.D. Stahl and D. Hodzic. 2006. Coupling of the nucleus and cytoplasm: role of the LINC complex. *J. Cell. Sci.* 172(1): 41–53.

Cui, Y. and C. Bustamante. 2000. Pulling a single chromatin fiber reveals the forces that maintain its higher-order structure. *P. Natl. Acad. Sci. USA* 97(1): 127–32.

Dahl, K.N., S.M. Kahn, K.L. Wilson and D.E. Discher. 2004. The nuclear envelope lamina network has elasticity and a compressibility limit suggestive of a molecular shock absorber. *J. Cell Sci.* 117(Pt 20): 4779–86.

Dahl, K.N., P. Scaffidi, M.F. Islam, A.G. Yodh, K.L. Wilson and T. Misteli. 2006. Distinct structural and mechanical properties of the nuclear lamina in Hutchinson-Gilford progeria syndrome. *P. Natl. Acad. Sci. USA* 103(27): 10271–6.

Dahl, K.N., A.J. Ribeiro and J. Lammerding. 2008. Nuclear shape, mechanics, and mechanotransduction. *Circ. Res.* 102(11): 1307–18.

Dahl, K.N., A. Kalinowski and K. Pekkan. 2010. Mechanobiology and the microcirculation: cellular, nuclear and fluid mechanics. *Microcirculation* 17(3): 179–91.

Davies, P.F. 1995. Flow-mediated endothelial mechanotransduction. *Physiol. Rev.* 75(3): 519–60.

De Caterina, R., P. Libby, H.B. Peng, V.J. Thannickal, T.B. Rajavashisth, M.A. Gimbrone, Jr., W.S. Shin and J.K. Liao. 1995. Nitric oxide decreases cytokine-induced endothelial activation. Nitric oxide selectively reduces endothelial expression of adhesion molecules and proinflammatory cytokines. *J. Clin. Invest.* 96(1): 60–8.

De Sandre-Giovannoli, A., R. Bernard, P. Cau, C. Navarro, J. Amiel, I. Boccaccio, S. Lyonnet, C.L. Stewart, A. Munnich, M. Le Merrer and N. Levy. 2003. Lamin a truncation in Hutchinson-Gilford progeria. *Science* 300(5628): 2055.

Deguchi, S., K. Maeda, T. Ohashi and M. Sato. 2005. Flow-induced hardening of endothelial nucleus as an intracellular stress-bearing organelle. *J. Biomech.* 38(9): 1751–9.

Delbarre, E., M. Tramier, M. Coppey-Moisan, C. Gaillard, J.C. Courvalin and B. Buendia. 2006. The truncated prelamin A in Hutchinson-Gilford progeria syndrome alters segregation of A-type and B-type lamin homopolymers. *Hum. Mol. Genet.* 15(7): 1113–22.

Djonov, V., M. Schmid, S.A. Tschanz and P.H. Burri. 2000. Intussusceptive angiogenesis: its role in embryonic vascular network formation. *Circ. Res.* 86(3): 286–92.

Dundr, M. and T. Misteli. 2001. Functional architecture in the cell nucleus. *Biochem. J.* 356(Pt 2): 297–310.

Enrico, B., P. Suranyi, C. Thilo, L. Bonomo, P. Costello and U.J. Schoepf. 2009. Coronary artery plaque formation at coronary CT angiography: morphological analysis and relationship to hemodynamics. *Eur. Radiol.* 19(4): 837–44.

Eriksson, M., W.T. Brown, L.B. Gordon, M.W. Glynn, J. Singer, L. Scott, M.R. Erdos, C.M. Robbins, T.Y. Moses, P. Berglund, A. Dutra, E. Pak, S. Durkin, A.B. Csoka, M. Boehnke, T.W. Glover and F.S. Collins. 2003. Recurrent *de novo* point mutations in lamin A cause Hutchinson-Gilford progeria syndrome. *Nature* 423(6937): 293–8.

Esser, S., M.G. Lampugnani, M. Corada, E. Dejana and W. Risau. 1998. Vascular endothelial growth factor induces VE-cadherin tyrosine phosphorylation in endothelial cells. *J. Cell Sci.* 111(Pt 13): 1853–65.

Ferrara, N. and T. Davis-Smyth. 1997. The biology of vascular endothelial growth factor. *Endocr. Rev.* 18(1): 4–25.

Gant, T.M. and K.L. Wilson. 1997. Nuclear assembly. *Annu. Rev. Cell. Dev. Biol.* 13: 669–95.

Georgatos, S.D., Y. Markaki, A. Christogianni and A.S. Politou. 2009. Chromatin remodeling during mitosis: a structure-based code? *Front. Biosci.* 14: 2017–27.

Gilbert, N., S. Boyle, H. Fiegler, K. Woodfine, N.P. Carter and W.A. Bickmore. 2004. Chromatin architecture of the human genome: gene-rich domains are enriched in open chromatin fibers. *Cell* 118(5): 555–66.

Goldberg, M.W., I. Huttenlauch, C.J. Hutchison and R. Stick. 2008. Filaments made from A- and B-type lamins differ in structure and organization. *J. Cell Sci.* 121(Pt 2): 215–25.

Goldman, R.D., D.K. Shumaker, M.R. Erdos, M. Eriksson, A.E. Goldman, L.B. Gordon, Y. Gruenbaum, S. Khuon, M. Mendez, R. Varga and F.S. Collins. 2004. Accumulation of mutant lamin A causes progressive changes in nuclear architecture in Hutchinson-Gilford progeria syndrome. *P. Natl. Acad. Sci. USA* 101(24): 8963–8.

Goldman, R.D., B. Grin, M.G. Mendez and E.R. Kuczmarski. 2008. Intermediate filaments: versatile building blocks of cell structure. *Curr. Opin. Cell Biol.* 20(1): 28–34.

Gong, C., K.V. Stoletov and B.I. Terman. 2004. VEGF treatment induces signaling pathways that regulate both actin polymerization and depolymerization. *Angiogenesis* 7(4): 313–21.

Griffin, G.K., G. Newton, M.L. Tarrio, D.X. Bu, E. Maganto-Garcia, V. Azcutia, P. Alcaide, N. Grabie, F.W. Luscinskas, K.J. Croce and A.H. Lichtman. 2012. IL-17 and TNF-alpha Sustain Neutrophil Recruitment during Inflammation through Synergistic Effects on Endothelial Activation. *J. Immunol.*

Gruenbaum, Y., K.L. Wilson, A. Harel, M. Goldberg and M. Cohen. 2000. Review: nuclear lamins—structural proteins with fundamental functions. *J. Struct. Biol.* 129(2–3): 313–23.

Gruenbaum, Y., A. Margalit, R.D. Goldman, D.K. Shumaker and K.L. Wilson. 2005. The nuclear lamina comes of age. *Nat. Rev. Mol. Cell Biol.* 6(1): 21–31.

Harborth, J., S.M. Elbashir, K. Bechert, T. Tuschl and K. Weber. 2001. Identification of essential genes in cultured mammalian cells using small interfering RNAs. *J. Cell Sci.* 114(Pt 24): 4557–65.

Harrison, D.G., J. Widder, I. Grumbach, W. Chen, M. Weber and C. Searles. 2006. Endothelial mechanotransduction, nitric oxide and vascular inflammation. *J. Intern. Med.* 259(4): 351–63.

Hazel, A.L. and T.J. Pedley. 2000. Vascular endothelial cells minimize the total force on their nuclei. *Biophys. J.* 78(1): 47–54.

Hegner, M. and W. Grange. 2002. Mechanics and imaging of single DNA molecules. *J. Muscle Res. Cell Motil.* 23(5–6): 367–75.

Heitlinger, E., M. Peter, A. Lustig, W. Villiger, E.A. Nigg and U. Aebi. 1992. The role of the head and tail domain in lamin structure and assembly: analysis of bacterially expressed chicken lamin A and truncated B2 lamins. *J. Struct. Biol.* 108(1): 74–89.

Herrmann, H. and R. Foisner. 2003. Intermediate filaments: novel assembly models and exciting new functions for nuclear lamins. *Cell Mol. Life Sci.* 60(8): 1607–12.

Herrmann, H., M. Hesse, M. Reichenzeller, U. Aebi and T.M. Magin. 2003. Functional complexity of intermediate filament cytoskeletons: from structure to assembly to gene ablation. *Int. Rev. Cytol.* 223: 83–175.

Hillen, F. and A.W. Griffioen. 2007. Tumour vascularization: sprouting angiogenesis and beyond. *Cancer Metastasis Rev.* 26(3–4): 489–502.

Himburg, H.A., D.M. Grzybowski, A.L. Hazel, J.A. LaMack, X.M. Li and M.H. Friedman. 2004. Spatial comparison between wall shear stress measures and porcine arterial endothelial permeability. *Am. J. Physiol. Heart Circ. Physiol.* 286(5): H1916–22.

Hohenadl, M., T. Storz, H. Kirpal, K. Kroy and R. Merkel. 1999. Desmin filaments studied by quasi-elastic light scattering. *Biophys. J.* 77(4): 2199–209.

Houchmandzadeh, B., J.F. Marko, D. Chatenay and A. Libchaber. 1997. Elasticity and structure of eukaryote chromosomes studied by micromanipulation and micropipette aspiration. *J. Cell Biol.* 139(1): 1–12.

Hu, Y., I. Kireev, M. Plutz, N. Ashourian and A.S. Belmont. 2009. Large-scale chromatin structure of inducible genes: transcription on a condensed, linear template. *J. Cell Biol.* 185(1): 87–100.

Huang, Z. and S.D. Bao. 2004. Roles of main pro-and anti-angiogenic factors in tumor angiogenesis. *World J. Gastroentero.* 10(4): 463–470.

Ketema, M., K. Wilhelmsen, I. Kuikman, H. Janssen, D. Hodzic and A. Sonnenberg. 2007. Requirements for the localization of nesprin-3 at the nuclear envelope and its interaction with plectin. *J. Cell Sci.* 120(Pt 19): 3384–94.

Khakpour, M. and K. Vafai. 2008. Critical assessment of arterial transport models. *International Journal of Heat and Mass Transfer* 51: 807–822.

Kiec-Wilk, B., J. Grzybowska-Galuszka, A. Polus, J. Pryjma, A. Knapp and K. Kristiansen. 2010. The MAPK-dependent regulation of the Jagged/Notch gene expression by VEGF, bFGF or PPAR gamma mediated angiogenesis in HUVEC. *J. Physiol. Pharmacol.* 61(2): 217–25.

Kliche, S. and J. Waltenberger. 2001. VEGF receptor signaling and endothelial function. *IUBMB Life* 52(1–2): 61–6.

Korf, B. 2008. Focus on research: Hutchinson-Gilford progeria syndrome, aging, and the nuclear lamina. *New Engl. J. Med.* 358(6): 552–555.

Kudlow, B.A., B.K. Kennedy and R.J. Monnat, Jr. 2007. Werner and Hutchinson-Gilford progeria syndromes: mechanistic basis of human progeroid diseases. *Nat. Rev. Mol. Cell Biol.* 8(5): 394–404.

Lamalice, L., F. Le Boeuf and J. Huot. 2007. Endothelial cell migration during angiogenesis. *Circ. Res.* 100(6): 782–94.

Lammerding, J., P.C. Schulze, T. Takahashi, S. Kozlov, T. Sullivan, R.D. Kamm, C.L. Stewart and R.T. Lee. 2004. Lamin A/C deficiency causes defective nuclear mechanics and mechanotransduction. *J. Clin. Invest.* 113(3): 370–8.

Lammerding, J., L.G. Fong, J.Y. Ji, K. Reue, C.L Stewart, S.G. Young and R.T. Lee. 2006. Lamins A and C but not lamin B1 regulate nuclear mechanics. *J. Biol. Chem.* 281(35): 25768–80.

Lanctot, C., T. Cheutin, M. Cremer, G. Cavalli and T. Cremer. 2007. Dynamic genome architecture in the nuclear space: regulation of gene expression in three dimensions. *Nat. Rev. Genet.* 8(2): 104–15.

Lehoux, S., Y. Castier and A. Tedgui. 2006. Molecular mechanisms of the vascular responses to haemodynamic forces. *J. Intern. Med.* 259(4): 381–92.

Leung, D.W., G. Cachianes, W.J. Kuang, D.V. Goeddel and N. Ferrara. 1989. Vascular Endothelial Growth-Factor Is a Secreted Angiogenic Mitogen. *Science* 246(4935): 1306–1309.

Libby, P. 2002. Inflammation in atherosclerosis. *Nature* 420(6917): 868–74.

Libby, P. 2006. Inflammation and cardiovascular disease mechanisms. *Am. J. Clin. Nutr.* 83(2): 456S–460S.

Liu, Y., B.P. Chen, M. Lu, Y. Zhu, M.B. Stemerman, S. Chien and J.Y. Shyy. 2002. Shear stress activation of SREBP1 in endothelial cells is mediated by integrins. *Arterioscler. Thromb. Vasc. Biol.* 22(1): 76–81.

Liu, Z., J.L. Tan, D.M. Cohen, M.T. Yang, N.J. Sniadecki, S.A. Ruiz, C.M. Nelson and C.S. Chen. 2010. Mechanical tugging force regulates the size of cell-cell junctions. *P. Natl. Acad. Sci. USA* 107(22): 9944–9.

Lutomska, A., A. Lebedev, K. Scharffetter-Kochanek and S. Iben. 2008. The transcriptional response to distinct growth factors is impaired in Werner syndrome cells. *Exp. Gerontol.* 43(9): 820–6.

Mammoto, A., K.M. Connor, T. Mammoto, C.W. Yung, D. Huh, C.M. Aderman, G. Mostoslavsky, L.E. Smith and D.E. Ingber. 2009. A mechanosensitive transcriptional mechanism that controls angiogenesis. *Nature* 457(7233): 1103–8.

Maniotis, A.J., C.S. Chen and D.E. Ingber. 1997. Demonstration of mechanical connections between integrins, cytoskeletal filaments, and nucleoplasm that stabilize nuclear structure. *P. Natl. Acad. Sci. USA* 94(3): 849–54.

Marko, J.F. and M.G. Poirier. 2003. Micromechanics of chromatin and chromosomes. *Biochem. Cell Biol.* 81(3): 209–20.

Matsumoto, T. and L. Claesson-Welsh. 2001. VEGF receptor signal transduction. *Sci. STKE* (112): re21.

Matsumoto, T. and H. Mugishima. 2006. Signal transduction via vascular endothelial growth factor (VEGF) receptors and their roles in atherogenesis. *J. Atheroscler. Thromb.* 13(3): 130–5.

Mattout-Drubezki, A. and Y. Gruenbaum. 2003. Dynamic interactions of nuclear lamina proteins with chromatin and transcriptional machinery. *Cell Mol. Life Sci.* 60(10): 2053–63.

McCormick, S.M., S.G. Eskin, L.V. McIntire, C.L. Teng, C.M. Lu, C.G. Russell and K.K. Chittur. 2001. DNA microarray reveals changes in gene expression of shear stressed human umbilical vein endothelial cells. *P. Natl. Acad. Sci. USA* 98(16): 8955–60.

Misteli, T. 2007. Beyond the sequence: cellular organization of genome function. *Cell* 128(4): 787–800.

Moir, R.D., A.D. Donaldson and M. Stewart. 1991. Expression in Escherichia coli of human lamins A and C: influence of head and tail domains on assembly properties and paracrystal formation. *J. Cell Sci.* 99(Pt 2): 363–72.

Moir, R.D., M. Yoon, S. Khuon and R.D. Goldman. 2000. Nuclear lamins A and B1: different pathways of assembly during nuclear envelope formation in living cells. *J. Cell Biol.* 151(6): 1155–68.

Moir, R.D., M. Yoon, S. Khuon and R.D. Goldman. 2000. Nuclear lamins A and B1: different pathways of assembly during nuclear envelope formation in living cells. *J. Cell Biol.* 151(6): 1155–68.

Mounkes, L.C., S. Kozlov, L. Hernandez, T. Sullivan and C.L. Stewart. 2003. A progeroid syndrome in mice is caused by defects in A-type lamins. *Nature* 423(6937): 298–301.

Nelson, C.M. and C.S. Chen. 2003. VE-cadherin simultaneously stimulates and inhibits cell proliferation by altering cytoskeletal structure and tension. *J. Cell Sci.* 116(Pt 17): 3571–81.

Nelson, C.M., D.M. Pirone, J.L. Tan and C.S. Chen. 2004. Vascular endothelial-cadherin regulates cytoskeletal tension, cell spreading, and focal adhesions by stimulating RhoA. *Mol. Biol. Cell* 15(6): 2943–53.

Ng, E.W.M. and A.P. Adamis. 2005. Targeting angiogenesis, the underlying disorder in neovascular age-related macular degeneration. *Can. J. Opthalmol.* 40(3): 352–368.

Ohura, N., K. Yamamoto, S. Ichioka, T. Sokabe, H. Nakatsuka, A. Baba, M. Shibata, T. Nakatsuka, K. Harii, Y. Wada, T. Kohro, T. Kodama and J. Ando. 2003. Global analysis of shear stress-responsive genes in vascular endothelial cells. *J. Atheroscler. Thromb.* 10(5): 304–13.

Orr, A.W., J.M. Sanders, M. Bevard, E. Coleman, I.J. Sarembock and M.A. Schwartz. 2005. The subendothelial extracellular matrix modulates NF-kappaB activation by flow: a potential role in atherosclerosis. *J. Cell Biol.* 169(1): 191–202.

Ostashevsky, J. 2002. A polymer model for large-scale chromatin organization in lower eukaryotes. *Mol. Biol. Cell* 13(6): 2157–69.

Papadaki, M. and S.G. Eskin. 1997. Effects of fluid shear stress on gene regulation of vascular cells. *Biotechnol. Prog.* 13(3): 209–21.

Petzold, T., A.W. Orr, C. Hahn, K. Jhaveri, J.T. Parsons and M.A. Schwartz. 2009. Focal adhesion kinase modulates activation of NF-{kappa}B by flow in endothelial cells. *Am. J. Physiol. Cell Physiol.*

Philip, J.T. and K.N. Dahl. 2008. Nuclear mechanotransduction: response of the lamina to extracellular stress with implications in aging. *J. Biomech.* 41(15): 3164–70.

Plank, M.J., D.J. Wall and T. David. 2007. The role of endothelial calcium and nitric oxide in the localisation of atherosclerosis. *Math Biosci.* 207(1): 26–39.

Poepping, T.L., R.N. Rankin and D.W. Holdsworth. 2010. Flow patterns in carotid bifurcation models using pulsed Doppler ultrasound: effect of concentric vs. eccentric stenosis on turbulence and recirculation. *Ultrasound Med. Biol.* 36(7): 1125–34.

Poirier, M., S. Eroglu, D. Chatenay and J.F. Marko. 2000. Reversible and irreversible unfolding of mitotic newt chromosomes by applied force. *Mol. Biol. Cell* 11(1): 269–76.

Poirier, M.G., T. Monhait and J.F. Marko. 2002. Reversible hypercondensation and decondensation of mitotic chromosomes studied using combined chemical-micromechanical techniques. *J. Cell Biochem.* 85(2): 422–34.

Qazi, Y., S. Maddula and B.K. Ambati. 2009. Mediators of ocular angiogenesis. *J. Genet.* 88(4): 495–515.

Razafsky, D. and D. Hodzic. 2009. Bringing KASH under the SUN: the many faces of nucleo-cytoskeletal connections. *J. Cell Biol.* 186(4): 461–72.

Reinhart-King, C.A., M. Dembo and D.A. Hammer. 2005. The dynamics and mechanics of endothelial cell spreading. *Biophys. J.* 89(1): 676–89.

Resnick, N., H. Yahav, A. Shay-Salit, M. Shushy, S. Schubert, L.C. Zilberman and E. Wofovitz. 2003. Fluid shear stress and the vascular endothelium: for better and for worse. *Prog. Biophys. Mol. Biol.* 81(3): 177–99.

Richmond, T.J., J.T. Finch, B. Rushton, D. Rhodes and A. Klug. 1984. Structure of the nucleosome core particle at 7 A resolution. *Nature* 311(5986): 532–7.

Risau, W. 1997. Mechanisms of angiogenesis. *Nature* 386(6626): 671–4.

Rouleau, L., M. Farcas, J.C. Tardif, R. Mongrain and R.L. Leask. 2010. Endothelial cell morphologic response to asymmetric stenosis hemodynamics: effects of spatial wall shear stress gradients. *J. Biomech. Eng.* 132(8): 081013.

Rousseau, S., F. Houle, J. Landry and J. Huot. 1997. p38 MAP kinase activation by vascular endothelial growth factor mediates actin reorganization and cell migration in human endothelial cells. *Oncogene* 15(18): 2169–77.

Roux, K.J., M.L. Crisp, Q. Liu, D. Kim, S. Kozlov, C.L Stewart and B. Burke. 2009. Nesprin 4 is an outer nuclear membrane protein that can induce kinesin-mediated cell polarization. *P. Natl. Acad. Sci. USA* 106(7): 2194–9.

Segura-Totten, M., A.K. Kowalski, R. Craigie and K.L. Wilson. 2002. Barrier-to-autointegration factor: major roles in chromatin decondensation and nuclear assembly. *J. Cell Biol.* 158(3): 475–85.

Shen, C.J., S. Raghavan, Z. Xu, J.D. Baranski, X. Yu, M.A. Wozniak, J.S. Miller, M. Gupta, L. Buckbinder and C.S. Chen. 2011. Decreased cell adhesion promotes angiogenesis in a Pyk2-dependent manner. *Exp. Cell Res.* 317(13): 1860–71.

Shiu, Y.T., J.A. Weiss, J.B. Hoying, M.N. Iwamoto, I.S. Joung and C.T. Quam. 2005. The role of mechanical stresses in angiogenesis. *Crit. Rev. Biomed. Eng.* 33(5): 431–510.

Sims, J.R., S. Karp and D.E. Ingber. 1992. Altering the cellular mechanical force balance results in integrated changes in cell, cytoskeletal and nuclear shape. *J. Cell Sci.* 103(Pt 4): 1215–22.

Stierle, V., J. Couprie, C. Ostlund, I. Krimm, S. Zinn-Justin, P. Hossenlopp, H.J. Worman, J.C. Courvalin and I. Duband-Goulet. 2003. The carboxyl-terminal region common to lamins A and C contains a DNA binding domain. *Biochem.* 42(17): 4819–28.

Stoletov, K.V., C. Gong and B.I. Terman. 2004. Nck and Crk mediate distinct VEGF-induced signaling pathways that serve overlapping functions in focal adhesion turnover and integrin activation. *Exp. Cell Res.* 295(1): 258–68.

Stroka, K.M. and H. Aranda-Espinoza. 2011. Effects of Morphology vs. Cell-Cell Interactions on Endothelial Cell Stiffness. *Cell Mol. Bioeng.* 4(1): 9–27.

Taddei, A., C. Giampietro, A. Conti, F. Orsenigo, F. Breviario, V. Pirazzoli, M. Potente, C. Daly, S. Dimmeler and E. Dejana. 2008. Endothelial adherens junctions control tight junctions by VE-cadherin-mediated upregulation of claudin-5. *Nat. Cell Biol.* 10(8): 923–34.

Thomas, C.H., J.H. Collier, C.S. Sfeir and K.E. Healy. 2002. Engineering gene expression and protein synthesis by modulation of nuclear shape. *P. Natl. Acad. Sci. USA* 99(4): 1972–7.

Traub, O. and B.C. Berk. 1998. Laminar shear stress: mechanisms by which endothelial cells transduce an atheroprotective force. *Arterioscler. Thromb. Vasc. Biol.* 18(5): 677–85.

Tricot, O., Z. Mallat, C. Heymes, J. Belmin, G. Leseche and A. Tedgui. 2000. Relation between endothelial cell apoptosis and blood flow direction in human atherosclerotic plaques. *Circulation* 101(21): 2450–3.

Tzima, E., M.A. Del Pozo, W.B. Kiosses, S.A. Mohamed, S. Li, S. Chien and M.A. Schwartz. 2002. Activation of Rac1 by shear stress in endothelial cells mediates both cytoskeletal reorganization and effects on gene expression. *Embo J.* 21(24): 6791–800.

Tzima, E., W.B. Kiosses, M.A. del Pozo and M.A. Schwartz. 2003. Localized cdc42 activation, detected using a novel assay, mediates microtubule organizing center positioning in endothelial cells in response to fluid shear stress. *J. Biol. Chem.* 278(33): 31020–3.

Vasava, P., P. Jalali, M. Dabagh and P.J. Kolari. 2012. Finite element modelling of pulsatile blood flow in idealized model of human aortic arch: study of hypotension and hypertension. *Comput. Math Methods Med.* 2012: 861837.

Vasir, B., J.C. Jonas, G.M. Steil, J. Hollister-Lock, W. Hasenkamp, A. Sharma, S. Bonner-Weir and G.C. Weir. 2001. Gene expression of VEGF and its receptors Flk-1/KDR and Flt-1 in cultured and transplanted rat islets. *Transplantation* 71(7): 924–35.

Versaevel, M., T. Grevesse and S. Gabriele. 2012. Spatial coordination between cell and nuclear shape within micropatterned endothelial cells. *Nat. Commun.* 3: 671.

Verstraeten, V.L., J.L. Broers, F.C. Ramaekers and M.A. van Steensel. 2007. The nuclear envelope, a key structure in cellular integrity and gene expression. *Curr. Med. Chem.* 14(11): 1231–48.

Vestweber, D., M. Winderlich, G. Cagna and A.F. Nottebaum. 2009. Cell adhesion dynamics at endothelial junctions: VE-cadherin as a major player. *Trends Cell Biol.* 19(1): 8–15.

Wary, K.K., G.D. Thakker, J.O. Humtsoe and J. Yang. 2003. Analysis of VEGF-responsive genes involved in the activation of endothelial cells. *Mol. Cancer* 2: 25.

Webber, M.J., J. Tongers, C.J. Newcomb, K.T. Marquardt, J. Bauersachs, D.W. Losordo and S.I. Stupp. 2011. Supramolecular nanostructures that mimic VEGF as a strategy for ischemic tissue repair. *P. Natl. Acad. Sci. USA* 108(33): 13438–13443.

Weston, G.C., I. Haviv and P.A. Rogers. 2002. Microarray analysis of VEGF-responsive genes in myometrial endothelial cells. *Mol. Hum. Reprod.* 8(9): 855–63.

White, C.R., M. Haidekker, X. Bao and J.A. Frangos. 2001. Temporal gradients in shear, but not spatial gradients, stimulate endothelial cell proliferation. *Circulation* 103(20): 2508–13.

Worman, H.J. 2005. Components of the nuclear envelope and their role in human disease. *Novart. Fdn. Symp.* 264: 35–42; discussion 42–50, 227–30.

Worman, H.J. and G. Bonne. 2007. "Laminopathies": a wide spectrum of human diseases. *Exp. Cell Res.* 313(10): 2121–33.

Yang, S., K. Toy, G. Ingle, C. Zlot, P.M. Williams, G. Fuh, B. Li, A. de Vos and M.E. Gerritsen. 2002. Vascular endothelial growth factor-induced genes in human umbilical vein endothelial cells: relative roles of KDR and Flt-1 receptors. *Arterioscler. Thromb. Vasc. Biol.* 22(11): 1797–803.

Ye, Q., I. Callebaut, A. Pezhman, J.C. Courvalin and H.J. Worman. 1997. Domain-specific interactions of human HP1-type chromodomain proteins and inner nuclear membrane protein LBR. *J. Biol. Chem.* 272(23): 14983–9.

Zarins, C.K., D.P. Giddens, B.K. Bharadvaj, V.S. Sottiurai, R.F. Mabon and S. Glagov 1983. Carotid bifurcation atherosclerosis. Quantitative correlation of plaque localization with flow velocity profiles and wall shear stress. *Circ. Res.* 53(4): 502–14.

Zastrow, M.S., S. Vlcek and K.L. Wilson. 2004. Proteins that bind A-type lamins: integrating isolated clues. *J. Cell Sci.* 117(Pt 7): 979–87.

Ziegler, T., K. Bouzourene, V.J. Harrison, H.R. Brunner and D. Hayoz. 1998. Influence of oscillatory and unidirectional flow environments on the expression of endothelin and nitric oxide synthase in cultured endothelial cells. *Arterioscler. Thromb. Vasc. Biol.* 18(5): 686–92.

Cancer Metastasis and Biomechanics of the Endothelium

Claudia T. Mierke

ABSTRACT

The malignancy of tumors and subsequently, most cancer-related deaths are caused by the capability of cancer cells to metastasize. The process of metastasis includes the dissemination of cancer cells from the primary tumor and their migration through the extracellular matrix to targeted organ sites. During the migration of cancer cells through the connective tissue microenvironment, which consists of endothelial cells and extracellular matrix components, cellular and extracellular biomechanical properties are crucial for the efficiency and speed of cancer cell invasion and subsequently, metastases formation. Biomechanical properties may enable cancer cells to migrate through tissue, transmigrate through basement membranes or endothelial monolayers and form metastases in targeted organs.

The current focus of cancer research still lies on the investigation of cancer cell's biochemical and molecular capabilities such as molecular genetics and gene signaling, but unfortunately these approaches

Faculty of Physics and Earth Science, Institute of Experimental Physics I, Soft Matter Physics Division, University of Leipzig, Linnéstr. 5, 04103 Leipzig, Germany.
Email: claudia.mierke@t-online.de

ignore the mechanical nature of the invasion process of cancer cells. In addition, even the role of the endothelium during the transmigration and invasion of cells remains unclear, as it has only been seen as a passive barrier, which could not explain all recent novel findings. This chapter discusses how cancer cells alter the structural, biochemical and mechanical properties of the endothelium to regulate their own invasiveness through extracellular matrices and hence, through the tissue microenvironment. Finally, this chapter sheds light on the mechanical properties of cancer cells as well as the interacting endothelium and points out the importance of the mechanical properties as a critical determinant for the efficiency of cancer cell invasion and the overall progression of cancer. Taken together, the regulation of the endothelial cell's biomechanical properties by cancer cells is a critical determinant of cancer cell invasiveness and may affect the future development of new cancer treatments.

Keywords: contractile forces, focal adhesions, cellular stiffness, cytoskeletal remodeling, cellular deformability, cell–cell adhesions

Introduction

The malignancy of the cancer disease depends mainly on the size of the primary tumor, its location, and the capability of single or clustered cancer cells to spread from the primary tumor, invade into the surroundings and finally to metastasize, which is indicated by the formation of secondary tumors in targeted organs. This process of metastasis formation has long been seen as a variety of genetic alterations that had occurred during the progression of cancer disease. The focus of molecular and cellular biological research was on the investigation of numerous mutants that showed altered cell motility in artificial extracellular matrices. The first motility assays have been performed on pure plastic or glass substrates, and they have been further improved to assays using thin layers of extracellular matrix (ECM) proteins as a surface-coating on top of the planar substrates. Then, motility of cells through a mesh of defined "undeformable" stabile pores (named transwell assays or Boyden Chamber assays) has been studied. As these artificial *in vitro* matrices are still far away from mimicking *in vivo* ECMs or connective tissue perfectly, they are still useful to investigate the mechanical properties, which are adjustable within these ECMs to obtain the mechanical properties of different tissue types. These in two dimensions performed motility assays have been further developed by using three-dimensional (3D) ECMs, which contain mainly collagen fibers or matrigel constituents (ECM extracted from tumor lesions) and have "deformable or bendable" pores with adjustable size suitable for each cell type studied. These artificial *in vitro* ECMs are still far away from rebuilding *in vivo* ECMs or connective tissue, but they are a useful tool to investigate the mechanical properties of the cell's surroundings, which are adjustable within these ECMs over a wide range to mimic the mechanical properties of different tissue types or even the enhanced stiffness of tumor tissue. Unfortunately, the mechanical properties of these ECM substrates have over a long time period been simply ignored. Moreover, even the mechanical properties of migratory cells or cells, which represent a barrier for migratory cells such as endothelial cells building the inner lining of blood or lymph vessels have not been studied. In the last few years, the mechanical properties of the (cancer) cells have become the focus of recent research studies (Zaman et al. 2006, Mierke et al. 2008a, Fritsch et al. 2010, Yu et al. 2011, Mierke et al. 2011a, Guck et al. 2005). Several studies analyzed (cancer) cells in connective tissue where they are exposed to forces such as protrusive forces, contractile forces, compressive forces, tensile forces, shear forces of the vessel stream (Ingber 2006, Paszek et al. 2009, Weaver et al. 2002). These forces fulfill prominent functions in cellular and tissue shaping, organ development as well as in maintenance and homeostasis of the tissue (Paszek et al. 2005, DuFort et al. 2011). The focus of these studies was on how (cancer) cells sense and respond to these forces, which seems to be set by the biomechanical properties of the (cancer) cells, their adjacent cells and the ECM (Ingber 2006, Paszek et al. 2009, Du Fort et al. 2011, Miteva et al. 2010). In a recent study, the mechanical properties of human endothelial cells have been investigated (Mierke 2011a). In more detail, alterations of endothelial mechanical properties by certain breast, bladder and kidney cancer cells have been observed (Mierke 2011a). Additionally the mechanical properties of the ECM and ECM-embedded cells such as endothelial cells play a prominent role in malignant cancer progression. As the surrounding tissue of neoplasms and tumors is altered in terms of the mechanical properties compared to "normal" tissue of healthy species, the mechanical properties of cancer cells such as the stiffness (the inverse of the compliance) and contractile forces may also be affected and become the focus of biophysical research groups.

In summary, this chapter points out whether the process of metastasis formation and especially the the transendothelial migration process of cancer cells depends on alterations of mechanical properties. In addition, it discusses how the cancer cells regulate the mechanical properties of endothelial cells during co-culture. Finally this chapter focusses on the biomechanics of the endothelium and its impact on the transendothelial migration, the invasiveness of cancer cells, and subsequently the capability of cancer cells to metastasize.

Cancer Metastasis

The steps of cancer metastasis

Many cancer-related deaths are due to the ability of the tumors to form metastasis rather than due to invasive growth of the primary tumor, which is often detected and removed by surgery. The problem is to decide whether metastasis formation will occur after surgery or not. There are several steps during the metastasis process which are known. These steps are the dissemination of single cancer cells from the primary tumor, their migration into connective tissue, the transmigration of these cancer cells into blood or lymph vessels (intravasation), their transport through the whole vascular or lymphoid system to targeted organs (Eger and Mikulitis 2005), possibly their adhesion to endothelial cells and growth inside the vessels (Al-Mehdi et al. 2000) or their adhesion and transmigration through the endothelial cell lining of the vessels (extravasation, Chambers et al. 2002), their invasion through the basal membrane (underneath the endothelium) into the connective tissue and the formation and growth of a secondary tumor in a targeted organ (Fig. 1). As many of these steps require mechanical properties of invasive cancer cells and the endothelium, the investigation of these biophysical parameters seems to be a key step to understand what is going on during the process of metastasis formation (Fig. 2).

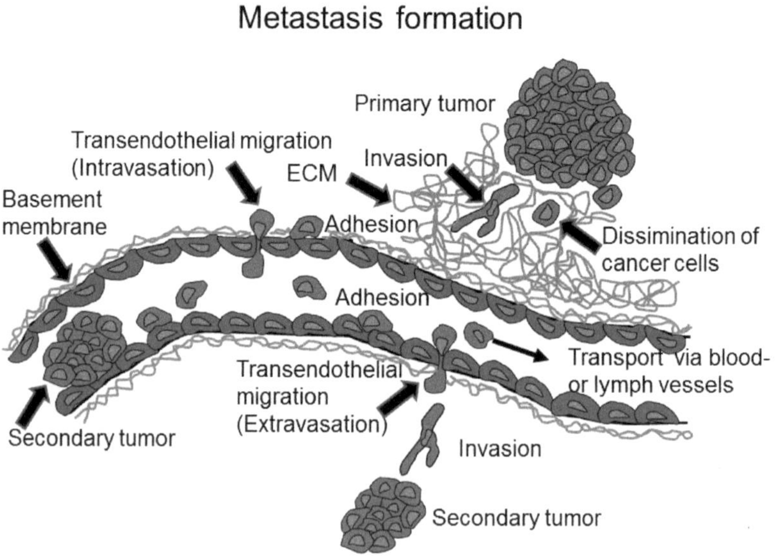

Metastasis formation

Figure 1. Process of metastasis formation.

Color image of this figure appears in the color plate section at the end of the book.

How do cancer cells transmigrate?

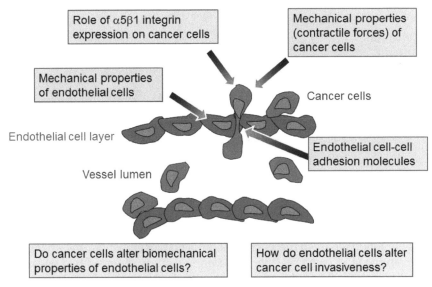

Figure 2. Mechanical aspects of cancer cell transendothelial migration.

Color image of this figure appears in the color plate section at the end of the book.

The invasion modes regulate cancer cell motility

During the progression of tumors, some cancer cells of the primary tumor as well as cells of the tumor microenvironment undergo characteristic molecular changes and biochemical changes, which lead to the dissemination of cancer cells. Due to the conditions of their surrounding microenvironment (cancer), cells use different migration modes and can migrate collectively in clusters or as single cells. Cancer cells migrating in cell clusters maintain their cell–cell adhesions and remain epithelial. When these cells start to migrate as single cells, they lose their cell–cell adhesions and thus their epithelial markers (Fig. 3). This process is called the epithelial-mesenchymal transition (EMT) (Kopfstein et al. 2006, Wolf et al. 2003). Cancer cell-intrinsic alterations include the loss of the cell polarity, alterations in cell–cell and cell-matrix adhesions, and the deregulation of the Src receptor kinase signaling, which all support the dissemination and hence, the invasion of cancer cells through tissue (Weis et al. 2004).

In this chapter, the focal lies on single cell migration. Individual migration strategies of (cancer) cells include the mesenchymal and amoeboid modes of invasion (Wolf et al. 2003). The cells can reversibly

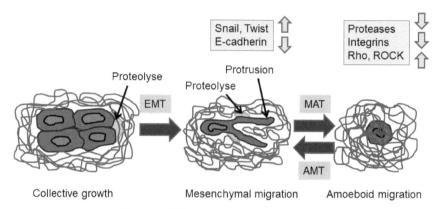

Figure 3. EMT and AMT or MAT transition.

Color image of this figure appears in the color plate section at the end of the book.

switch between both modes of cell invasion, which depends on their biochemical and mechanical properties as well as that of the ECM. This phenomenon is named mesenchymal-amoeboid or amoeboid-mesenchymal transition (MAT or AMT) (Fig. 3). Recently, a study reported that one main regulatory protein of this transition is the transcription factor snail that represses epithelial specific gene expression and thereby, induces epithelial-mesenchymal transition and subsequently cell invasion (Stemmer et al. 2008). Furthermore, this study showed that the Wnt/β-catenin pathway is also involved in epithelial-mesenchymal transition by the β-catenin medicated activation of snail.

The transition of tumors plays an important role in different stages of tumor progression. While the epithelial-mesenchymal transition requires the transcriptional regulation of gene expression, the mesenchymal-amoeboid or amoeboid-mesenchymal transition can react much faster to alterations of the microenvironment, which then evokes both morphological and functional adaption of cancer cells to facilitate their invasion in the ECM (Mishima et al. 2010). In particular, the mesenchymal-amoeboid transition is induced by overexpression of the LIM domain kinase 1 (LIMK1) gene that induced cell rounding. Hence, these results suggest that LIMK1 plays an important role in the Rho/ROCK-induced mesenchymal-to-amoeboid cell morphological transition of human fibrosarcoma HT1080 cells cultured in 3D collagen matrices (Mishima et al. 2010).

Nevertheless, the invasion of cancer cells (mesenchymal or amoeboid) remodels the tumor stroma that consists of embedded endothelial cells, fibroblasts and immune cells such as macrophages. The association of fibroblasts to cancer cells has been reported to be correlated with a poor prognosis in rectal, breast and pancreatic cancer (Saigusa et al. 2011, De

Monte et al. 2011). Cancer cells interact with endothelial cells to induce angiogenic processes such as blood and lymph vessel angiogenesis within tumors, because it is necessary to vascularize the primary tumor to further increase its tumor size (Reid et al. 2009). In addition, cancer cells induce inflammatory processes, immune-suppressive responses and growth factors which can induce the proliferation of adjacent cells (De Monte et al. 2011). For example, tumor-associated macrophages are key modulators of the tumor microenvironment, as they directly affect neoplastic cell growth, induce neoangiogenesis, and restructure or remodel the extracellular matrix (Solinas et al. 2010). Taken together, cancer cells simply stimulate their surroundings. This behavior has then in turn an impact on the tumor growth, on the invasiveness of cancer cells (Solinas et al. 2010) and subsequently, on the process of metastasis formation.

The mesenchymal type of cell invasion is determined by fibroblast-like motility used by a large number of cancer cell types (Wolf et al. 2003). The cells that migrate with a mesenchymal type exhibit an elongated spindle-like morphology with one or more protrusions of the pseudopod kind at the leading edge and the dragged cell body at the trailing edge (Fig. 4). The mesenchymal movement of cells is driven by Rac1, Cdc42 and other Rho-family dependent smallGTPases that initiate the formation of protrusions such as filopodia and pseudopodia at the leading edge of the invasive cell (Nobes et al. 2010).

After pharmacological inhibition of matrix metalloproteinases as well as siRNA-mediated knock-down of the matrix-metalloproteinsae MT1-MMP, the cells squeeze through the ECM by deforming their cytoskeleton and even their nucleus pronouncedly, instead of degrading the ECM to migrate without drastic cellular deformation (Wolf et al. 2007). The cellular

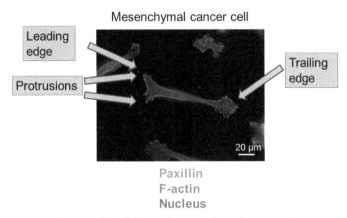

Figure 4. Morphology of a mesenchymal cancer cell.

Color image of this figure appears in the color plate section at the end of the book.

deformations is necessitate large reorganization of the cell's cytoskeleton and organelles. Thus, the mechanical properties of cells seem to be crucial for cell migration through connective tissue and ECMs (Kumar and Weaver 2009). For example, the alteration of the human pancreatic adenocarcinoma cell's keratin cytoskeleton influences their cellular mechanics through the modification of the acto-myosin cytoskeleton indicating that the mechanical properties are highly dependent on the cell's cytoskeletal organization (Seufferlein and Rosengurt 1995). In particular, these modification results lead to decreased cellular stiffness, which enables the cells to squeeze through the pores of the membranes in a Boyden chamber assay. These results lead to the suggestion that the disruption of the cell's keratin cytoskeleton increases the metastatic potential of pancreatic cancer cells (Suresh et al. 2005).

Mechanical properties of the (cancer) cells regulate their invasiveness

The mechanical properties of the cancer cells and of other tissue cells such as endothelial cells determine the ability of cancer cells to migrate through connective tissue. For example, the cell stiffness affects together with the cytoskeletal remodeling dynamics and the ability to generate contractile forces the invasiveness of cells into 3D ECMs (Mierke et al. 2008, Mierke et al. 2011a, Mierke et al. 2010) (Fig. 2).

Role of contractile forces in cell invasion

Cells were able to adhere to the ECM through cell matrix adhesion molecules on their cell surface. During the past two decades, adhesion sites of cells with mesenchymal origin have been extensively investigated on flat and rigid 2D substrates. At least, three different types of adhesion sites can be distinguished: focal complexes, focal adhesions (focal contacts) and fibrillar adhesions (Geiger et al. 2001). All cell matrix contacts contain most popular integrins as their major transmembrane receptors (Geiger et al. 2001, Zaidel-Bar et al. 2003). As these integrins anchor the cell to the adjacent microenvironment, thus, the cell matrix adhesions are able to transmit forces derived from the ECM to the interior of the cell. In turn, cytoskeletal generated forces of the cell are transmitted to the exterior. In stationary cells that are firmly adhered to the ECM, the external and internal forces cancel out each other. In contrast, in motile cells the force balance is shifted towards one direction that results in cell contraction, extension and translocation of the cell body including the nucleus (Ingber 2003, Ingber 2006). These forces are applied through adhesion molecules such as integrins towards the ECM and to other neighboring cells. Indeed, cells continuously respond by exerting reciprocal contractile forces to external applied forces that are generated by the restructuring of the ECM or by surrounding cells such as endothelial cells which withstand shear forces of the blood stream or adhesive forces from immune cells (Ingber 2003).

The contractile forces cause cytoskeletal tension maintaining the shape of the cell. In particular, the traction forces are generated by the non-muscle cytoskeletal motor protein myosin II (Cai et al. 2006, Sims et al. 1992) and are usually time and spatial dependent except in the rapid and dynamically restructuring of actin filaments (Galbraith and Sheetz 1997). Due to the long distances between cell–cell adhesions of adjacent cells, the stability of the cell bridges relies not only in these cell adhesions. In order to maintain the structure and morphology, these bridges are composed of acto-myosin networks that produce cytoskeletal tension to provide a structured cytoplasm, the so-called cytoskeletal scaffold or cytoskeleton, between adhesion contacts (Thery et al. 2006).

The anisotropy of the cell's adhesive microenvironment controls the intracellular organization and regulates the polarity of the cell (Thery et al. 2006). Indeed, the microenvironment of cells may influence their motility within 3D ECMs. In particular, the ECM stiffness alters the expression of integrin subunits, enhances focal adhesion assembly and composition, disrupts the cellular architecture and facilitates the invasiveness of cells into 3D ECMs (Paszek et al. 2005). Recently, we investigated whether the mechanical properties of 3D ECMs hinder or enable cells to invade. In this study, we found that invasive cells need to overcome the steric hindrance of the surrounding microenvironment through the transmission and generation of contractile forces (Mierke et al. 2010). In more molecular detail, the increased contractility may result from elevated Rho GTPase activity and its downstream effectors such as phosphorylated myosin light chain (MLC) (Rosel et al. 2008, Mierke et al. 2008b) or by increased levels of growth factor mediated ERK activity (Sawhney et al. 2006). In addition, the importance of growth factor receptors that may cooperate with integrins through connecting intercellular molecules such as PINCH and integrin-linked kinase (ILK) in the process of cell invasion has the potential to become a new focus of motility research (Fig. 5).

Role of cell stiffness in cell invasion

Recently, we reported that stiffer cells migrate at higher numbers and further into 3D ECMs (Mierke et al. 2010). The difference in stiffness is due to a difference in vinculin gene expression: vinculin expressing cells are stiffer compared to their wildtype counterparts. Functional knock-down of vinculin using a siRNA-mediated approach indicated that the effect on

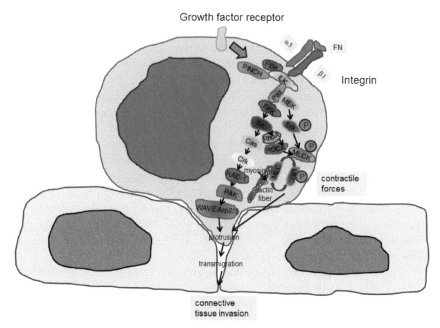

Figure 5. Signal transduction pathway during transmigration of cancer cells.
Color image of this figure appears in the color plate section at the end of the book.

cellular stiffness is indeed regulated by vinculin expression (Mierke et al. 2010). Consistently, we have found that even breast cancer cells with at least a 2-fold increased stiffness migrate at higher degrees and deeper into 3D ECMs (Mierke et al. 2011a). When the stiffness is reduced by using actin-depolymerizing agents such as latrunculin A, the migration into 3D ECMs is decreased as indicated by both a reduced number of cells that are capable to invade into these ECMs and reduced mean invasion depth of these cells (Mierke et al. 2011a). Since strain and stress are related, the cell stiffness cannot be seen separated from the generation of contractile forces, which plays an important role in the invasiveness of (cancer) cells.

Role of cytoskeletal remodeling dynamics in cell invasion

In addition to cell stiffness and the generation of contractile forces, the cytoskeletal remodeling dynamics of cells may play a role in cell invasion. Indeed, the cells that were capable to invade into 3D ECMs (highly invasive cells) showed a higher migration speed and higher cytoskeletal remodeling dynamics compared to cells that were less invasive cells (Mierke et al. 2010). When a cell can restructure its cytoskeleton, as it is appropriate as for amoeboid migration or maybe during cell division, the cell is then subsequently able to squeeze through the pores of 3D matrices or exchange cytoplasmic material to get two daughter cells of appropriate size, respectively.

The Role of the Endothelium during Cancer Disease

The role of endothelial cells in promoting tumor growth has for a long time been seen as well defined. In nearly all kinds of tumors, endothelial cells are needed as a vascularization system for a certain tumor to grow larger than a certain critical tumor size and to become finally malignant (Folkman 1992). Only a few years ago, the role of endothelial cells in cancer metastasis was unquestionably perfectly defined, as the endothelium has been regarded as a barrier for the invasiveness of cancer cells (Weis et al. 2004, Bauer et al. 2007). Similarly it has been reported for diseases involving tissue injury or immune-response reactions that the endothelium represents a barrier for immune cells such as macrophages, T-cells and eosinophil granulocytes (Wittchen et al. 2005). Now, the view of endothelial cells regarding their function in metastasis formation has been changed dramatically, as endothelial cells have been reported to promote and even increase the invasiveness of certain cancer cell lines into dense 3D ECMs (Mierke et al. 2008a). In the following parts the old and new role of endothelial cells is discussed in more detail.

Old Role: The endothelium acts a barrier for (cancer) cell invasion

The endothelium is more than a mere coating or passive lining of the vessel walls (Fig. 6). In particular, the endothelium is involved in a variety of different physiological processes and plays a role in several diseases (Ruoslahti et al. 2010, Scotland et al. 2005). For example, the endothelial monolayer regulates as a "selective" barrier the exchange of substances or cells such as immune cells between the tissue and the blood or the lymphoid system (Miteva et al. 2010, Belvitch et al. 2012, van Hinsbergh et al. 1997, Algars et al. 2011). To regulate the blood pressure, the endothelium can produce substances such as nitric oxide, which serves in the regulation of tone (stress) of the vascular muscle in the cardiovascular system (Scotland et al. 2005). In addition, the endothelium regulates the fluidity of the blood system by inhibition or activation of the coagulation process within the blood vessels (Achneck et al. 2008). Moreover, the endothelium fulfills an important role as it functions in immunologically reactions during acute and chronic inflammations. In particular, various endogenous cytokines, chemokines, growth factors or microbial substances are able to activate the endothelium locally (Zhang et al. 2011, Bates et al. 2002, Suprapitichat et al.

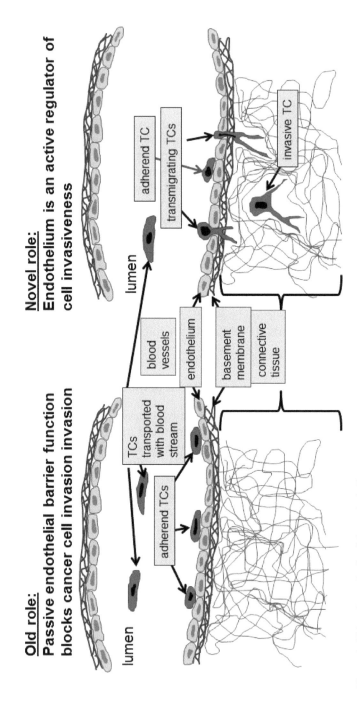

Figure 6. Old and novel role of the endothelium.

Color image of this figure appears in the color plate section at the end of the book.

2007, Steinberg et al. 2012). The activation of the endothelium promotes the adhesion of white blood cells such as neutrophil granulocytes, monocytes, macrophages and T cells to the endothelium and helps these immune cells to transmigrate through the endothelial monolayer into the underlying tissue, where they are able to fight an infection (Wittchen et al. 2005, Hashimoto et al. 2011, Muller et al. 2011, Moser et al. 1992). The transmigration process of immune cells evoked by tissue injury activates the endothelium as well as the transmigrating white blood cells, which deform themselves to squeeze through the cell–cell contacts between neighboring endothelial cells.

Another function of the endothelium is the formation of new vessels by sprouting of new blood vessels from existing vessels after vessel injury. This sprouting process is facilitated by growth factors or cytokines produced by fibroblasts or adjacent endothelial cells (Muller et al. 2011). This "normal" function of the endothelium is called angiogenesis and is similar to tumor angiogenesis, which occurs in growing neoplasms developing into a tumor (Amini et al. 2012). The process of tumor angiogenesis is the formation of a tumor-associated vasculature, which is essential for malignant tumor progression (Folkman 1971), otherwise the primary tumor grows only to a certain size of 1–2 mm^3 (Folkman 1974). In more detail, tumor vessels promote the growth of the primary tumor by providing oxygen as well as nutrients and thus promote metastasis formation by mediating the entry of cancer cells into the blood vessel system (Padua e t al. 2008). In particular, cancer metastasis depends on the dissemination of cancer cells from the primary tumor side and their adaptation to distant targeted organs (Chiang et al. 2008, Langley et al. 2007, Scheel et al. 2007). The shedding of cancer cells into the circulation of blood or lymphoid vessels may occur early in the tumor development, at several time points or stages and involves large numbers of cancer cells (Husemann et al. 2008, Pantel and Brakenhoff 2004, Stoecklein et al. 2008). In contrast, cancer metastasis is evoked solely by a certain set of these dispersed cancer cells displaying specific biomechanical properties. The tightness of the vessel walls represents a barrier for cancer cell migration, unfavorable conditions for the survival of cancer cells in distant "non-targeted" organs, and a limited number of circulating cancer cells which possess the ability to colonize targeted organs impair the formation of distant metastasis (Nguyen et al. 2009). These criteria seem to be less stringent compared to the ability of circulating cancer cells to re-infiltrate their tumors of origin (the primary tumor). This hypothesis is supported by the leakiness of the neovasculature of tumors (Carmeliet et al. 2000, Rafii et al. 2003), which favors not only the transmigration (intravasation) of cancer cells into the circulation but also the re-entry of dispersed cancer cells from the circulation back into the primary tumor (Kim et al. 2009). In addition the circulating cancer cells need no further adaptation to enter the microenvironment of their source tumor. This re-infiltration of the primary tumor by circulating cancer cells may lead to an enrichment of aggressive cancer cells, which underwent alterations during the "vessel journey" after their dissemination from the primary tumor. This process is called "tumor self-seeding", and may have indeed implications on tumor growth and the development of metastatic cancer cell progenitors (Kim et al. 2009, Norton and Massague 2008). During the development of tissues, the vascular endothelial growth factor (VEGF) induces the proliferation and subsequently the differentiation of endothelial cells from endothelial progenitor stem cells to differentiated endothelial cell which form a vascular plexus. This process is called vasculogenesis. Angiopoietin-1 (Ang-1) and proteins such as ephrins induce the restructuring of the whole vascular plexus into a clearly restructured mature vascular scaffold system by endothelial cell-facilitated sprouting, endothelial differentiation and recruitment of pericytes to the vessels (Sato et al. 1995, Mansson-Broberg 2008). This process is called angiogenesis. In addition, during the process of tumor angiogenesis, angiopoietin-2 (Ang-2) destabilizes the mature vessel walls (Fagiani et al. 2011). In particular, Ang-1 reduces the permeability of the vascular endothelium and enhances the vascular stabilization through the recruitment of pericytes and smooth muscle cells to novel growing blood vessels, whereas Ang-2 supports the sprouting of blood vessels and the vascular regression (Fagiani et al. 2011). Taken together, these findings may explain the leakiness of tumor vessels of certain tumor types.

There exist three fundamentally different types of endothelium in different tissues, which are named continuous, fenestrated and discontinuous endothelium. The main difference between the three types of endothelium is their permeability for various substances of the blood stream (Aurrand-Lions 2001). The continuous endothelium such as the endothelium of the blood-brain boundary is a strong barrier for cells and substances, because each endothelial cell forms tight junctions with the neighboring endothelial cells (Aurrand-Lions 2001). However, there is still a remaining transfer between the tissue microenvironment and blood system, which is regulated and restricted by highly selective transport mechanisms (Miteva et al. 2010, Algars et al. 2011, Bates et al. 2002). In contrast, the fenestrated endothelium is much more permeable compared to the continuous endothelium: larger molecules can pass through the 'window' in the endothelium without any problem (Braet et al. 2009). These windows in the endothelial layer have normally a diameter of 70 nm and possess in humans—with one exception in the kidney—always diaphragms, which can be thought of as spokes limiting the permeability for large molecules and cells (Braet et al. 2009). The fenestrated endothelium has in all cases a continuous endothelial basement membrane. It has been found in the kidney glomerulus, the intestine and endocrine glands. In particular, the endothelial basement of the fenestrated endothelium membrane is an additional barrier for the migration of (cancer) cells into the connective tissue, which needs to be overcome. In contrast to the fenestrated endothelium, discontinuous endothelium is highly permeable, even whole cells can easily transmigrate through it to invade the tissue microenvironment (Burns et al. 2000). Additionally, the neighboring endothelial cells are in several cases not in close contact with each other, so that a large "hole" is build up within the endothelial monolayer (Li and Zhu 1999). Moreover, the basement membrane partially or completely is even not existent (an example is the endothelium of the liver), which then enhances the permeability of the vessel lining. In particular, even parts of the basement membrane can stimulate the transendothelial migration of cells and hence enhance the transmigration of cells (Dennis et al. 2010). These three examples of different endothelial layer types clearly demonstrate that the type of endothelial cells is critical for the investigation of the endothelial barrier function and may affect the outcome of the transendothelial migration of cancer cells as well as the endothelial cell induced effects. Therefore, it has to be clearly decided which type of endothelium can be chosen for the investigation of endothelial mediated effects on the process of cancer cell invasion.

Novel role: The endothelium increases cancer cell invasion

The novel role of the endothelium in cancer metastasis is that it facilitates the invasion of several types of cancer cells into dense 3D ECMs. Moreover, it even promotes or increases the invasiveness of certain cancer cells (Mierke et al. 2008a). These findings have altered dramatically the canonical view (old view) of the endothelium as a passive barrier, which inhibits the invasiveness of cancer cells. The novel view of the endothelium is as an active cell layer that acts as an active modulator or enhancer of cancer cell invasion (Fig. 6).

Regarding the novel role of the endothelium, a lot questions can be raised: what happened on the mechanical level during the migration of cancer cells through the endothelium? Do cancer cells affect the mechanical properties of the endothelium during adhesion and transmigration? Do endothelial cells, in turn, alter the mechanical properties of transmigrating cancer cells? Do mechanical alterations in the endothelium disrupt its barrier function by down-regulation of cell–cell adhesion molecules such as the platelet endothelial cell adhesion molecule (PECAM-1)? Do cancer cells actively use the cytoskeletal scaffold of the endothelium to transmigrate through it? Do cancer cells apply mechanical forces to the endothelium in order to transmigrate? Several recent findings have proposed a novel paradigm in which endothelial cells regulate the invasiveness of certain cancer cells by increasing their dissemination through blood or lymphoid vessels (Kedrin et al. 2008) or even by enhancing the invasive capability of certain cancer cells to migrate into 3D ECMs (Mierke et al. 2008a). Moreover, evidence has been provided that the mechanical properties of cancer cells such as the transmission and generation of contractile forces are involved in the invasion process of cancer cells (Zaman et al. 2006, Mierke et al. 2008c). These observations are consistent with confocal images showing that the endothelial monolayer is pressed deeply into the dense 3D-ECM after 16 hours of tumor-endothelial cell co-culture. In addition, even the involvement of the contractile acto-myosin cytoskeleton of endothelial cells, which are adjacent to cancer cells, has been investigated in terms of increasing the invasiveness of interacting cancer cells (Khuon et al. 2010). In particular, it has been reported that cancer cells transmigrate directly through the endothelial cell body without any support of other interacting cell types by generating a hole in the membrane of underneath located endothelial cells. Then, subsequently this hole enables cancer cells to move along the endothelial acto-myosin contractile filaments, migrate through the endothelial cell monolayer and further invade in the connective tissue microenvironment (Khuon et al. 2010). Whether the endothelial cell dies (Khuon et al. 2010) or not needs to be still investigated in more detail (Mierke et al. 2008a, Weis et al. 2004). However, the transcellular migration of cancer cells through the endothelium requires an actin cytoskeleton of cancer cells that is tensioned by myosin-II acting on actin fibers. In addition, an intact contractile apparatus of endothelial cells is necessary to provide a cytoskeletal scaffold that is strong enough to withstand the external forces applied by transmigrating cancer cells towards the endothelium without the rupture of endothelial cells from the whole endothelial lining.

Biomechanics of the Endothelium

The following paragraph discusses the biomechanical function of the endothelium during the progression of cancer disease. The focus is on the adhesion and transendothelial migration of cancer cells during the step of cancer cell intra- or extravasation. For the malignancy of cancers, the process of metastasis formation is mainly responsible. Hence, metastasis formation is a main cause for most cancer-related deaths. In particular, a benign tumor becomes malignant when cancer cells disseminate and thus, spread from the primary tumor and finally form metastases in targeted organs (Frixen et al. 1991, Batle et al. 2000, Cano et al. 2000, De Crane et al. 2005). The process of metastasis can be divided into several steps which involve the dissemination and spreading of cancer cells from the primary tumor site into the surrounding ECM, the invasion of cancer cells through connective tissue, the adhesion of cancer cells to the endothelium lining blood or lymph vessels, possibly the transmigration of cancer cells through the endothelium (intravasation and/or extravasation), and subsequently, the formation of a secondary tumor in a distant targeted organ (Steep 2006, Al-Mehdi et al. 2000). The impact of the endothelium on the regulation of cancer cell invasiveness in 3D-ECMs is still largely unknown and hence, under investigated. In particular, even the regulation of cancer cell invasiveness regardless of the endothelial role is a complex scenario, which is not fully characterized yet (Frield and Wolf 2003, Wolf et al. 2003, Discher et al. 2005, Mierke et al. 2008c). In many previous studies, the endothelium acts as a strong barrier against the invasion of cancer cells, which is here described as the "old" classical role of the endothelium (Al-Medhi et al. 2000, Zijlstra et al. 2008). In most cases, the endothelium gets leaky after stimulation with certain substances such as cytokines, chemokines or peptides, which may additionally crosslink endothelial cell surface receptors to increase their leakiness (Pober et al. 2005, Ferrara et al. 1999, van Sluis et al. 2009). Moreover in these studies, the endothelium inhibits pronouncedly the invasion of cancer cells and hence the formation of metastases (Kollmannsberger et al. 2011). However, several recent reports propose a "novel" barrier break-down role of the endothelium in which endothelial cells regulate the invasiveness of several invasive cancer cells by increasing their dissemination through vessels (Kedrin et al. 2008) or by increasing the invasive capability of cancer cells to migrate into the ECM (Mierke et al. 2008a). Among these cancer cells are the human breast cancer cell line MDA-MB-231, human kidney cancer cell line 786-O and the human bladder cancer cell line T24 (Mierke et al. 2008a).

Biomechanical interactions between cancer and endothelial cells

Numerous cell surface receptors such as adhesion molecules have been identified, which play a role in the tumor–endothelial cell interactions and, hence metastasis formation, but the mechanical properties of endothelial cells co-cultured with cancer cells, are still elusive. In this part of the chapter, the focus lies on the biomechanical properties of endothelial cells supporting cancer progression. In particular, there are certain biomechanical properties of endothelial cells that determine whether the endothelium acts as a barrier or as an enhancer for the invasiveness of cancer cells.

The Endothelium enhances the invasiveness of cancer cells

The endothelium increases the invasiveness of invasive cancer cells (Fig. 6). Recently, it has been investigated how the invasive behavior of highly-invasive and weakly-invasive breast cancer cells is affected by the co-culture of these cancer cells with an endothelial cell monolayer on top of 3D-ECMs (Mierke 2011a). To investigate the effect of the endothelium on cancer cell invasiveness, two established human breast cancer cell lines, the highly-invasive MDA-MB-231 cells and the weakly-invasive MCF-7 cells, were cultured in the presence or absence of macro- (human umbilical vein endothelial cells, HUVECs) and microvascular (human pulmonary or dermal microvascular endothelial cells, HPMECs or HDMECs) endothelial cells. Indeed, the invasiveness of highly-invasive MDA-MB-231 cells has been increased in the presence of a closed endothelial cell monolayer on top of dense (1.3 mm average pore-size) 3D ECMs (Mierke 2011a). In contrast, the invasiveness of the weakly-invasive MCF-7 cells has not been affected neither by macro- nor microvascular endothelial cells (Mierke 2011a). These results show that the endothelial cell monolayer does not increase the invasiveness of all breast cancer cell lines or even of all cancer cell lines investigated, which is consistent with another previous study (Mierke et al. 2008a). These findings indicate that the endothelial-facilitated mechanism to increase the invasiveness of cancer cells is cancer cell-specific and may additionally depend on the biomechanical properties of cancer cells. In turn, endothelial cells do not invade significantly into 3D ECMs when co-cultured with highly-invasive cancer cells indicating that highly-invasive cancer cells cannot induce motility within the endothelial cell lining (Mierke 2011a).

Invasive Cancer cells regulate Biomechanical Properties of Endothelial Cells

Do Invasive Cancer Cells alter the Stiffness of Endothelial Cells?

To investigate whether highly-invasive human breast cancer cells facilitate the transmigration through the endothelium by altering biomechanical properties of endothelial cells such as cell stiffness, the cell stiffness of co-cultured and mono-cultured endothelial cells was determined using magnetic tweezer microrheology (Kollmannsberger et al. 2011). Indeed, the cellular stiffness is a measure of the resistance by an elastic body such as an endothelial cell to external applied forces, which cause non-permanent deformation. The stiffness of cells seems to be determined by their ability to generate and transmit contractile forces (Mierke et al. 2011a). The stiffness of endothelial cells has been measured by applying force to fibronectin (FN)-coated beads that are bound through integrin cell–matrix adhesion receptors such as $\alpha 5\beta 1$ integrins to the acto-myosin cytoskeleton of mono-cultured and with MDA-MB-231 cancer cells co-cultured endothelial cells, respectively. The bead displacement can be tracked during force application and fitted to a power-law. The pre-factor a of the power-law equation stands for the compliance of the cells and the inverse of the compliance corresponds to the cell stiffness (Rauapch et al. 2007). In order to show that the bead binding is to endothelial cells, the FN-bead-binding towards endothelial cells has been analyzed by staining of endothelial cells with CyTrack orange and cancer cells with carboxyfluorescein diacetate (CFDA). As expected, almost all FN-beads bind to endothelial cells, which have been incubated with the FN-beads two hours before co-culture start, whereas the FN-bead binding to cancer cells during the co-culture is below 1%. Indeed, the highly-invasive MDA-MB-231 cells have decreased the cellular stiffness in co-cultured HPMECs compared to mono-cultured HPMECs, whereas the co-culture of weakly-invasive MCF-7 cells with HPMECs has not affected the endothelial cell stiffness (Mierke 2011a).

In summary, the weakly-invasive MCF-7 breast cancer cells are not able to regulate the biomechanical properties of co-cultured endothelial cells suggesting that only highly-invasive MDA-MB-231 breast cancer cells are able to decrease the mechanical stiffness of endothelial cells. These results suggest that the reduced endothelial cell stiffness can be described as an increased deformability of endothelial cells, which facilitates the transmigration of cancer cells through the endothelium. Furthermore, this behavior may be a prerequisite to overcome and break-down the endothelial barrier and may indeed be a critical parameter for the aggressiveness of cancer cells.

Do Invasive Cancer Cells alter the Cytoskeletal Remodeling Dynamics of Endothelial Cells?

Structural or dynamical alterations in the acto-myosin cytoskeleton of endothelial cells are supposed to play a role in transendothelial migration (Mierke 2008b). Thus, it has been investigated whether the remodeling of cytoskeletal structures such as the formation or degradation of actin stress fibers in endothelial cells affects their barrier function against cellular transmigration in the presence and absence of the invasive MDA-MB-231 breast cancer cells. The nanoscale particle tracking method has been used to analyze whether the acto-myosin cytoskeletal remodeling dynamics are altered in co-cultured endothelial cells compared to mono-cultured endothelial cells (Fig. 7). Therefore, the displacement of FN-cell-bound beads, which were connected to the cell's acto-myosin cytoskeleton through FN-binding integrin receptors expressed on the surface of endothelial cells, was analyzed. In summary, this method allows directly the visualization of the cytoskeletal-bound bead movement, which is evoked by ATP-driven cytoskeletal rearrangements (Fig. 7). In particular, the nanoscale particle tracking method records and analyzes the random walk of unforced, spontaneously diffusing beads. Cytoskeletal-bound beads do not change their position until the focal adhesions and/or the actin structures (microfilaments) to which they are firmly connected reorganize or reassemble (Raupach et al. 2007, Metzner et al. 2010). The

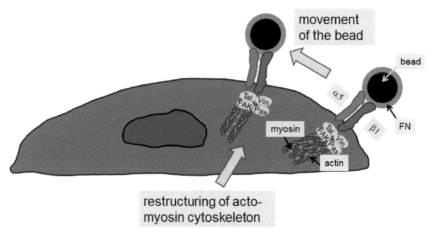

Figure 7. Bead binding to cells.

Color image of this figure appears in the color plate section at the end of the book.

evolution of the cytoskeletal-bound beads's mean squared displacement (MSD) over time (t) can be described by a diffusion coefficient (D) and the direction of bead-motion as an index of persistence (b) according to MSD = D $(t/t_0)^b$ + c (Raupach et al. 2007). The MSD of endothelial cell bound beads which are co-cultured with highly invasive cancer cells has been shown to be significantly higher compared to mono-cultured endothelial cells or co-cultures with weakly invasive cancer cells (Mierke 2011a). The diffusion coefficient D of FN-beads bound to co-cultured endothelial cells (HDMECs) has been reported to be pronouncedly higher compared to mono-cultured endothelial cells, indicating that the speed of cytoskeletal remodeling and restructuring processes, which can be obtained by the square-root of the diffusion coefficient D, is increased in endothelial cells co-cultured with the highly invasive MDA-MB-231 cells (Mierke 2011a). These findings demonstrate that the highly invasive MDA-MB-231 cells induce a dynamic remodeling of the actin cytoskeleton within co-cultured endothelial cells compared to mono-cultured endothelial cells, whereas the weakly invasive MCF-7 cells do not have such a dramatic effect on the mechanical properties of co-cultured endothelial cells, on which they are still able to adhere. Due to the values of the power-law exponent b, which is an index of persistence, the bead motion can be described as Brownian or diffusive motion for b = 1, subdiffusive motion for b < 1, superdiffusive motion (called persistent motion or directed motion) for b > 1, and ballistic motion for b = 2 (Raupach et al. 2007). When the FN-beads are bound to the endothelial cell's acto-myosin cytoskeleton through cell-surface expressed integrins, the persistent motion of these FN-beads reflects the cytoskeletal remodeling dynamics such as the reorganization or assembly/disassembly of actin microfilaments. During co-culture with the highly-invasive MDA-MB-231 cells the persistence of the endothelial FN-bead movement, which can be regarded as a measure of the directionality, has not been altered compared with mono-cultured or with weakly-invasive MCF-7 cells co-cultured microvascular endothelial cells (Mierke 2011a).

Role of the α5β1 integrin expression on cancer cells in endothelial facilitated cancer cell invasion

Do cancer cells with high α5β1 expression transmigrate more efficiently compared to cancer cells with low α5β1 expression?

In a recent study, it has been investigated whether the increased invasiveness of MDA-MB-231 cells is facilitated by increased expression of the α5β1 integrin on their cell surface. Thus, the invasiveness of highly and lowly α5β1 integrin expressing subcell lines derived from parental MDA-MB-231 cells has been determined using a 3D-ECM invasion assay. Indeed, the study shows that the α5β1^{high} cells have migrated more numerous and deeper into 3D ECMs during co-culture with HDMECs or HPMECs compared to mono-cultured α5β1^{high} cells, whereas the invasiveness of α5β1^{low} cells has not been significantly affected (Mierke et al. 2011a). Knock-down of the α5 integrin subunit in α5β1^{high} cells has decreased the endothelial-facilitated increased invasiveness to the level of mono-cultured α5β1^{high} cells as shown by the decreased numbers of invasive cells and their decreased invasion depth (Mierke 2011a). These results demonstrate that the knock-down of the α5 integrin subunit abolishes the endothelial-facilitated increased invasiveness of α5β1^{high} cells. These findings suggest an involvement of α5β1 integrin signaling in the endothelial-facilitated increased invasiveness, which may additionally be stimulated by the endothelial cells produced and secreted α5β1 integrin ligand FN (Mierke et al. 2011a).

Do contractile forces alter the endothelial-facilitated invasiveness of cancer cells expressing high amounts of the α5β1 integrin?

Several studies reported that the transmission and generation of contractile forces play a prominent role during the invasion of cell in 3D-ECMs (Zaman et al. 2006, Mierke et al. 2008a, Mierke et al. 2011a, Stemmer et al. 2008, Mierke et al. 2010, Mierke et el. 2011b, Mierke 2011b). In order to analyze whether the inhibition of contractile forces can affect the endothelial-facilitated increased invasiveness of α5β1^{high} cells, the myosin light chain kinase (MLCK) and the Rho kinase (ROCK) have been inhibited by adding the inhibitor ML-7 and Y27632, respectively, to α5β1^{high} cells prior to co-culture and invasion assay start. The inhibition of contractile forces has been reported to decrease in both cases the invasiveness of HPMEC-co-cultured α5β1^{high} cells, which has been shown by the decreased percentage of invasive cells and their decreased invasion depth. Finally, these results indicate that the endothelial-facilitated increased invasiveness of α5β1^{high} cells depends on the generation or transmission of contractile forces within the cancer cells. Taken together, both the mechanical properties of endothelial and cancer cells play an important role for the invasiveness of cancer cells, and the physical properties of both cell types should be analyzed together in a co-culture experiment as well as in monocultures.

Does the inhibition of small GTPase signaling reduce the endothelial-facilitated invasiveness of α5β1^{high} cells?

The mechanism by which co-cultured endothelial cells increase the invasiveness of cancer cells is currently still elusive and hence under investigation. Recently, the transendothelial migration and invasion assay have been performed in the presence of embedded FN in 3D-FN-ECMs. Indeed, the results show indeed that embedded FN enhances the endothelial-facilitated invasiveness of α5β1^{high} cells. Furthermore, as expected, the number of transmigrating and invasive α5β1^{high} cells and their invasion depths have been increased in the presence of an HPMEC cell monolayer, whereas the invasiveness of α5β1^{low} cells has not been affected. These results suggest that FN activates α5β1 integrins and subsequently, increases the transmission and generation of contractile forces in α5β1^{high} cells, which further increases their ability to transmigrate and finally to invade in connective tissue. These findings show that the activation of FN-binding receptors with embedded FN, but not with soluble FN, may facilitate the increased invasiveness of highly invasive and highly α5β1 integrin expressing cancer cells through enhanced contractile force transmission and generation. Additionally, endothelial cells initiate and enhance the invasiveness of certain cancer cells by producing IL-8 and Gro-β that bind to the CXCR2 receptor expressed on cancer cells and induce increased motility of cancer cells. Indeed, this terminal step of the α5β1 integrin-dependent inside–out signaling may be triggered by IL-8 and Gro-β signaling through their binding to the CXCR2 receptor. Taken together, this observation of chemokine-induced triggering of integrin-dependent cell motility seems to be comparable to the chemokine-dependent integrin inside–out signaling of leukocytes (Ley et al. 2007, Shimonaka et al. 2003).

Furthermore, to investigate whether small GTPase signaling pathways play a role in endothelial cell facilitated increased invasion of MDA-MB-231 cells, inhibitors against Rho kinase (Y27632), Rac-1 (Rac Inh), MEK (PD98059) and PI3 kinase (LY294002) have been added prior to co-culture invasion assay start. Indeed, the results show reduced numbers of invasive cancer cells as well as reduced invasion depths in the presence of HPMECs and after addition of Rho kinase inhibitor (Y27632), Rac-1 inhibitor (Rac Inh), MEK inhibitor (PD98059) or PI3 kinase inhibitor (LY294002). These findings indicate that small GTPases are involved in the endothelial-facilitated MDA-MB-231 cancer cell invasion. In addition, the endothelial-facilitated increased invasiveness of $\alpha5\beta1^{high}$ cells is not restricted to human breast cancer cells, as the invasiveness of $\alpha5\beta1^{high}$ cells derived from human bladder and kidney cancer cells has been shown to be similarly increased by endothelial cells. In contrast, it has been shown that these inhibitors have no effect on weakly-invasive $\alpha5\beta1^{low}$ cells derived from breast cancer cells. These results indicate that there seems to be a general mechanism that enables integrin $\alpha5\beta1$ highly expressing aggressive cancer cells to invade more pronouncedly in ECMs of connective tissue compared to integrin $\alpha5\beta1$ lowly expressing and less aggressive cells. These biomechanical results are consistent with an up-regulation of the Erk phosphorylation reported in HUVECs during co-culture with colon cancer cells (Tremblay et al. 2006). Moreover, this study reported that the inhibition of the upstream regulatory protein of Erk, called MEK, abolished the endothelial-facilitated increased invasiveness of $\alpha5\beta1^{high}$ breast cancer cells. These results are also consistent with the inhibition of cancer cell transmigration and invasiveness by addition of the MEK inhibitor. Moreover, these findings support the idea that cancer cells may apply forces towards the endothelial cell monolayer to alter endothelial mechanical properties and hence, facilitate cancer cell transendothelial migration. Indeed, contractile forces are necessary for highly-invasive MDA-MB-231 cells to overcome the endothelial barrier and invade connective tissue, because inhibition of the MLCK and subsequently, the transmission or generation of contractile forces reduces the number of transmigrating and invading cells. Finally, these results indicate that the endothelial facilitated increased invasiveness depends on the $\alpha5\beta1$ integrin expression, suggesting that $\alpha5\beta1$ integrins connect the mechanical properties of the cancer cell's acto-myosin cytoskeleton with the mechanical properties of the microenvironment. Alterations of endothelial cell's mechanical properties by highly-invasive cancer cells may also depend on contractile forces, as $\alpha5\beta1^{high}$ cells are able to transmit and generate 7-fold higher contractile forces compared to $\alpha5\beta1^{low}$ cells (Mierke et al. 2011a). To reveal the mechanism through which highly-invasive cancer cells transmit forces to endothelial cells, disrupt the cell–cell adhesion contacts and finally reduce the mechanical stiffness within endothelial cells, further traction force measurements are needed. In future traction force measurements the forces towards endothelial cells could be measured in the presence and absence of highly invasive cancer cells.

Invasive cancer cells regulate the expression cell–cell adhesion molecules on endothelial cells

Do invasive cancer cells alter cell–cell adhesion molecule expression on endothelial cells?

Until now it is still controversially discussed how cancer cells transmigrate through an endothelial cell monolayer lining blood or lymph vessels. Recently, it has been investigated whether highly-invasive cancer cells can alter the cell–cell adhesion molecule expression on human microvascular endothelial cells during co-culture. Therefore primary endothelial cells derived from the lung HPMECs have been co-cultured for 16 hours with highly-invasive MDA-MB-231 and weakly-invasiveMCF-7 cancer cells. Flow cytometric analysis has been shown that microvascular endothelial cells co-cultured with highly-invasive MDA-MB-231 breast cancer cells decreased the cell surface expression of the platelet endothelial cell adhesion molecule-1 (PECAM-1) and the vascular endothelial cadherin (VE-cadherin) compared to monocultured endothelial cells (Mierke 2011a). In particular, the downregulation of PECAM-1 and VE-cadherin has solely been observed during co-culture with highly-invasive MDA-MB-231 cells, but not during the co-culture with weakly-invasive MCF-7 cells. These findings indicate that the down-regulation of cell–cell adhesion molecules on the endothelium seems to be cancer cell-specific and may depend on the individual invasive potential and the mechanical properties of the co-cultured cancer cells. In summary, these results show that the co-culture of endothelial cells with highly-invasive cancer cells has led to a down-regulation in the cell surface expression of the cell–cell adhesion molecules vascular endothelial (VE)-cadherin and platelet endothelial cell adhesion molecule-1 (PECAM-1), altered biomechanical properties of endothelial cells and subsequently, the breakdown of the endothelial barrier function. To investigate which mechanism leads to reduced endothelial cell–cell adhesion molecule expression during co-culture with MDA-MB-231 cells, the membrane-shedding of these adhesion molecules has been inhibited during the transendothelial migration of the MDA-MB-231 cells. In particular, it has been analyzed whether the decreased expression of PECAM-1 and VE-cadherin receptors on endothelial cells (HPMECs) during co-culture with MDA-MB-231 cells is due to membrane shedding of these receptors. Thus the co-culture has been performed in the presence and absence of the broad matrix-metallo-proteinase inhibitor GM6001. Indeed, the reduction of PECAM-1 and VE-cadherin receptor expression on endothelial cells during co-culture with MDA-MB-231 cells has been reported to depend on enhanced membrane shedding of both cell–cell adhesion receptors.

Adhesive cells transmit external forces to the endothelium

The endothelium is exposed to high external forces and hence, reacts to them by disrupting cell–cell adhesions between neighboring endothelial cells. Thus, the whole endothelial monolayer explores now the new internal forces, which are no longer balanced through endothelial cell–cell adhesions, which finally disturbs the force homeostasis within the endothelial monolayer. The whole endothelial monolayer is exposed to external forces by neutrophils or lymphocytes adhering to the endothelium and causing the rupture of cell–cell adhesions (Cinamon et al. 2001, Radodzy et al. 2008). In addition, the endothelium is also exposed to shear stresses, when neutrophil granulocytes and lymphocytes transmit (shear) forces to transmigrate through the endothelium and finally, into the connective tissue (Cinamon et al. 2001, Radodzy et al. 2008). It has been reported that the expression of platelet-endothelial cell adhesion molecule-1 (PECAM-1) and vascular endothlelial cadherin (VE-cadherin) on the endothelial cell surface is down-regulated in co-culture with adhesive neutrophils or lymphocytes compared to monocultured endothelial cells (Ionescu et al. 2003, Su et al. 2002, Dejana et al. 2008, Shaw et al. 2001, Woodfin et al. 2003). These findings indicate that it might be possible to compare the transmigration of neutrophils or lymphocytes with the transmigration of cancer cells on biochemical and biomechanical levels. In more detail, whether the reduction of cell–cell adhesion molecules on endothelial cells and the decrease in the stiffness of endothelial cells are linked by co-regulation or are otherwise related has to be further investigated. In summary, both the biochemical and the biomechanical properties of the endothelium determine the maintenance or the reduction of endothelial cell barrier function for transmigrating certain cancer cells. Moreover, for cancer cell transmigration, a dual function of the endothelium has been suggested which can act either as an enhancer or inhibitor of cancer cell invasion depending on the cancer cell type co-cultured (Mierke 2011a).

Adhesive and transmigrating cells down-regulate endothelial focal adhesion molecules

During microbial inflammation (for example a bacterial infection), the endothelium undergoes structural and morphological alterations, which enables the transmigration of neutrophils from the blood stream to the sites of tissue injury or primary infection (Lee and Liles 2011, Yuan et al. 2012). During inflammatory processes endothelial cell–cell junctions and the endothelial cytoskeleton reorganize, but still less is known about the role of the cellular structures forming cell–cell contacts such as the focal adhesions. The selective loss of the focal adhesion protein focal adhesion kinase (FAK) from endothelial focal adhesions in close proximity to transmigrating neutrophils has been recently reported, whereas the levels of the β1 integrin subunit and the focal adhesion protein vinculin were unaffected (Parsons et al. 2012). Another focal adhesion protein paxillin, which is able to bind to vinculin, was not found in focal adhesions during neutrophil transmigration. As paxillin interacts with FAK, the effect of FAK during the transmigration of neutrophil granulocytes has been investigated in more detail. The disruption of the FAK protein by using a siRNA-mediated approach or blockage of FAK signaling by using a kinase-deficient FAK has been reported to decrease in approaches the transendothelial migration of neutrophils. Taken together, these findings postulate a novel role for paxillin and FAK in the endothelial monolayer during the regulation of the transmigration of neutrophils (Parsons et al. 2012). Furthermore the down-regulation of FAK may give an explanation why the expression of cell–cell adhesion molecules such as PECAM-1 and VE-cadherin in the endothelium is downregulated (Shaw et al. 2001, Woodfin et al. 2003). In addition, FAK seems to be a mechano-sensory protein, and hence, a biomechanical mechanism may be responsible for its down-regulation in endothelial focal adhesions during transmigration of neutrophils.

Conclusions

This chapter discussed the structural and mechanical properties of the endothelium that impact the invasiveness of cancer cells as well as the metastasis formation. An agreement has been reached about the importance of the endothelium to promote tumor progression either by enhancing the growth of the primary tumor or by targeting cancer cells to blood or lymph vessels. In conclusion, this review pointed out that the endothelium is no passive barrier for cancer cell invasion. Instead, the endothelium seems to act cancer cell-type-specific, as it can increase or decrease cancer cell transmigration and invasion of cancer cells. In addition, biomechanical properties of endothelial cells are altered by a certain type of cancer cells with distinct biomechanical properties indicating that these biomechanical alterations may play a role in the transendothelial migration process of cancer cells. In summary, biomechanical measurements may shed light on mechanisms that facilitate the transendothelial migration of cancer cells. Furthermore, this article discussed the possibility that aggressive cancer cells with a distinct set of biomechanical properties such as increased cell stiffness and contractile force generation break-down the endothelial barrier function by lowering cellular stiffness through remodeling of the actin cytoskeleton. In addition, endothelial cells may secrete FN which then further activates α5β1 integrins on cancer cells and enables cancer cells to transmit and generate increased contractile forces and subsequently, to increase their invasiveness. Finally, this article suggests that biomechanical alterations in endothelial cells evoked by certain cancer cells may provide a novel biomechanical selection process towards higher invasiveness of cancer cells. In conclusion, biomechanical interactions between highly-invasive cancer cells and endothelial cells can facilitate the transmigration of cancer cells, further enhance their invasion into connective tissue and subsequently, may determine the malignancy of tumors.

Acknowledgements

This work was supported by the Deutsche Krebshilfe (109432). I thank Thomas Mierke for excellent proof-reading and editing of this chapter.

References

Achneck, H.E., B. Sileshi and J.H. Lawson. 2008. Review of the biology of bleeding and clotting in the surgical patient. *Vascular* 16: S6–S13.

Algars, A., M. Karikoski, G.G. Yegutkin, P. Stoitzner, J. Niemela, M. Salmi and S. Jalkanen. 2011. Different role of CD73 in leukocyte trafficking via blood and lymph vessels. *Blood* 117(16): 4387–4393. Epub 2011 Feb 23.

Al-Mehdi, A.B., K. Tozawa, A.B. Fisher, L. Shientag, A. Lee and R.J. Muschel. 2000. Intravascular origin of metastasis from the proliferation of endothelium-attached tumor cells: a new model for metastasis. *Nat. Med.* 6: 100–102.

Amini, A., S.M. Moghaddam, D.L. Morris and M.H. Pourgholami 2012. The Critical Role of Vascular Endothelial Growth Factor in Tumor Angiogenesis. *Curr. Cancer Drug Targets* 12: 23–43.

Aurrand-Lions, M., C. Johnson-Leger, C. Wong, L. Du Pasquier and B.A. Imhof. 2001. Heterogeneity of endothelial junctions is reflected by differential expression and specific subcellular localization of the three JAM family members. *Blood* 98: 3699–3707.

Bates, D.O., N.J. Hillman, B. Williams, C.R. Neal and T.M. Pocock. 2002. Regulation of microvascular permeability by vascular endothelial growth factors. *J. Anat.* 200: 581–597.

Batlle, E., E. Sancho, C. Franci, D. Dominguez, M. Monfar, J. Baulida and A. Garcia De Herreros. 2000. The transcription factor snail is a repressor of E-cadherin gene expression in epithelial tumour cells. *Nat. Cell Biol.* 2: 84–89.

Bauer, K., C. Mierke and J. Behrens. 2007. Expression profiling reveals genes associated with transendothelial migration of tumor cells: a functional role for alphavbeta3 integrin. *Int. J. Cancer.* 121: 1910–8.

Belvitch, P. and S.M. Dudek. 2012. Role of FAK in S1P-regulated endothelial permeability. *Microvasc. Res.* 83(1): 22–30. Epub 2011 Sep 5.

Braet, F., J. Riches, W. Geerts, K.A. Jahn, E. Wisse and P. Frederik. 2009. Three-dimensional organization of fenestrae labyrinths in liver sinusoidal endothelial cells. *Liver Int.* 29: 603–613.

Burns, A.R., R.A. Bowden, S.D. MacDonell, D.C. Walker, T.O. Odebunmi, E.M. Donnachie, S.I. Simon, M.L. Entman and C.W. Smith. 2000. Analysis of tight junctions during neutrophil transendothelial migration. *J. Cell Sci.* 113: 45–57.

Cai, Y., N. Biais, G. Giannone, M. Tanase, G. Jiang, J.M. Hofman, C.H. Wiggins, P. Silberzan, A. Buguin, B. Ladoux and M.P. Sheetz. 2006. Nonmuscle myosin IIA-dependent force inhibits cell spreading and drives F-actin flow. *Biophys. J.* 91: 3907–20.

Cano, A., M.A. Perez-Moreno, I. Rodrigo, A. Locascio, M.J. Blanco, M.G. del Barrio, F. Portillo and M.A. Nieto. 2000. The transcription factor snail controls epithelial–mesenchymal transitions by repressing E-cadherin expression. *Nat. Cell Biol.* 2: 76–83.

Carmeliet, P. and R.K. Jain. 2000. Angiogenesis in cancer and other diseases. *Nature* 407: 249–257.

Chambers, A.F., A.C. Groom and I.C. MacDonald. 2002. Dissemination and growth of cancer cells in metastatic sites. *Nat. Rev. Cancer* 2: 563–572.

Chiang, A.C. and J. Massague. 2008. Molecular basis of metastasis. *N. Engl. J. Med.* 359: 2814–2823.

Cinamon, G., V. Shinder and R. Alon. 2001. Shear forces promote lymphocyte migration across vascular endothelium bearing apical chemokines. *Nat. Immunol.* 2: 515–522.

De Craene, B., B. Gilbert, C. Stove, E. Bruyneel, F. van Roy and G. Berx. 2005. The transcription factor snail induces tumor cell invasion through modulation of the epithelial cell differentiation program. *Cancer Res.* 65: 6237–6244.

De Monte, L., M. Reni, E. Tassi, D. Clavenna, I. Papa, H. Recalde, M. Braga, V. Di Carlo, C. Doglioni and M.P. Protti. 2011. Intratumor T helper type 2 cell infiltrate correlates with cancer-associated fibroblast thymic stromal lymphopoietin production and reduced survival in pancreatic cancer. *J. Exp. Med.* 208: 469–479.

Dejana, E., F. Orsenigo and M.G. Lampugnani. The role of adherens junctions and VE-cadherin in the control of vascular permeability. *J. Cell Sci.* 121(Pt 13): 2115–2122.

Dennis, J., D.T. Meehan, D. Delimont, M. Zallocchi, G.A. Perry, S. O'Brien, H. Tu, T. Pihlajaniemi and D. Cosgrove. 2010. Collagen XIII induced in vascular endothelium mediates alpha1beta1 integrindependent transmigration of monocytes in renal fibrosis. *Am. J. Pathol.* 177: 2527.7.

Discher, D.E., P. Janmey and Y.L. Wang. 2005. Tissue cells feel and respond to the stiffness of their substrate. *Science* 310: 1139–1143.

DuFort, C.C., M.J. Paszek and V.M. Weaver. 2011. Balancing forces: architectural control of mechanotransduction, *Nat. Rev. Mol. Cell Biol.* 12: 308–319.

Eger, A. and W. Mikulits. 2005. Models of epithelial–mesenchymal transition. *Drug Discov. Today: Dis. Models* 2: 57–63.

Fagiani, E., P. Lorentz, L. Kopfstein and G. Christofori. 2011. Angiopoietin-1 and -2 exert antagonistic functions in tumor angiogenesis, yet both induce lymphangiogenesis. *Cancer Res.*71: 5717–5727.

Ferrara, N. 1999. Vascular endothelial growth factor: molecular and biological aspects. *Curr. Top. Microbiol. Immunol.* 237: 1–30.

Folkman, J. 1971. Tumour angiogenesis: therapeutic implications, *N. Engl. J. Med.* 285: 1182–1186.

Folkman, J. 1974. Tumour angiogenesis. *Adv. Cancer Res.* 19: 331–358.

Folkman, J. 1992. The role of angiogenesis in tumor growth. *Semin. Cancer Biol.* 3: 65–71.

Friedl, P. and K. Wolf. 2003. Tumour-cell invasion and migration: diversity and escape mechanisms. *Nat. Rev. Cancer* 3: 362–374.

Fritsch, A., M. Höckel, T. Kiessling, K.D. Nnetu, F. Wetzel, M. Zink and J.A. Käs. 2010. Are biomechanical changes necessary for tumour progression? *Nature Physics* 6: 730–732.

Frixen, U.H., J. Behrens, M. Sachs, G. Eberle, B. Voss, A. Warda, D. Lochner and W. Birchmeier. 1991. E-cadherin-mediated cell–cell adhesion prevents invasiveness of human carcinoma cells. *J. Cell Biol.* 113: 173–185.

Galbraith, C.G. and M.P. Sheetz. 1997. A micromachined device provides a new bend on fibroblast traction forces. *Proc. Natl. Acad. Sci. USA* 94: 9114–8.

Geiger, B., A. Bershadsky, R. Pankov and K.M. Yamada. 2001. Transmembrane crosstalk between the extracellular matrix—cytoskeleton crosstalk. *Nat. Rev. Mol. Cell Biol.* 2: 793–805.

Guck, J., S. Schinkinger, B. Lincoln, F. Wottawah, S. Ebert, M. Romeyke, D. Lenz, H.M. Erickson, R. Ananthakrishnan, D. Mitchell, J. Käs, S. Ulvick and C. Bilby. 2005. *Biophys. J.* 88: 3689–3698.

Hashimoto, K., N. Kataoka, E. Nakamura, K. Hagihara, M. Hatano, T. Okamoto, H. Kanouchi, Y. Minatogawa, S. Mohri, K. Tsujioka and F. Kajiya. 2010. Monocyte trans-endothelial migration augments subsequent transmigratory activity with increased PECAM-1 and decreased VE-cadherin at endothelial junctions. *Int. J. Cardiol.* Dec 28. [Epub ahead of print].

Husemann, Y., J.B. Geigl, F. Schubert, P. Musiani, M. Meyer, E. Burghart, G. Forni, R. Eils, T. Fehm, G. Riethmuller and C.A. Klein. 2008. Systemic spread is an early step in breast cancer. *Cancer Cell* 13: 58–68.

Ingber, D.E. 2003. Mechanobiology and diseases of mechanotransduction. *Ann. Med.* 35: 564–77.

Ingber, D.E. 2006. Cellular mechanotransduction: putting all the pieces together again. *Faseb J.* 20: 811–27.

Ionescu, C.V., G. Cepinskas, J. Savickiene, M. Sandig and P.R. Kviets. 2003. Neutrophils induce sequential focal changes in endothelial adherens junction components: role of elastase. *Microcirculation* 10(2): 205–220.

Kedrin, D., B. Gligorijevic, J. Wyckoff, V.V. Verkhusha, J. Condeelis, J.E. Segall and J. van Rheenen. 2008. Intravital imaging of metastatic behavior through a mammary imaging window. *Nat. Methods* 5: 1019–21.

Khuon, S., L. Liang, R.W. Dettman, P.H. Sporn, R.B. Wysolmerski and T.L. Chew. 2010. Myosin light chain kinase mediates transcellular intravasation of breast cancer cells through the underlying endothelial cells: a three-dimensional FRET study. *J. Cell Sci.* 123: 431–40.

Kim, M.Y., T. Oskarsson, S. Acharyya, D.X. Nguyen, X.H. Zhang, L. Norton and J. Massague´.2009. Tumor Self-Seeding by Circulating Cancer Cells. *Cell* 139: 1315–1326.

Kollmannsberger, P., C.T. Mierke and B. Fabry. 2011. Nonlinear viscoelasticity of adherent cells is controlled by cytoskeletal tension. *Soft Matter.* 7: 3127–3132.

Kopfstein, L. and G. Christofori. 2006. Metastasis: cell-autonomous mechanisms versus contributions by the tumor microenvironment. *Cell Mol. Life Sci.* 63: 449–68.

Kumar, S. and V.M. Weaver. 2009. Mechanics, malignancy, and metastasis: The force journey of a tumor cell. *Cancer and Metastasis Reviews* 28: 113–127.

Langley, R.R. and I.J. Fidler.2007. Tumor cell–organ microenvironment interactions in the pathogenesis of cancer metastasis. *Endocr. Rev.* 28: 297–321.

Lee W.L. and W.C. Liles. 2011. Endothelial activation, dysfunction and permeability during severe infections. *Curr. Opin. Hematol.* 18(3): 191–196.

Ley, K., C. Laudanna, M.I. Cybulsky and S. Nourshargh. 2007. Getting to the site of inflammation: the leukocyte adhesion cascade updated. *Nat. Rev. Immunol.* 7: 678–689.

Li Y.H. and C. Zhu. 1999. A modified Boyden chamber assay for tumor cell transendothelial migration *in vitro. Clin. Exp. Metastasis* 17: 423–429.

Månsson-Broberg, A., A.J. Siddiqui, M. Genander, K.H. Grinnemo, X. Hao, A.B. Andersson, E. Wä rdell, C. Sylve´n and M. Corbascio. 2008. Modulation of ephrinB2 leads to increased angiogenesis in ischemic myocardium and endothelial cell proliferation. *Biochem. Biophys. Res. Commun.* 373: 355–359.

Metzner, C., C. Raupach, C.T. Mierke and B. Fabry. 2010. Fluctuations of cytoskeleton-bound microbeads—effect of bead-receptor binding dynamics. *J. Phys.: Condens. Matter.* 22: 194105.

Mierke, C.T. 2008. Role of the Endothelium during Tumor Cell Metastasis: Is the Endothelium a Barrier or a Promoter for Cell Invasion and Metastasis? *J. Biophys.* 200813, ID 183516.

Mierke, C.T. 2011. The Biomechanical Properties of 3d Extracellular Matrices and Embedded Cells Regulate the Invasiveness of Cancer Cells. *Cell Biochem. Biophys.* 61: 217–236.

Mierke, C.T. 2011. Cancer cells regulate biomechanical properties of human microvascular endothelial cells. *J. Biol. Chemisty* 286: 40025–40037.

Mierke, C.T., B. Frey, M. Fellner, M. Herrmann and B. Fabry. 2011. Integrin α5β1 facilitates cancer cell invasion through enhanced contractile forces. *J. Cell Science* 124: 369–83.

Mierke, C.T., D. Rosel, B. Fabry and J. Brabek. 2008. Contractile forces in tumor cell migration. *Eur. J. Cell Biol.* 87: 669–76.

Mierke, C.T., D. Rosel, B. Fabry and J. Brabek. 2008. Contractile forces in tumor cell migration. *Eur. J. Cell Biol.* 87: 669–676.

Mierke, C.T., D.P. Zitterbart, P. Kollmannsberger, C. Raupach, U. Schlotzer-Schrehardt, T.W. Goecke, J. Behrens and B. Fabry. 2008. Breakdown of the endothelial barrier function in tumor cell transmigration. *Biophys. J.* 94: 2832–46.

Mierke, C.T., N. Bretz and P. Altevogt. 2011. Contractile forces contribute to increased GPI-anchored receptor CD24 facilitated cancer cell invasion. *J. Biol. Chem.* 286: 34858–34871.

Mierke, C.T., P. Kollmannsberger, D.P. Zitterbart, G. Diez, T.M. Koch, S. Marg, W.H. Ziegler, W.H. Goldmann and B. Fabry. 2010. Vinculin facilitates cell invasion into three-dimensional collagen matrices. *J. Biol. Chem.* 285: 13121–30.

Mishima, T., M. Naotsuka, Y. Horita, M. Sato, K. Ohashi and K. Mizuno. 2010. LIM-kinase is critical for the mesenchymal-to-amoeboid cell morphological transition in 3D matrices. *Biochem. Biophys. Res. Commun.* 392: 577–81.

Miteva, D.O., J.M. Rutkowski, J.B. Dixon, W. Kilarski, J.D. Shields and M.A. Swartz. 2010. Transmural flow modulates cell and fluid transport functions of lymphatic endothelium. *Circ. Res.* 106: 920–931.

Moser, R., J. Fehr, L. Olgiati and P.L. Bruijnzeel. 1992. Migration of primed human eosinophils across cytokine-activated endothelial cell monolayers. *Blood* 79: 2937–45.

Muller, W.A. 2011. Mechanisms of leukocyte transendothelial migration. *Annu. Rev. Pathol.* 6: 323–44.

Nguyen, D.X., P.D. Bos and J. Massague. 2009. Metastasis: from dissemination to organ-specific colonization. *Nat. Rev. Cancer* 9: 274–284.

Nobes, C.D. and A. Hall. 1995. Rho, rac, and cdc42 GTPases regulate the assembly of multimolecular focal complexes associated with actin stress fibers, lamellipodia, and filopodia. *Cell* 81: 53–62.

Norton, L. and J. Massague. 2006. Is cancer a disease of self-seeding? *Nat. Med.* 12: 875–878.

Padua, D., X.H. Zhang, Q. Wang, C. Nadal, W.L. Gerald, R.R. Gomis and J. Massague. 2008. TGFbeta primes breast tumors for lung metastasis seeding through angiopoietin-like 4. *Cell* 133: 66–77.

Pantel, K. and R.H. Brakenhoff. 2004. Dissecting the metastatic cascade. *Nat. Rev. Cancer* 4: 448–456.

Parsons, S.A., R. Sharma, D.L. Roccamatisi, H. Zhang, B. Petri, P. Kubes, P. Colarusso and K.D. Patel. 2012. Endothelial paxillin and focal adhesion kinase play a critical role in neutrophil transmigration. *Eur. J. Immunol.* 42: 436–446.

Paszek, M.J., D. Boettiger, V.M. Weaver and D.A. Hammer. 2009. Integrin clustering is driven by mechanical resistance from the Glycocalyx and the substrate. *PLoS Comput. Biol.* 5: e1000604.

Paszek, M.J., N. Zahir, K.R. Johnson, J.N. Lakins, G.I. Rozenberg, A. Gefen, C.A. Reinhart-King, S.S Margulies, M. Dembo, D. Boettiger, D.A. Hammer and V.M. Weaver. 2005. Tensional homeostasis and the malignant phenotype. *Cancer Cell* 8: 241–54.

Pober, J.S. 1988. Cytokine-mediated activation of vascular endothelium. *Am. J. Pathol.* 133: 426–433.

Rabodzey, A., P. Alcaide, F.W. Luscinskas and B. Ladoux. 2008. Mechanical forces induced by the transendothelial migration of human neutrophils. *Biophys. J.* 95: 1428–1438.

Rafii, S., S.T. Avecilla and D.K. Jin. 2003. Tumor vasculature address book: identification of stage-specific tumor vessel zip codes by phage display. *Cancer Cell* 4: 331–333.

Raupach, C., D. Paranhos-Zitterbart, C. Mierke, C. Metzner, A.F. Müller and B. Fabry. 2007. Stress fluctuations and motion of cytoskeletal-bound markers. *Phys. Rev. E: Stat. Phys., Plasmas, Fluids, Relat. Interdiscip. Top.* 76011918.

Reid, P.E., N.J. Brown and I. Holen. 2009. Breast cancer cells stimulate osteoprotegerin (OPG) production by endothelial cells through direct cell contact. *Mol. Cancer.* 15, 8: 49.

Rosel, D., J. Brabek, O. Tolde, C.T. Mierke, D.P. Zitterbart, C. Raupach, K. Bicanova, P. Kollmannsberger, D. Pankova, P. Vesely, P. Folk and B. Fabry. 2008. Up-regulation of Rho/ROCK Signaling in Sarcoma Cells Drives Invasion and Increased Generation of Protrusive Forces. *Mol. Cancer Res.* 6: 1410–20.

Ruoslahti, E., S.N. Bhatia and M.J. Sailor. 2010. Targeting of drugs and nanoparticles to tumors. *J. Cell Biol.* 188: 759–768.

Saigusa, S., Y. Toiyama, K. Tanaka, T. Yokoe, Y. Okugawa, H. Fujikawa, K. Matsusita, M. Kawamura, Y. Inoue, C. Miki and M. Kusunoki. 2011. Cancer-associated fibroblasts correlate with poor prognosis in rectal cancer after chemoradiotherapy. *Int. J. Oncol.* Jan 14. doi: 10.3892/ijo.2011.906. [Epub ahead of print].

Sato, T.N., Y. Tozawa, U. Deutsch, K. Wolburg-Buchholz, Y. Fujiwara, M. Gendron-Maguire, T. Gridley, H. Wolburg, W. Risau and Y. Qin. 1995. Distinct roles of the receptor tyrosine kinases Tie-1 and Tie-2 in blood vessel formation. *Nature* 376: 70–74.

Sawhney, R.S., M.M. Cookson, Y. Omar, J. Hauser and M.G. Brattain. 2006. Integrin alpha2-mediated ERK and calpain activation play a critical role in cell adhesion and motility via focal adhesion kinase signaling: identification of a novel signaling pathway. *J. Biol. Chem.* 281: 8497–510.

Scheel, C., T. Onder, A. Karnoub and R.A. Weinberg. 2007. Adaptation versus selection: the origins of metastatic behavior. *Cancer Res.* 67: 11476–11479.

Scotland, R.S., M. Madhani, S. Chauhan, S. Moncada, J. Andresen, H. Nilsson, A.J. Hobbs and A. Ahluwalia. 2005. Investigation of vascular responses in endothelial nitric oxide synthase/cyclooxygenase-1 double-knockout mice: key role for endothelium-derived hyperpolarizing factor in the regulation of blood pressure *in vivo*. *Circulation* 111: 796–803.

Seufferlein, T. and E. Rozengurt. 1995. Sphingosylphosphorylcholine rapidly induces tyrosine phosphorylation of p125FAK and paxillin, rearrangement of the actin cytoskeleton and focal contact assembly. Requirement of p21rho in the signaling pathway. *J. Biol. Chem.* 270: 24343–24351.

Shaw, S.K., B.N. Perkins, Y.C. Lim, Y. Liu, A. Nusrat, F.J. Schnell, C.A. Parkos and F.W. Luscinskas. 2001. Reduced expression of junctional adhesion molecule and platelet/endothelial cell adhesion molecule-1 (CD31) at human vascular endothelial junctions by cytokines tumor necrosis factor-alpha plus interferongamma does not reduce leukocyte transmigration under flow. *Am. J. Pathol.* 159: 2281–2291.

Shimonaka, M., K. Katagiri, T. Nakayama, N. Fujita, T. Tsuruo, O. Yoshie and T. Kinashi. 2003. Rap1 translates chemokine signals to integrin activation, cell polarization, and motility across vascular endothelium under flow. *J. Cell Biol.* 161: 417–427.

Sims, J.R., S. Karp and D.E. Ingber. 1992. Altering the cellular mechanical force balance results in integrated changes in cell, cytoskeletal and nuclear shape. *J. Cell Sci.* 103(Pt 4): 1215–22.

Solinas, G., S. Schiarea, M. Liguori, M. Fabbri, S. Pesce, L. Zammataro, F. Pasqualini, M. Nebuloni, C. Chiabrando, A. Mantovani and P. Allavena. 2010. Tumor-conditioned macrophages secrete migration-stimulating factor: a new marker for M2-polarization, influencing tumor cell motility. *J. Immunol.* 185: 642–52.

Steeg, P.S. 2006. Tumor metastasis: mechanistic insights and clinical challenges. *Nat. Med.* 12: 895–904.

Steinberg, B.E., N.M. Goldenberg and W.L. Lee. 2012. Do viral infections mimic bacterial sepsis? The role of microvascular permeability: a review of mechanisms and methods. *Antiviral Res.* 93: 2–15.

Stemmer, V., B. de Craene, G. Berx and J. Behrens. 2008. Snail promotes Wnt target gene expression and interacts with beta-catenin. *Oncogene* 27: 5075–80.

Stoecklein, N.H., S.B. Hosch, M. Bezler, F. Stern, C.H. Hartmann, C. Vay, A. Siegmund, P. Scheunemann, P. Schurr, W.T. Knoefel, P.E. Verde, U. Reichelt, A. Erbersdobler, R. Grau, A. Ullrich, J.R. Izbicki and C.A. Klein. 2008. Direct genetic analysis of single disseminated cancer cells for prediction of outcome and therapy selection in esophageal cancer. *Cancer Cell* 13: 441–453.

Su, W.H., H.I. Chen and C.J. Jen. 2002. Differential movements of VEcadherin and PECAM-1 during transmigration of polymorphonuclear leukocytes through human umbilical vein endothelium. *Blood* 100: 3597–3603.

Surapisitchat, J., K.I. Jeon, C. Yan and J.A. Beavo. 2007. Differential regulation of endothelial cell permeability by cGMP via phosphodiesterases 2 and 3. *Circ. Res.* 101: 811–818.

Suresh, S., J. Spatz, J.P. Mills, A. Micoulet, M. Dao, C.T. Lim, M. Beil and T. Seufferlein. 2005. Connections between single-cell biomechanics and human disease states: gastrointestinal cancer and malaria. *Acta Biomaterialia* 1: 15–30.

Thery, M., A. Pepin, E. Dressaire, Y. Chen and M. Bornens. 2006. Cell distribution of stress fibres in response to the geometry of the adhesive environment. *Cell Motil. Cytoskeleton* 63: 341–55.

Tremblay, P.-L., F.A. Auger and J. Huot. 2006. Regulation of transendothelial migration of colon cancer cells by E-selectin-mediated activation of p38 and ERK MAP kinases *Oncogene* 25: 6563–6573.

van Hinsbergh, W.M. 1997. Endothelial permeability for macromolecules. Mechanistic aspects of pathophysiological modulation. *Arterioscler. Thromb. Vasc. Biol.* 17: 1018–1023.

Van Sluis, G.L., T.M. Niers, C.T. Esmon, W. Tigchelaar, D.J. Richel, H.R. Buller, C.J. Van Noorden and C.A. Spek. 2009. Endogenous activated protein C limits cancer cell extravasation through sphingosine-1-phosphate receptor 1-mediated vascular endothelial barrier enhancement. *Blood* 114: 1968–1973.

Weaver, V. M., S. Lelièvre, J.N. Lakins, M.A. Chrenek, J.C.R. Jones, F. Giancotti, Z. Werb and M.J. Bissell. 2002. b4 integrindependent formation of polarized three-dimensional architecture confers resistance to apoptosis in normal and malignant mammary epithelium. *Cancer Cell* 2: 205–216.

Weis, S., J. Cui, L. Barnes and D. Cheresh. 2004. Endothelial barrier disruption by VEGF-mediated Src activity potentiates tumor cell extravasation and metastasis. *J. Cell Biol.* 167: 223–9.

Wittchen, E.S., R.A. Worthylake, P. Kelly, P.J. Casey, L.A. Quilliam and K. Burridge. 2005. Rap1 GTPase inhibits leukocyte transmigration by promoting endothelial barrier function. *J. Biol. Chem.* 280: 11675–82.

Wolf, K., I. Mazo, H. Leung, K. Engelke, U.H. von Andrian, E.I. Deryugina, A.Y. Strongin, E.B. Brocker and P. Friedl. 2003. Compensation mechanism in tumor cell migration: mesenchymal-amoeboid transition after blocking of pericellular proteolysis. *J. Cell Biol.* 160: 267–77.

Wolf, K., Y.I. Wu, Y. Liu, J. Geiger, E. Tam, C. Overall, M.S. Stack and P. Friedl. 2007. Multi-step pericellular proteolysis controls the transition from individual to collective cancer cell invasion. *Nat. Cell Biol.* 9: 893–904.

Woodfin, A., M.B. Voisin, B.A. Imhof, E. Dejana, B. Engelhardt and S. Nourshargh. 2009. Endothelial cell activation leads to neutrophil transmigration as supported by the sequential roles of ICAM-2, JAM-A, and PECAM-1. *Blood* 113: 6246–6257.

Yu, H., J.K. Mouw and V.M. Weaver. 2011. Forcing form and function: biomechanical regulation of tumor evolution. *Trends Cell Biol.* 21: 47–56.

Yuan, S.Y., Q. Shen, R.R. Rigor and M.H. Wu. 2012. Neutrophil transmigration, focal adhesion kinase and endothelial barrier function. *Microvasc. Res.* 83(1): 82–88 Epub 2011 Aug 16.

Zaidel-Bar, R., C. Ballestrem, Z. Kam and B. Geiger. 2003. Early molecular events in the assembly of matrix adhesions at the leading edge of migrating cells. *J. Cell Sci.* 116: 4605–13.

Zaman, M.H., L.M. Trapani, A. Siemeski, D. Mackellar, H. Gong, R.D. Kamm, A. Wells, D.A. Lauffenburger and P. Matsudaira. 2006. Migration of tumor cells in 3D matrices is governed by matrix stiffness along with cell-matrix adhesion and proteolysis. *Proc. Natl. Acad. Sci. USA* 103: 10889–94.

Zhang, J., P. Alcaide, L. Liu, J. Sun, A. He, F.W. Luscinskas and G.P. Shi. 2011. Regulation of endothelial cell adhesion molecule expression by mast cells, macrophages, and neutrophils. *PLoS One* Jan 14; 6(1): e14525.

Zijlstra, A., J. Lewis, B. Degryse, H. Stuhlmann and J.P. Quigley. 2008. The inhibition of tumor cell intravasation and subsequent metastasis via regulation of *in vivo* tumor cell motility by the tetraspanin CD151. *Cancer Cell* 13: 221–234.

4

Valvular Endothelial Mechanobiology

Suthan Srigunapalan[1] and Craig A. Simmons[1],*

Introduction

The human heart contains four valves that ensure unidirectional blood flow throughout the cardiac cycle. The semilunar valves (pulmonary on the right side and aortic on the left) prevent retrograde flow back into the ventricles during diastole. The atrio-ventricular valves (tricuspid on the right and mitral on the left) prevent retrograde flow from the ventricle to the atrium when the heart contracts during systole. All four valves function passively, opening and closing due to inertial forces exerted by the flowing blood. Although their function is seemingly simple, the biomechanical demands placed on the cardiac valves are substantial, and specialized valve tissue and cells have evolved to ensure proper function under conditions of extreme multi-modal loading repeated with every heartbeat.

Remarkably, the majority of people will never suffer valve malfunction in their lifetime, a testament to the robustness of the valve tissue and its homeostatic mechanisms. Nonetheless, valvular diseases are among the most common heart diseases and are associated with significant

[1]Department of Mechanical and Industrial Engineering and Institute of Biomaterials and Biomedical Engineering, University of Toronto, Toronto, ON, Canada, M5S 3G8.
*Corresponding author: c.simmons@utoronto.ca

cardiovascular morbidity (Rajamannan et al. 2011). The most common valvular disease is calcific aortic valve disease (CAVD), which describes a spectrum of disease from initial alterations in valve cell biology to end-stage calcification resulting in valve malfunction (Rajamannan et al. 2011). Because of the prevalence of CAVD, the aortic valve is the most studied and best understood in terms of its mechanical and biological characteristics, and is the focus of this review. The hypothesized etiology of CAVD has changed considerably in recent years. Originally thought to be a passive disease associated with aging and long-term wear and tear, CAVD is now recognized to be an active process that cannot be described simply as senile or degenerative (Rajamannan et al. 2011).

Although CAVD is associated epidemiologically with systemic risk factors (Stewart et al. 1997), lesions do not form uniformly throughout the aortic valve, but rather develop focally and preferentially in the fibrosa, the interstitial layer on the outflow side of the leaflet closest to the aorta (Otto et al. 1994, Obrien et al. 1996) (Fig. 1). Side-specific pathosusceptibility suggests that factors local to the fibrosa promote, or at least permit, pathological development. While several biochemical, cellular, and extracellular matrix factors may define the fibrosa microenvironment to contribute to CAVD (Chen and Simmons 2011, Yip and Simmons 2011), local hemodynamic forces stand out as a likely contributor. Altered hemodynamics likely also explains accelerated CAVD in patients with congenital valve malformations, such as bicuspid aortic valves (Beppu et al. 1993). The hypothesized connection between hemodynamics and valve function is supported by

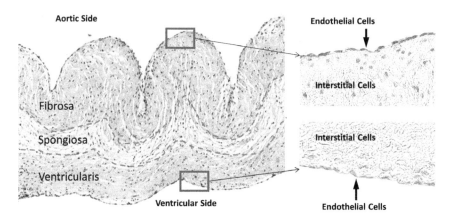

Figure 1. Cross-section of a porcine aortic valve leaflet demonstrating the striated three layer structure, which contains valvular interstitial cells and is lined by the valvular endothelial cells. Adapted from Simmons et al. (2005).

Color image of this figure appears in the color plate section at the end of the book.

emerging evidence of the role of mechanical forces in regulating aortic valve cell biology and pathological responses, including those of the aortic valvular endothelium, the focus of this chapter.

The Structure and Composition of the Aortic Valve

The aortic valve is situated between the left ventricle and the aorta. Its primary function is to maintain unidirectional blood flow during systole to prevent regurgitation of blood from the aorta back into the left ventricle during diastole. The aortic valve is composed of three thin semilunar cusps (or leaflets) that attach at their base to the aortic root. During diastole, the free edges of the leaflets coapt to prevent retrograde flow. The leaflets, which are normally less than 1 mm in thickness in humans (Otto et al. 1994), are stratified into three distinct layers: the fibrosa, spongiosa, and ventricularis (Fig. 1). The fibrosa layer is closest to the aortic root on the outflow side of the valve and is composed of dense, circumferentially-oriented type I and III collagen fibres that provide structural integrity to the leaflets and bear much of the load placed on the valve. The middle layer, the spongiosa, is rich in proteoglycans and gylcosaminoglycans with scattered collagen fibres. It links and lubricates the fibrosa and ventricularis layers as they shear and deform relative to each other during leaflet movement and pressurization (Sacks et al. 2009). The ventricularis is on the ventricular (inflow) surface of the leaflets and is a thin layer composed of collagen and elastin fibres. The radial orientation of the elastin is thought to reduce radial strains that occur during systole and to aid in leaflet recoil (Schoen 2008).

Cells of the Aortic Valve

The cellular components of the aortic valve include valvular interstitial cells (VICs) that populate the extracellular matrix of the leaflet tissue and a monolayer of valvular endothelial cells (VECs) that line the outer blood-contacting surfaces of the leaflets (Fig. 1).

Valvular Interstitial Cells

In healthy adult aortic valves, VICs are a heterogeneous population of fibroblasts (>95% of total population (Rabkin-Aikawa et al. 2004, Pho et al. 2008)), myofibroblasts, and smooth muscle cells. VICs maintain normal valve structure and function through matrix synthesis and remodeling activities. As maladaptive remodeling of valve tissue is a defining feature of many valve diseases, dysregulation of VICs is likely responsible for disease progression. In swine (Chen et al. 2009), and likely in humans (Liu et al. 2007,

Schoen 2008), there is a sizeable subpopulation of mesenchymal progenitor/
stem cells with differentiation potential to the myofibrogenic, osteogenic,
chondrogenic, and adipogenic lineages, pathological phenotypes that are
observed to contribute to ectopic mesenchymal tissue formation in diseased
valves. Another source of pathological VICs may be from the endothelium
through endothelial-to-mesenchymal transformation (EMT) (Bischoff and
Aikawa 2011). In valve development, VICs arise from endocardial cells
of the endocardial cushion that undergo EMT (Lincoln et al. 2004) and as
discussed below, the capacity for EMT persists into adulthood.

Valvular endothelial cells

VECs line the endocardial cushion during development and also contribute
to the VIC population through EMT during development (Lincoln et al.
2004). VECs from adult valves also can be differentiated to VIC-like cells
that express α-smooth muscle actin (SMA) when treated with transforming
growth factor (TGF)-ß1, an effect that is opposed by vascular endothelial
growth factor (VEGF) (Paranya et al. 2001) (Fig. 2). Additionally, cells
co-expressing the endothelial marker CD31 and SMA are observed on the
endothelial surface and in subendothelial regions of human aortic valves
(Paranya et al. 2001). Notably, mitral VECs demonstrate osteogenic and
chondrogenic differentiation potential *in vitro*, and osteocalcin, a marker
of osteogenic differentiation, was detected *in vivo* along the endothelium
of mechanically-stressed mitral valves (Wylie-Sears et al. 2011). These
intriguing findings, which are not observed in non-valvular endothelial
cells, demonstrate the potential for VECs to directly contribute to valve
pathology through EMT and ectopic tissue formation. It is not known,
however, how frequently this occurs in humans *in vivo*, and therefore the
significance and impact of EMT to CAVD has yet to be determined.

Mature VECs have a limited ability to proliferate, and thus endothelial
progenitor cells (EPCs) may be important to maintaining valvular
endothelial monolayer integrity and function (Matsumoto et al. 2009).
Patients with severe aortic valve stenosis have a reduction in the number
and function of EPCs and increased EC senescence on the aortic side, where
lesions are most likely to form (Matsumoto et al. 2009). Dysfunction or
disruption of the endothelial monolayer on the surface of the aortic valve
may contribute to disease initiation and progression by compromising
its barrier function and its roles in maintenance of a non-thrombogenic
environment, mediation of inflammation, regulation of permeability, and
regulation of VIC function.

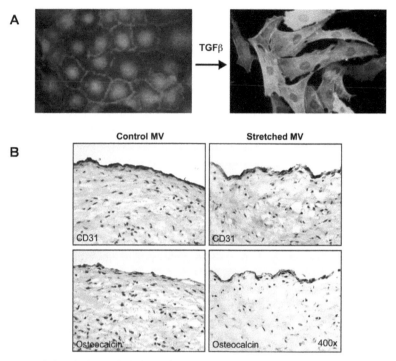

Figure 2. Adult valvular endothelial cells (VECs) can undergo endothelial-to-mesenchymal transformation (EMT). (A) Ovine aortic VEC clone treated without (left) or with (right) transforming growth factor-β1 for 5 days. The loss of endothelial marker expression (CD31 in red), the loss of cell-cell contacts, and the induction of α-smooth muscle actin expression (green) are hallmarks of EMT. Adapted from Bischoff and Aikawa (2011). (B) In ovine mitral valves subjected to elevated mechanical stress *in vivo*, VECs co-express CD31 and osteocalcin, suggesting that they have osteogenic differentiation potential that can be induced mechanically. Adapted from Wylie-Sears et al. (2011).

Color image of this figure appears in the color plate section at the end of the book.

Regulation of coagulation

Similar to vascular endothelial cells, VECs express nitric oxide (Siney and Lewis 1993), prostacyclin (Manduteanu et al. 1988, Pompilio et al. 1998), von Willebrand factor (Lester et al. 1993), and tissue factor (Drake and Pang 1989) to regulate coagulation. VECs are normally non-thrombogenic, and even under conditions that favor platelet adhesion and thrombosis in the vasculature (e.g., high shear rates that occur on the inflow surface of the aortic valve (Tsai 2003)), thrombosis of the aortic valve is rare (Barandon et al. 2004). This suggests that valvular and vascular endothelial cells differ in their abilities to resist the denudation and coagulation that accompanies high shear stresses (Butcher and Nerem 2007).

Regulation of inflammation

Activation of the vascular endothelium to express adhesion receptors and recruit monocytes is an initiating event in atherosclerosis. Similarly, VEC dysfunction and activation is thought to play a role in initiating valvular diseases, particularly calcific aortic valve disease (CAVD). In support of this hypothesis, valve disease is associated with systemic endothelial dysfunction (Poggianti et al. 2003) and VECs in diseased valves have increased expression of VCAM-1, ICAM-1, and E-selectin (Ghaisas et al. 2000, Muller et al. 2000). Notably, VECs on the disease-prone aortic side of normal aortic valves and aortic valves in swine fed a high cholesterol diet for two weeks demonstrate anti-inflammatory and anti-oxidative transcriptional profiles (Simmons et al. 2005, Guerraty et al. 2010), which may balance the aortic side's apparent susceptibility to calcification (Simmons et al. 2005) and provide acute protection against systemic pro-pathological insult.

Regulation of permeability

VECs play an important role in regulating transport of macromolecules between circulating blood and the valve interstitium. Tomkins et al. (1989) quantified the uptake of labeled low-density lipoprotein (LDL) and albumin in valve leaflets of squirrel monkeys and rabbits. Using a simplified model they determined that the aortic valve was much more permeable than major arteries. Additionally, there were large variations in concentration profiles measured across individual leaflets suggesting that the valve endothelium contains focal regions of high and low permeability. This non-uniformity in macromolecular transport has also been observed in rat aortic valves (Zeng et al. 2007) and may be responsible for focal lesion development during early stages of disease (Guerraty et al. 2010). In advanced aortic valve disease, the leaflet endothelium is damaged at lesion sites, which occur primarily on the aortic side of the valve (Mirzaie et al. 2002). In addition to a loss of endothelial cells, stenotic aortic valves contain endothelial cells with abnormal morphologies and microvilli and increased intercellular separation (Riddle et al. 1980), which likely increases permeability (Tompkins et al. 1989).

Regulation of VICs

Endothelial cells are believed to play a significant role in regulating VIC function. In addition to their barrier function, which regulates the penetration of biochemical factors and inflammatory cells and shields VICs from hemodynamic shear stress, VECs regulate the function of VICs and other local cell types through paracrine signaling.

Just as vascular ECs regulate blood vessel compliance, VECs have been shown to regulate aortic valve stiffness *ex vivo* (El-Hamamsy et al. 2009). Using intact porcine aortic valves, El-Hamamsy et al. measured valve tissue elastic modulus and contraction force in response to serotonin, which induces VIC contraction (Chester et al. 2000). Tissue mechanical properties were measured with the valve leaflet held in tension, stretched equally in the radial and circumferential directions to mimic the physiological loading due to diastolic transvalvular pressure. In this basal state, VIC-mediated contraction contributed to tissue stiffness, but the endothelium did not. Stiffness and contraction were *decreased* in intact tissue samples when treated with serotonin, likely due to endothelial release of nitric oxide (NO) and its induction of VIC relaxation, as inhibition of NO synthase or denudation of the endothelium reversed the effect of serotonin, causing an increase in tissue stiffness (and presumably VIC contractility). VECs have been shown to secrete NO (Siney and Lewis 1993) and NO has also been implicated in regulating dilation of the AV leaflets through interaction with neurons (Chester et al. 2008). NO donors have also been shown to inhibit calcified aggregate formation by VICs *in vitro* (Kennedy et al. 2009) (a model of VIC pathological differentiation), and consistent with this, removal of VECs from aortic valve explants promotes calcific nodule formation (Mohler et al. 1999).

Importantly, these studies did not include fluid flow-induced shear stress applied to the endothelium, which is known in the vasculature to modulate EC secretion of vasoactive substances like NO (Cooke 2003). As discussed in a later section, shear stress may play an important role in regulating VEC-VIC communication and the focal nature of valve disease (Yip and Simmons 2011). For example, VECs on the disease-prone aortic side of aortic valve leaflets (where shear stresses are low in magnitude) have reduced expression of several proteins that putatively inhibit local fibrosis and calcification by VICs via paracrine signaling (Simmons et al. 2005).

Forces Experienced by the Aortic Valve

The aortic valve opens and closes over 30 million times each year, exposing the leaflets and cells of the valve to some of the most complex and greatest stresses within the body. Aortic valve tissue mechanics has been well studied in the context of the design and failure analysis of mechanical and bioprosthetic valves. Increased appreciation of the association between hemodynamics and focal valve disease development along with the putative role for mechanics in regulating valve cell (dys)function has motivated similar studies aimed at characterizing the forces experienced by native valves, as summarized below.

Hemodynamic shear stress

The aortic valve opens during systole and closes during diastole in response to contraction and relaxation of the left ventricle, respectively. The movement of blood from the left ventricle into the aorta induces shear stresses on the surfaces of the valve leaflets. Along the aortic wall, low inertial flow develops and results in vortices in the sinuses of the aortic root that help close the aortic valve during diastole (Sacks and Yoganathan 2007). As a consequence, the ventricular and aortic surfaces of the aortic valve (and the VECs on them) experience very different hemodynamic environments (Fig. 3).

Characterization of shear stresses on the surface of aortic valve leaflets is challenging. Non-invasive medical imaging modalities such as ultrasound and magnetic resonance imaging can provide general information about flow through the valve, but these techniques do not have sufficient spatial or temporal resolution to accurately quantify fluid shear stresses on the surface of moving leaflets. A variety of approaches have been used to

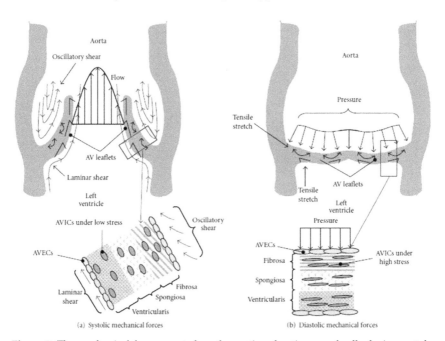

Figure 3. The mechanical forces exerted on the aortic valve tissue and cells during systole (left) and diastole (right). From Balachandran et al. (2011).

characterize shear stresses on or around artificial and native valves, with estimates of peak wall shear stresses ranging from 20 dynes/cm² to over 1000 dynes/cm² (Butcher and Nerem 2007). The best estimates to date come from recent experimental studies that used laser Doppler velocimetry to estimate fluid shear stresses on artificial (Weston et al. 1999) or natural valves (Yap et al. 2011, Yap et al. 2011) mounted in *ex vivo* flow loops that replicate the geometries of the aortic valve environment. Using physiological pressures and flow waveforms that mimic the complex fluid environment *in vivo*, Yap et al. (Yap et al. 2011, Yap et al. 2011) estimated peak wall shear stress of 64–71 dynes/cm² and 18–20 dynes/cm² on the ventricular (inflow) and aortic (outflow) surfaces of native aortic valves, respectively (Fig. 4). The difference in shear stress magnitude on opposite sides of the leaflets is intuitive and consistent with previous studies (Kilner et al. 2000, Ge and Sotiropoulos 2010, Balachandran et al. 2011). However, contrary to expectations that the ventricular side of the valve experiences unidirectional

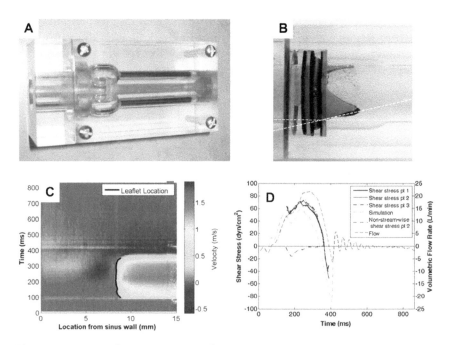

Figure 4. Experimental system to estimate shear stresses on the surface of the aortic valve. (A) Acrylic chamber used to measure shear stresses on polymeric aortic valve leaflets. (B) Image of polymeric valve during opening. Dotted lines depict how opening angles were quantified. (C) Velocity map obtained using laser Doppler velocimetry in stream wise direction. (D) Plot of fluid shear stress measured on the ventricular surface of the valve leaflets at three separate points. Adapted from Yap et al. (2011).

Color image of this figure appears in the color plate section at the end of the book.

shear stress and the aortic side is subjected to recirculating flow, Yap et al. found that opposite: the ventricular surface of the aortic valve experienced oscillatory shear stress and the aortic surface experienced shear stresses that were mostly unidirectional (Yap et al. 2011, Yap et al. 2011). The relative sensitivity of VECs to shear stress of different magnitudes and patterns has not been studied, so the implications of these findings for valve biology are still unclear. Furthermore, the authors noted that *in vitro* models lack features of the *in vivo* environment that influence valve dynamics, including compliance and geometry of the aortic root, and calculated shear stresses are sensitive to the positioning of the valve within the flow loop. Also, shear stresses were measured at only one location on the leaflet surface (Yap et al. 2011). Thus, more accurate characterization of the *in vivo* hemodynamic environment will require improved models and methods to spatially map shear stress profiles. Advanced high-resolution computational models that account for fluid-structure interactions may be a valuable tool to that end (De Hart et al. 2003, De Hart et al. 2004, Weinberg and Mofrad 2008, Ge and Sotiropoulos 2010, Chandra et al. 2011).

Other mechanical stresses

In addition to fluid shear stress, the aortic valve experiences mechanical stresses in the forms of pressures, strains, and bending forces (Fig. 3). Pressure on the aortic valve varies over a cardiac cycle. Under normal conditions, the peak transvalvular pressure is ~80 mm Hg. The pressure load is carried completely by the valve during diastole, and decreases to zero during systole (Thubrikar 1990). Pressure across the valve cause the leaflets to curve and bend, inducing both compressive and tensile stresses. Pressure also causes the leaflets to increase in length thus inducing a circumferential and radial strain on the tissue and residing cells. *In vivo* studies on canine aortic valves determined that the circumferential stress is 2.4 kPa during diastole compared to 0.167 kPa during systole (Thubrikar 1990). These results are comparable to a maximum principal stress of 2.19 kPa estimated by computational modeling of a human aortic valve leaflets (Cataloglu et al. 1976). Pressure loading during diastole results in elongation of the canine leaflets by 31% in the radial direction and only 11% in the circumferential direction (Thubrikar 1990). The anisotropic response of the leaflet tissue to pressure loading is explained by the arrangement of collagen and elastin within the valve tissue: the circumferential organization of collagen fibers make valve leaflets less compliant and stronger in the circumferential direction compared to the radial direction.

The Response of the Valvular Endothelium to Forces

While the mechanobiology of vascular ECs has a rich history, valvular EC mechanobiology is comparably understudied. However, with the recognition that CAVD and other valvular diseases are active processes with endothelial involvement, interest in VEC mechanobiology and its role in valve function and disease has increased. Historical and recent investigations into how mechanical forces influence the biology of VECs and their specific role in valve (patho) biology are reviewed below.

Valve endothelial responses to fluid shear stress

While not studied nearly as extensively as the vascular endothelium, several studies clearly demonstrate the shear stress sensitivity of VEC morphology and gene and protein expression; suggest differences in how VECs and vascular endothelial cells respond to shear stress; and implicate important roles for shear stress in regulating valve development, homeostasis, and disease.

Valve development

Morphogenesis of the developing embryonic heart is influenced by biomechanical and hemodynamic forces (Santhanakrishnan and Miller 2011). Valvulogenesis is particularly sensitive to mechanical stimuli, including flow-induced shear stress (Freund et al. 2012). Manipulation of intracardiac blood flow in embryonic zebrafish hearts (which are transparent and therefore amenable to flow visualization *in vivo*) can disturb or arrest valve growth. Hove et al. (2003) used small beads to obstruct flow into or out of the heart tube to reduce wall shear stress by ~10-fold, and demonstrated disrupted valvulogenesis and cardiac function (Fig. 5). However, this approach also affected myocardial function, making it difficult to tease out the relative contributions of shear stress versus contraction-induced cell and tissue deformation to valve formation. Indeed, inhibition of myocardial contraction genetically or pharmacologically also results in defective valvulogenesis (Bartman et al. 2004), although this study could not discount the influence of shear stress. Flow patterns rather than shear stress magnitude may be a more potent signal for valvulogenesis: for example, valve dysgenesis in zebrafish is more sensitive to reduced oscillatory (reversing) flow than changes in shear stress magnitude (Vermot et al. 2009). Further, reducing flow oscillations decreased expression of the endothelial shear sensitive gene *klf2a*, and knockdown of *klf2a* with morpholine antisense oligonucleotides resulted in valve dysfunction, suggesting shear-dependent regulation of klf2a is required for normal valve

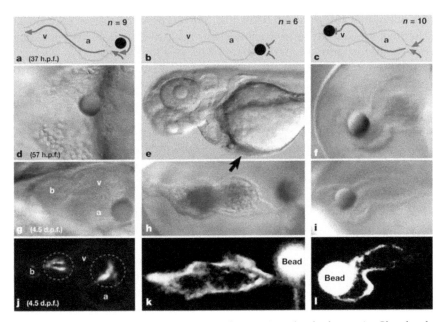

Figure 5. Impaired blood flow influences cardiogenesis and valvulogenesis. Glass beads (50 µm diameter) were inserted into zebrafish embryos in one of three positions: (a) close to the entrance of the primitive heart tube, but not blocking flow (sham); (b) in front of the heart tube, blocking inflow; or (c) into the outflow tract to block blood efflux. (d-f) Blockage was successful as evidenced by the lack of blood in the atrium when the inflow was blocked (e) and the accumulation of blood when the outflow was blocked (f). The heart (g) and valves (j) developed normally in the sham operated embryos, but cardiogenesis (h, i) and valve formation (k, l) were severely disrupted independent of blockage location. From Hove et al. (2003).

Color image of this figure appears in the color plate section at the end of the book.

development and subtle flow perturbations can impair the process (Vermot et al. 2009). Current approaches aimed at more detailed and quantitative understanding of hemodynamic patterning of heart development include the use of advanced imaging techniques and computational models (Yalcin et al. 2011).

Cell morphology

The most notable response of endothelial cells to fluid flow-induced shear stress is their change in morphology. Under static conditions, endothelial cells appear cobblestone-like and rounded; with application of shear stress, endothelial cells become elongated with the direction of flow and resemble ellipsoids. Parallel alignment of vascular endothelial cells to the direction of flow is observed in many arterial locations (Nerem et al. 1981), but not in regions of disturbed flow such as bifurcations, where endothelial cells are polygonal and rounded (Gau et al. 1980). These patterns of endothelial alignment are highly reproducible in vitro using either unidirectional flow (for parallel alignment) or oscillatory flow (for no alignment).

Based on vascular EC morphological responses to differing hemodynamics environments in the arterial system, one would expect differences in VEC morphology on opposite sides of valve leaflets owing to the distinct flow patterns. Indeed, side-specific VEC morphologies are evident early in human valve development, when distinct shear stresses on opposite sides of the valves are established (Maron and Hutchins 1974). Consistent with expectations from vascular EC response, VECs on the ventricular, high shear outflow side of the developing valve are elongated and flattened in morphology, while VECs on the low shear aortic side display cuboidal morphology. These observations suggest that side-dependent morphological differences are established early and rapidly as a result of side-specific shear stresses. However, in post-natal valves, VECs are oriented circumferentially on the valve leaflet concomitant with underlying collagen fibrils (Deck 1986). Therefore, unlike vascular endothelial cells that align parallel to the direction of flow *in vivo* (Reidy and Langille 1980, Langille and Adamson 1991, Noria et al. 1999), there exist regions on valve leaflets in which VECs are elongated parallel to blood flow, but also regions where the cell alignment and elongation is perpendicular to flow. It is notable, however, that actin stress fibres in bovine VECs align parallel to the direction of flow *in vivo*, independent of cell shape (Wong et al. 1983). For example, VECs at the base of both the aortic and ventricular sides of the leaflets are spindle shaped with actin bundles aligning parallel to blood flow, whereas VECs in the center of the valve leaflets are larger and more oval but still have stress fibers oriented parallel to flow. Thus, f-actin alignment in VECs follows flow direction and does not always coincide with cell elongation *in vivo*.

The first study of VEC morphological responses to shear stress *in vitro* reported *perpendicular* alignment of porcine VECs to the direction of flow (Butcher et al. 2004). In this study, VECs demonstrated the same arrangement of focal adhesion complexes as vascular endothelial cells (i.e., at the ends of the long axis), but the signaling pathways involved in shear-mediated alignment differed: Rho kinase signaling was required for alignment of both VECs and vascular ECs, but vascular ECs also required phosphatidylinositol 3-kinase. These initial observations suggested that VECs are phenotypically different from vascular endothelial cells, specifically in their response to shear stress. Other differences between VECs and vascular endothelial cells include transcriptional profiles *in vitro* (Farivar et al. 2003) and adhesion strength to different extracellular matrices (Young et al. 2007). However, recent results indicate that human VECs align parallel to the direction of flow *in vitro* (Holliday et al. 2011), and work from our own lab suggests this is also the case for pure populations of post-confluent porcine VECs (S. Srigunapalan, E.W.K. Young, and C.A. Simmons, unpublished data). Differences in alignment between the first and subsequent studies may be a result of cell type, purity of the EC population, cell density, and/or the matrix on which the cells were grown (Holliday et al. 2011).

Gene and protein expression

VEC mechanobiological responses to shear stress have only recently been studied. By analogy with the vasculature endothelium, VEC mRNA and protein expression were expected to be shear-sensitive, with distinct expression patterns in response to different hemodynamic environments at multiple length scales (Davies et al. 2004). Indeed, we showed by global transcriptional profiling that VECs were phenotypically different on the aortic versus ventricular side of normal porcine aortic valves, with 584 genes being differentially expressed (Simmons et al. 2005). In normal valves that showed no signs of inflammation, the transcriptional profiles suggested that the disease-prone aortic surface is permissive to calcification but protected by an enhanced antioxidative state (Fig. 6). In a follow up study, pigs were fed a high cholesterol diet for two weeks to induce early aortic valve disease (Guerraty et al. 2010). The aortic side endothelium was more responsive than the ventricular side endothelium to the high cholesterol diet, as measured by changes in side-specific global gene expression. Unexpectedly, however, systemic insult resulted in induction and persistence of a *protective* endothelial phenotype on the pathosusceptible aortic side. Micro RNA-370, which is upregulated during endothelial-to-mesenchymal transdifferentiation (EMT), is also expressed in a side-dependent manner, with increased expression on the ventricular side of aortic valves from human patients undergoing heart transplantation (Holliday et al. 2011).

While these studies were the first to confirm side-dependent VEC phenotypes that correlate with distinct hemodynamic environments, they did not demonstrate a causal relationship between shear stress and VEC phenotypes. Butcher et al. (2006) were the first to investigate the effects of fluid shear stress on VEC mRNA expression. Porcine aortic valve ECs and vascular (aortic) ECs were subjected to 20 dynes/cm^2 unidirectional steady shear stress for 48 hours or maintained under static conditions. Transcriptional profiles were compared between the different conditions. Over 400 genes were differentially expressed between valvular and vascular ECs under static conditions, supporting other evidence that these EC populations, while anatomically close, display phenotypic differences. Fluid shear stress differentially regulated hundreds of genes in both cell

Differentially expressed genes related to skeletal development and vascular calcification

Gene*	Lower Expression on Aortic Side			Higher Expression on Aortic Side			
	Accession No.	A/V Fold Change†	Putative Effect‡	Gene*	Accession No.	A/V Fold Change†	Putative Effect‡
TNFRSF11B	U94332	−3.53	+	BMP4	NM_001202	1.57	+
NPPC	D90337	−3.12	+	PTN	AU120808	1.53	+
CHRD	AF209928	−1.37	+	HAPLN1	U43328	1.49	+
PTH	V00597	−1.31	−	FBN1	X63556	1.39	+
COL11A1	J04177	−1.44	+	CHAD	AF371328	1.37	+
BMP1	NM_006129	−1.52	?	OSTF1	BC007459	1.24	?
BMP6	AA426596	−1.29	?				

*CHAD indicates chondroadherin; CHRD, chordin; COL3A1, collagen type III α1; COL11A1, collagen type XI α1; FBN1, fibrillin 1; HAPLN1, hyaluronan and proteoglycan link protein 1; NPPC, natriuretic peptide precursor C (CNP); OSTF1, osteoclast-stimulating factor 1; PTN, pleiotrophin; TNFRSF11B, tumor necrosis factor receptor superfamily member 11b (osteoprotegerin).
†Aortic side to ventricular side fold change by microarray analysis; ‡ + procalcific on aortic side; −anticalcific on aortic side; ?unknown; refer to supplemental Table 3.

Differentially expressed genes related to inflammation and oxidation

Gene*	Lower Expression on Aortic Side			Higher Expression on Aortic Side			
	Accession No.	A/V Fold Change†	Putative Effect‡	Gene*	Accession No.	A/V Fold Change†	Putative Effect‡
ALOX12	M62982	−4.53	−	MGST2	U77604	1.82	−
NPPC	D90337	−3.12	−	NOS3	BG741096	1.44	−
LGALS1	BC001693	−2.22	−	PRDX2	BC000452	1.28	+
FY	AF030521	−1.97	−	SELP	NM_003005	1.66	+
CCL13	U59808	−1.53	−	vWF	X04385	1.60	+
OLR1	AB017444	−1.32	−				
GPX3	D00632	−1.36	+				

*ALOX12 indicates arachidonate 12-lipoxygenase; CCL13, chemokine (C-C motif) ligand 13; FY, Duffy blood group; GPX3, glutathione peroxidase 3; LGALS1, galectin 1; MGST2, microsomal glutathione S-transferase 2; NPPC, natriuretic peptide precursor C; OLR1, oxidized low-density lipoprotein receptor 1; PRDX2, peroxiredoxin 2; SELP, selectin P.
†Aortic side to ventricular side fold change by microarray analysis; ‡ + proinflammatory/oxidative on the aortic side; −anti-inflammatory/oxidative on the aortic side; refer to supplemental Table 3.

Figure 6. The valvular endothelium in normal valves exhibits side-dependent phenotypic heterogeneity. Transcriptional profiling of valve endothelial cells from the aortic versus ventricular sides of normal porcine leaflets revealed side-specific expression patterns. The expression patterns suggested that the aortic side was pro-calcific (top table), but anti-inflammatory and anti-oxidative (bottom table). Side-dependent expression at the protein level was confirmed by immunohistochemical staining for (A) osteoprotegerin (expressed lower on the aortic surface; (B) than on the ventricular surface (C)); and (D) arachidonate 12-lipoxygenase (expressed higher on the aortic surface, (F)). Adapted from Simmons et al. (2005).

types, with each showing similar decreases in oxidative and inflammatory gene expression patterns with shear stress. Of putative relevance to valve calcification, shear stress inhibited expression of chondrogenic and osteogenic genes in VECs and vascular ECs, respectively.

It is notable that only 15% of the genes that were differentially expressed under static vs. shear conditions *in vitro* (Butcher et al. 2006) were also differentially expressed on aortic vs. ventricular sides of the aortic valve *in vivo* (Simmons et al. 2005). Of these genes, 64% of those expressed at higher levels on the ventricular side were also up-regulated by shear stress, and 44% of those expressed at higher levels on the fibrosa side were also down-regulated by shear stress (Butcher et al. 2008). Thus, side-dependent VEC phenotypes appear to be in part hemodynamically regulated, but there are clearly other factors at play. Some of the observed discrepancy may be attributable to limitations of the *in vitro* model and phenotypic drift that occurs with subculture of cells. But there also may be inherent differences in aortic versus ventricular side VECs that are controlled epigenetically, as has been suggested as a partial explanation for EC heterogeneity between different vascular beds (Aird 2007).

To test for differences in how aortic and ventricular side VECs respond to shear stress, Sucosky et al. (2009) subjected the aortic or ventricular surface of leaflets to unidirectional, high magnitude pulsatile flow (representing ventricular side flow) or bidirectional, low magnitude oscillatory flow (representing aortic side flow) *ex vivo*. They found that exposure of the aortic surface to altered shear stress (i.e., high magnitude flow) increased expression of the inflammatory proteins VCAM-1, ICAM-1, BMP-4, and TGF-β1. In contrast, exposure of the aortic surface to its physiological flow pattern or exposure of the ventricular surface to either flow pattern had no effect on inflammatory protein expression. Thus, aortic and ventricular side VECs exhibit different shear responsiveness. It has yet to be determined if these differences are determined developmentally and/or epigenetically, or if they are a result of hemodynamic conditioning.

Shear stress-dependent paracrine signaling

The endothelium plays a critical role in vascular homeostasis through shear stress-dependent secretion of paracrine factors that signal to circulating and mural cells. For example, vascular smooth muscle tone and vessel compliance are regulated by shear-dependent expression of nitric oxide, endothelin-1 and natriuretic peptides by the vascular endothelium.

Similarly, VEC communication with VICs via paracrine signals is likely shear-regulated. This was demonstrated *in vitro* by co-culturing a sub-confluent monolayer of VECs on a collagen hydrogel that encapsulated VICs, and subjecting the endothelium to shear stress in a parallel plate flow chamber (Butcher and Nerem 2006). Under static conditions, the presence of VECs reduced VIC expression of α-smooth muscle actin (SMA), a marker of activated myofibroblasts that is associated with remodeling and disease in the aortic valve. When the VECs were subjected to shear stress, VIC SMA expression decreased further. Compared to static conditions, fluid shear stress also increased protein content, reduced proliferation, and diminished loss of GAGs within the underlying VIC-laden hydrogel. While the factor or factors responsible for VEC regulation of VIC activation were not identified, nitric oxide may play a role as it is up-regulated in endothelial cells by shear stress and has been shown to inhibit VIC activation and calcification under static conditions *in vitro* (Kennedy et al. 2009).

Another intriguing candidate for shear stress-dependent VEC paracrine regulation of VIC phenotype is C-type natriuretic peptide (CNP). CNP in the heart and vasculature is a 53 amino acid peptide that binds natriuretic peptide receptor (NPR)-B, a membrane-bound guanylylcyclase (Potter et al. 2006). CNP binding to NPR-B potentiates intracellular synthesis of cGMP, and thus CNP can elicit similar responses to NO. We first discovered that CNP was expressed in the aortic valve, with higher expression by ventricular side VECs (Simmons et al. 2005). It is likely that side-dependent expression of CNP in the aortic valve is hemodynamically-regulated, as CNP expression by vascular ECs is stimulated by shear stress (Zhang et al. 1999), but this has yet to be confirmed in VECs. We subsequently showed that CNP inhibits pathological differentiation of VICs to the myofibroblasts and osteoblast lineages *in vitro* (Yip et al. 2011). The implication from these studies is that paracrine factors expressed in a shear-dependent manner on one side of the leaflet may contribute to the local biochemical milieu to protect or promote pathological development. In the case of CNP, its higher expression in ventricular side VECs may protect the ventricularis, whereas the lack of expression on the aortic side may leave the fibrosa vulnerable to lesion formation (Yip and Simmons 2011, Yip et al. 2011). Additional similar examples include osteoprotegerin and chordin, both of which are putative inhibitors of calcification and are expressed at lower levels on the disease-prone aortic side than on the disease-protected ventricular side of normal aortic valves (Simmons et al. 2005).

Valve endothelial response to mechanical stress

VECs are likely subjected to mechanical strains as their substrate (the leaflet tissue) undergoes large deformation during the cardiac cycle. Vascular ECs align perpendicular to the major direction of stretch *in vitro* (Zhao et al. 1995, Moretti et al. 2004) and *in vivo*, where circumferential strain dominates and ECs align longitudinally. In contrast, VECs cultured isotropically on elastomeric substrates and stretched cyclically at either 10% or 20% aligned parallel to direction of strain (Balachandran et al. 2011). These results imply that circumferential VEC alignment *in vivo* is most likely due to topographical cues from circumferentially-oriented collagen fibres (Deck 1986) and not cyclic strains, which are dominant in the radial direction in the aortic valve.

A few recent studies suggest roles for mechanical stretch in valve pathobiology. VECs cultured *in vitro* and stretched cyclically display magnitude-dependent expression of the inflammatory proteins VCAM-1, ICAM-1, and E-selectin (Metzler et al. 2008). Aortic valve leaflets exposed to pathological levels of stretch (15% strain versus 10% strain for physiological loading) *ex vivo* had elevated VEC expression of BMP-2, primarily on the aortic side (Balachandran et al. 2010). Matrix remodeling, cell proliferation, and apoptosis are also elevated with stretch (reviewed in (Balachandran et al. 2011)), but the specific roles of VECs in these responses have not been investigated. Cyclic strain has also been shown to induce EMT in VEC monolayers *in vitro* (Balachandran et al. 2011). EMT in this culture system was TGF-ß-dependent under low (10%) strain conditions, but not high (20%) strain conditions. Instead, the wnt/ß-catenin pathway was activated under high strain. VECs co-expressing CD31 and the bone matrix protein osteocalcin have been detected *in vivo* in sheep mitral valves that had been surgically altered to mimic the pathological mechanical stretch that occurs in ischemic mitral regurgitation (Bischoff and Aikawa 2011). Thus, mechanical forces may regulate EMT to maintain the valve (perhaps by replenishing the VIC population (Bischoff and Aikawa 2011)) or contribute to disease development, although a significant role for EMT in adult valve (patho)biology has yet to be established.

Summary

Compared to other tissues in the cardiovascular system, heart valve biology is in its infancy. What has emerged from recent increased focus on native valves is that while the structure, function, cells, and (patho) physiology of heart valves share features with the vasculature, there are clear distinctions. This is perhaps most evident in the unique characteristics of the valvular endothelium, which have yet to be defined completely but are worthy of continued study. Open questions remain about the role of VECs in EMT and recapitulation of developmental programs in valve disease; the heterogeneity of VECs at multiple length scales and the impact on their regulation of valve homeostasis; the roles of VECs in disease progression and their potential as a target for therapies; and of course, the mechanisms by which mechanical forces regulate each of these aspects of VEC biology. Recent advances in the understanding of heart valve biology and pathogenesis have primed the field for the development of new therapies that will meet the current unmet need for medical treatments. To that end, an improved understanding of VEC biology, its role in homeostasis and pathogenesis, and the influence of the valvular microenvironment, including hemodynamics, will yield new insights into the complexities of valve pathobiology and will be essential to discovering effective therapies.

References

Aird, W.C. 2007. Phenotypic heterogeneity of the endothelium I. Structure, function, and mechanisms. *Circ. Res.* 100(2): 158–173.

Balachandran, K., P.W. Alford, J. Wylie-Sears, J.A. Goss, A. Grosberg, J. Bischoff, E. Aikawa, R.A. Levine and K.K. Parker. 2011. Cyclic strain induces dual-mode endothelial-mesenchymal transformation of the cardiac valve. *Proc. Natl. Acad. Sci. USA* 108(50): 19943–19948.

Balachandran, K., P. Sucosky, H. Jo and A.P. Yoganathan. 2010. Elevated Cyclic Stretch Induces Aortic Valve Calcification in a Bone Morphogenic Protein-Dependent Manner. *Am. J. Pathol.* 177(1): 49–57.

Balachandran, K., P. Sucosky and A.P. Yoganathan. 2011. Hemodynamics and mechanobiology of aortic valve inflammation and calcification. *Int. J. Inflam.* 263870.

Barandon, L., P. Clerc, C. Chauvel and P. Plagnol. 2004. Native aortic valve thrombosis: a rare cause of acute ischemia of the lower limb. *Interact. Cardiovasc. Thorac. Surg.* 3(4): 675–677.

Bartman, T., E.C. Walsh, K.K. Wen, M. McKane, J. Ren, J. Alexander, P.A. Rubenstein and D.Y. Stainier. 2004. Early myocardial function affects endocardial cushion development in zebrafish. *PLoS Biol.* 2(5): E129.

Beppu, S., S. Suzuki, H. Matsuda, F. Ohmori, S. Nagata and K. Miyatake. 1993. Rapidity of progression of aortic stenosis in patients with congenital bicuspid aortic valves. *Am. J. Cardiol.* 71(4): 322–327.

Bischoff, J. and E. Aikawa. 2011. Progenitor Cells Confer Plasticity to Cardiac Valve Endothelium. *J. Cardiovasc. Transl. Res.* 4(6): 710–719.

Butcher, J.T. and R.M. Nerem. 2006. Valvular endothelial cells regulate the phenotype of interstitial cells in co-culture: Effects of steady shear stress. *Tissue Eng.* 12(4): 905–915.

Butcher, J.T. and R.M. Nerem. 2007. Valvular endothelial cells and the mechanoregulation of valvular pathology. *Philos. Trans. R. Soc., B* 362(1484): 1445–1457.

Butcher, J.T., A.M. Penrod, A.J. Garcia and R.M. Nerem. 2004. Unique morphology and focal adhesion development of valvular endothelial cells in static and fluid flow environments. *Arterioscler., Thromb., Vasc. Biol.* 24(8): 1429–1434.

Butcher, J.T., C.A. Simmons and J.N. Warnock. 2008. Mechanobiology of the aortic heart valve. *J. Heart Valve Dis.* 17(1): 62–73.

Butcher, J.T., S. Tressel, T. Johnson, D. Turner, G. Sorescu, H. Jo and R.M. Nerem. 2006. Transcriptional profiles of valvular and vascular endothelial cells reveal phenotypic differences—Influence of shear stress. *Arterioscler., Thromb., Vasc. Biol.* 26(1): 69–77.

Cataloglu, A., P.L. Gould and R.E. Clark. 1976. Refined stress analysis of human aortic heart valves. *J. Eng. Mech. Div., Am. Soc. Civ. Eng.* 102(1): 135–150.

Chandra, S., N.M. Rajamannan and P. Sucosky. 2011. Computational assessment of bicuspid aortic valve wall-shear stress: implications for calcific aortic valve disease. *Biomech. Model Mechanobiol.* 11(7): 1085–1096.

Chen, J.H. and C.A. Simmons. 2011. Cell-matrix interactions in the pathobiology of calcific aortic valve disease: critical roles for matricellular, matricrine, and matrix mechanics cues. *Circ. Res.* 108(12): 1510–1524.

Chen, J.H., C.Y. Yip, E.D. Sone and C.A. Simmons. 2009. Identification and characterization of aortic valve mesenchymal progenitor cells with robust osteogenic calcification potential. *Am. J. Pathol.* 174(3): 1109–1119.

Chester, A.H., J.D.B. Kershaw, P. Sarathchandra and M.H. Yacoub. 2008. Localisation and function of nerves in the aortic root. *J. Mol. Cell. Cardiol.* 44(6): 1045–1052.

Chester, A.H., M. Misfeld and M.H. Yacoub. 2000. Receptor-mediated contraction of aortic valve leaflets. *J. Heart Valve Dis.* 9(2): 250–254; discussion 254–255.

Cooke, J.P. 2003. Flow, NO, and atherogenesis. *Proc. Natl. Acad. Sci. USA* 100(3): 768–770.

Davies, P.F., A.G. Passerini and C.A. Simmons. 2004. Aortic valve—Turning over a new leaf(let) in endothelial phenotypic heterogeneity. *Arterioscler., Thromb., Vasc. Biol.* 24(8): 1331–1333.

De Hart, J., G.W.M. Peters, P.J.G. Schreurs and F.P.T. Baaijens. 2003. A three-dimensional computational analysis of fluid-structure interaction in the aortic valve. *J. Biomech.* 36(1): 103–112.

De Hart, J., G.W.M. Peters, P.J.G. Schreurs and F.P.T. Baaijens. 2004. Collagen fibers reduce stresses and stabilize motion of aortic valve leaflets during systole. *J. Biomech.* 37(3): 303–311.

Deck, J.D. 1986. Endothelial-cell orientation on aortic-valve leaflets. *Cardiovasc. Res.* 20(10): 760–767.

Drake, T.A. and M. Pang. 1989. Effects of interleukin-1, lipopolysaccharide, and streptococci on procoagulant activity of cultured human cardiac valve endothelial and stromal cells. *Infect. Immun.* 57(2): 507–512.

El-Hamamsy, I., K. Balachandran, M.H. Yacoub, L.M. Stevens, P. Sarathchandra, P.M. Taylor, A.P. Yoganathan and A.H. Chester. 2009. Endothelium-Dependent Regulation of the Mechanical Properties of Aortic Valve Cusps. *J. Am. Coll. Cardiol.* 53(16): 1448–1455.

Farivar, R.S., L.H. Cohn, E.G. Soltesz, T. Mihaljevic, J.D. Rawn and J.G. Byrne. 2003. Transcriptional profiling and growth kinetics of endothelium reveals differences between cells derived from porcine aorta versus aortic valve. *Eur. J. Cardiothorac. Surg.* 24(4): 527–534.

Freund, J.B., J.G. Goetz, K.L. Hill and J. Vermot. 2012. Fluid flows and forces in development: functions, features and biophysical principles. *Development* 139(7): 1229–1245.

Gau, G.S., T.A. Ryder and M.L. Mackenzie. 1980. The effect of blood flow on the surface morphology of the human endothelium. *J. Pathol.* 131(1): 55–64.

Ge, L. and F. Sotiropoulos. 2010. Direction and magnitude of blood flow shear stresses on the leaflets of aortic valves: is there a link with valve calcification? *J. Biomech. Eng.* 132(1): 014505.

Ghaisas, N.K., J.B. Foley, D.S. O'Briain, P. Crean, D. Kelleher and M. Walsh. 2000. Adhesion molecules in nonrheumatic aortic valve disease: Endothelial expression, serum levels and effects of valve replacement. *J. Am. Coll. Cardiol.* 36(7): 2257–2262.

Guerraty, M.A., G.R. Grant, J.W. Karanian, O.A. Chiesa, W.F. Pritchard and P.F. Davies. 2010. Hypercholesterolemia Induces Side-Specific Phenotypic Changes and Peroxisome Proliferator-Activated Receptor-gamma Pathway Activation in Swine Aortic Valve Endothelium. *Arterioscler., Thromb., Vasc. Biol.* 30(2): 225–231.

Holliday, C.J., R.F. Ankeny, H. Jo and R.M. Nerem. 2011. Discovery of shear- and side-specific mRNAs and miRNAs in human aortic valvular endothelial cells. *Am. J. Physiol. Heart Circ. Physiol.* 301(3): H856–867.

Hove, J.R., R.W. Koster, A.S. Forouhar, G. Acevedo-Bolton, S.E. Fraser and M. Gharib. 2003. Intracardiac fluid forces are an essential epigenetic factor for embryonic cardiogenesis. *Nature* 421(6919): 172–177.

Kennedy, J.A., X. Hua, K. Mishra, G.A. Murphy, A.C. Rosenkranz and J.D. Horowitz. 2009. Inhibition of calcifying nodule formation in cultured porcine aortic valve cells by nitric oxide donors. *Eur. J. Pharmacol.* 602(1): 28–35.

Kilner, P.J., G.Z. Yang, A.J. Wilkes, R.H. Mohiaddin, D.N. Firmin and M.H. Yacoub. 2000. Asymmetric redirection of flow through the heart. *Nature* 404(6779): 759–761.

Langille, B.L. and S.L. Adamson. 1991. Relationship between blood flow direction and endothelial cell orientation at arterial branch sites in rabbits and mice. *Circ. Res.* 48(4): 481–488.

Lester, W.M., A.A. Damji, I. Gedeon and M. Tanaka. 1993. Interstitial cells from the atrial and ventricular sides of bovine mitral valve respond differently to denuding endocardial injury. *In Vitro Cell. Dev. Biol.: Anim.* 29A(1): 41–50.

Lincoln, J., C.M. Alfieri and K.E. Yutzey. 2004. Development of heart valve leaflets and supporting apparatus in chicken and mouse embryos. *Dev. Dyn.* 230(2): 239–250.

Liu, A.C., V.R. Joag and A.I. Gotlieb. 2007. The emerging role of valve interstitial cell phenotypes in regulating heart valve pathobiology. *Am. J. Pathol.* 171(5): 1407–1418.

Manduteanu, I., D. Popov, A. Radu and M. Simionescu. 1988. Calf cardiac valvular endothelial cells in culture: production of glycosaminoglycans, prostacyclin, and fibronectin. *J. Mol. Cell. Cardiol.* 20(2): 103–118.

Maron, B.J. and G.M. Hutchins. 1974. The development of semilunar valves in the human heart. *Am. J. Pathol.* 74(2): 331–344.

Matsumoto, Y., V. Adams, C. Walther, C. Kleinecke, P. Brugger, A. Linke, T. Walther, F.W. Mohr and G. Schuler. 2009. Reduced number and function of endothelial progenitor cells in patients with aortic valve stenosis: a novel concept for valvular endothelial cell repair. *Eur. Heart J.* 30(3): 346–355.

Metzler, S.A., C.A. Pregonero, J.T. Butcher, S.C. Burgess and J.N. Warnock. 2008. Cyclic Strain Regulates Pro-Inflammatory Protein Expression in Porcine Aortic Valve Endothelial Cells. *J. Heart Valve Dis.* 17(5): 571–578.

Mirzaie, M., T. Meyer, P. Schwarz, S. Lotfi, A. Rastan and F. Schondube. 2002. Ultrastructural alterations in acquired aortic and mitral valve disease as revealed by scanning and transmission electron microscopical investigations. *Ann. Thorac. Cardiovasc. Surg.* 8: 24–30.

Mohler, E.R., M.K. Chawla, A.W. Chang, N. Vyavahare, R.J. Levy, L. Graham and F.H. Gannon. 1999. Identification and characterization of calcifying valve cells from human and canine aortic valves. *J. Heart Valve Dis.* 8(3): 254–260.

Moretti, M., A. Prina-Mello, A.J. Reid, V. Barron and P.J. Prendergast. 2004. Endothelial cell alignment on cyclically-stretched silicone surfaces. *J. Mater. Sci.: Mater. Med.* 15(10): 1159–1164.

Muller, A.M., C. Cronen, L.I. Kupferwasser, H. Oelert, K.M. Muller and C.J. Kirkpatrick. 2000. Expression of endothelial cell adhesion molecules on heart valves: up-regulation in degeneration as well as acute endocarditis. *J. Pathol.* 191(1): 54–60.

Nerem, R.M., M.J. Levesque and J.F. Cornhill. 1981. Vascular endothelial morphology as an indicator of the pattern of blood flow. *J. Biomech. Eng.* 103(3): 172–176.

Noria, S., D.B. Cowan, A.I. Gotlieb and B.L. Langille. 1999. Transient and steady-state effects of shear stress on endothelial cell adherens junctions. *Circ. Res.* 85(6): 504–514.

Obrien, K.D., D.D. Reichenbach, S.M. Marcovina, J. Kuusisto, C.E. Alpers and C.M. Otto. 1996. Apolipoproteins B, (a), and E accumulate in the morphologically early lesion of 'degenerative' valvular aortic stenosis. *Arterioscler., Thromb., Vasc. Biol.* 16(4): 523–532.

Otto, C.M., J. Kuusisto, D.D. Reichenbach, A.M. Gown and K.D. Obrien. 1994. Characterization of the early lesion of degenerative valvular aortic stenosis haracterization of the early lesion of degenerative valvular aortic-stenosis—histological and immunohistochemical studies. *Circulation* 90(2): 844–853.

Paranya, G., S. Vineberg, E. Dvorin, S. Kaushal, S.J. Roth, E. Rabkin, F.J. Schoen and J. Bischoff. 2001. Aortic valve endothelial cells undergo transforming growth factor-beta-mediated and non-transforming growth factor-beta-mediated transdifferentiation *in vitro*. *Am. J. Pathol.* 159(4): 1335–1343.

Pho, M., W. Lee, D.R. Watt, C. Laschinger, C.A. Simmons and C.A. McCulloch. 2008. Cofilin is a marker of myofibroblast differentiation in cells from porcine aortic cardiac valves. *Am. J. Physiol.: Heart Circ. Physiol.* 294(4): H1767–H1778.

Poggianti, E., L. Venneri, V. Chubuchny, Z. Jambrik, L.A. Baroncini and E. Picano. 2003. Aortic valve sclerosis is associated with systemic endothelial dysfunction. *J. Am. Coll. Cardiol.* 41(1): 136–141.

Pompilio, G., G. Rossoni, A. Sala, G.L. Polvani, F. Berti, L. Dainese, M. Porqueddu and P. Biglioli. 1998. Endothelial-dependent dynamic and antithrombotic properties of porcine aortic and pulmonary valves. *Ann. Thorac. Surg.* 65(4): 986–992.

Potter, L.R., S. Abbey-Hosch and D.M. Dickey. 2006. Natriuretic peptides, their receptors, and cyclic guanosine monophosphate-dependent signaling functions. *Endocr. Rev.* 27(1): 47–72.

Rabkin-Aikawa, E., M. Farber, M. Aikawa and F.J. Schoen. 2004. Dynamic and reversible changes of interstitial cell phenotype during remodeling of cardiac valves. *J. Heart Valve Dis.* 13(5): 841–847.

Rajamannan, N.M., F.J. Evans, E. Aikawa, K.J. Grande-Allen, L.L. Demer, D.D. Heistad, C.A. Simmons, K.S. Masters, P. Mathieu, K.D. O'Brien, F.J. Schoen, D.A. Towler, A.P. Yoganathan and C.M. Otto. 2011. Calcific Aortic Valve Disease: Not Simply a Degenerative Process A Review and Agenda for Research From the National Heart and Lung and Blood Institute Aortic Stenosis Working Group. *Circulation* 124(16): 1783–1791.

Reidy, M.A. and B.L. Langille. 1980. The effect of local blood flow patterns on endothelial cell morphology. *Exp. Mol. Pathol.* 32(3): 276–289.

Riddle, J.M., D.J. Magilligan and P.D. Stein. 1980. Surface topography of stenotic aortic valves by scanning electron microscopy. *Circulation* 61(3): 496–502.

Sacks, M.S., W. David Merryman and D.E. Schmidt. 2009. On the biomechanics of heart valve function. *J. Biomech.* 42(12): 1804–1824.

Sacks, M.S. and A.P. Yoganathan. 2007. Heart valve function: a biomechanical perspective. *Philos. Trans. R. Soc., B* 362(1484): 1369–1391.

Santhanakrishnan, A. and L.A. Miller. 2011. Fluid dynamics of heart development. *Cell Biochem. Biophys.* 61(1): 1–22.

Schoen, F.J. 2008. Evolving Concepts of Cardiac Valve Dynamics The Continuum of Development, Functional Structure, Pathobiology, and Tissue Engineering. *Circulation* 118(18): 1864–1880.

Simmons, C.A., G.R. Grant, E. Manduchi and P.F. Davies. 2005. Spatial heterogeneity of endothelial phenotypes correlates with side-specific vulnerability to calcification in normal porcine aortic valves. *Circ. Res.* 96(7): 792–799.

Siney, L. and M.J. Lewis. 1993. Nitric oxide release from porcine mitral valves. *Cardiovasc. Res.* 27(9): 1657–1661.

Stewart, B.F., D. Siscovick, B.K. Lind, J.M. Gardin, J.S. Gottdiener, V.E. Smith, D.W. Kitzman and C.M. Otto. 1997. Clinical factors associated with calcific aortic valve disease. Cardiovascular Health Study. *J. Am. Coll. Cardiol.* 29(3): 630–634.

Sucosky, P., K. Balachandran, A. Elhammali, H. Jo and A.P. Yoganathan. 2009. Altered Shear Stress Stimulates Upregulation of Endothelial VCAM-1 and ICAM-1 in a BMP-4-and TGF-beta 1-Dependent Pathway. *Arterioscler., Thromb., Vasc. Biol.* 29(2): 254–260.

Thubrikar, M.J. 1990. The Aortic Valve. Boca Raton, Florida, CRC Press.

Tompkins, R.G., J.J. Schnitzer and M.L. Yarmush. 1989. Macromolecular transport within heart valves. *Circ. Res.* 64(6): 1213–1223.

Tsai, H.M. 2003. Shear stress and von Willebrand factor in health and disease. *Semin. Thromb. Hemost.* 29(5): 479–488.

Vermot, J., A.S. Forouhar, M. Liebling, D. Wu, D. Plummer, M. Gharib and S.E. Fraser. 2009. Reversing blood flows act through klf2a to ensure normal valvulogenesis in the developing heart. *PLoS Biol.* 7(11): e1000246.

Weinberg, E.J. and M.R.K. Mofrad. 2008. A multiscale computational comparison of the bicuspid and tricuspid aortic valves in relation to calcific aortic stenosis. *J. Biomech.* 41(16): 3482–3487.

Weston, M.W., D.V. LaBorde and A.P. Yoganathan. 1999. Estimation of the shear stress on the surface of an aortic valve leaflet. *Ann. Biomed. Eng.* 27(4): 572–579.

Wong, A.J., T.D. Pollard and I.M. Herman. 1983. Actin Filament Stress Fibers in Vascular Endothelial Cells *in vivo*. *Science* 219(4586): 867–869.

Wylie-Sears, J., E. Aikawa, R.A. Levine, J.H. Yang and J. Bischoff. 2011. Mitral valve endothelial cells with osteogenic differentiation potential. *Arterioscler. Thromb. Vasc. Biol.* 31(3): 598–607.

Yalcin, H.C., A. Shekhar, T.C. McQuinn and J.T. Butcher. 2011. Hemodynamic patterning of the avian atrioventricular valve. *Dev. Dyn.* 240(1): 23–35.

Yap, C., N. Saikrishnan and A. Yoganathan. 2011. Experimental measurement of dynamic fluid shear stress on the ventricular surface of the aortic valve leaflet. *Biomech. Model Mechanobiol.* 11(1): 231–244.

Yap, C.H., N. Saikrishnan, G. Tamilselvan and A.P. Yoganathan. 2011. Experimental Technique of Measuring Dynamic Fluid Shear Stress on the Aortic Surface of the Aortic Valve Leaflet. *J. Biomech. Eng.* 133(6).

Yip, C.Y. and C.A. Simmons. 2011. The aortic valve microenvironment and its role in calcific aortic valve disease. *Cardiovasc. Pathol.* 20(3): 177–182.

Yip, C.Y.Y., M.C. Blaser, Z. Mirzaei, X. Zhong and C.A. Simmons. 2011. Inhibition of Pathological Differentiation of Valvular Interstitial Cells by C-Type Natriuretic Peptide. *Arterioscler., Thromb., Vasc. Biol.* 31(8): 1881–U1387.

Young, E.W.K., A.R. Wheeler and C.A. Simmons. 2007. Matrix-dependent adhesion of vascular and valvular endothelial cells in microfluidic channels. *Lab Chip* 7(12): 1759–1766.

Zeng, Z.Q., Y.Y. Yin, A.L. Huang, K.M. Jan and D.S. Rumschitzki. 2007. Macromolecular transport in heart valves. I. Studies of rat valves with horseradish peroxidase. *Am. J. Physiol.: Heart Circ. Physiol.* 292(6): H2664–H2670.

Zhang, Z.H., Z.S. Xiao and S.L. Diamond. 1999. Shear stress induction of C-type natriuretic peptide (CNP) in endothelial cells is independent of NO autocrine signaling. *Ann. Biomed. Eng.* 27(4): 419–426.

Zhao, S.M., A. Suciu, T. Ziegler, J.E. Moore, E. Burki, J.J. Meister and H.R. Brunner. 1995. Synergistic effects of fluid shear stress and cyclic circumferential stretch on vascular endothelial cell morphology and cytoskeleton. *Arterioscler., Thromb., Vasc. Biol.* 15(10): 1781–1786.

5

Environmental Factors that Influence the Response of the Endothelium to Flow

Tracy M. Cheung[1] and George A. Truskey[2,]*

Introduction

Complications from cardiovascular disease, such as heart attack and stroke, represent the leading cause of death in the United States and many developed and developing countries. In 2008, approximately 672,000 people died of heart disease or stroke, which accounts for about 27% of all deaths (Roger et al. 2012). While mortality and morbidity is decreasing in the United States (Roger et al. 2012), the incidence of the disease is increasing in developed countries such as China and India. Atherosclerosis is the primary pathology underlying heart disease and stroke. The disease is localized to large and medium sized arteries and involves a complex sequence of events in which the accumulation of lipids and cholesterol in the arterial intima initiates an inflammatory response. The result is a thickened intima consisting of lipids, macrophages and smooth muscle cells. The thickened intima can

[1]Department of Biomedical Engineering, Duke University, Durham, NC 27708.
Email: tracy.cheung@duke.edu
[2]Department of Biomedical Engineering, 136 Hudson Hall, CB 90281, Duke University, Durham, NC 27708.
*Corresponding author: george.truskey@duke.edu

compromise blood flow or rupture, causing a thrombus to form, blocking blood flow and precipitating a heart attack or stroke. The earliest events in the disease process involve changes to the function of endothelium leading to increased adhesion of leukocytes, altered permeability and decreased release of the vasodilator nitric oxide (NO). Risk factors for atherosclerosis include age, high cholesterol levels, diabetes, genetics, hypertension, and smoking (Chu and Peters 2008).

The endothelium plays an important role in regulating water and macromolecular transport, leukocyte adhesion and transmigration, vessel remodeling, apoptosis, generation and metabolism of biochemical substances, and modulation of smooth muscle cell (SMC) function and proliferation (Cines et al. 1998). Endothelial cells (ECs) respond to mechanical forces such as shear stress by modulating signaling pathways and overall cell function (Chien 2008). As a result, the local hemodynamics impact EC vascular homeostasis and influence the development of atherosclerosis. Transport of monocytes by arterial fluid dynamics and subsequent adhesion to activated endothelium may also influence the location of atherosclerotic lesions (Truskey et al. 2002).

Vascular geometry influences the local flow properties and the initiation of atherosclerosis (Nerem 1984). Atherogenesis tends to occur at specific locations, such as vessel branches and regions of curvature, in the blood vessel. In these areas, there is significant variation in flow fields and shear stress (Karino et al. 1987, Nerem 1984) and lesion formation and intimal thickening are correlated with low and oscillating shear stresses that change direction (Ku et al. 1985). ECs in this region are polygonal in shape and are prone to express adhesion molecules for leukocytes and release cytokines, which promotes the development of atherosclerosis (Chiu and Chien 2011). In contrast, regions away from branches or regions of curvature are exposed to a unidirectional pulsatile shear stress (Nerem 1984). The ECs are quiescent and express proteins that limit platelet and leukocyte adhesion and are anti-inflammatory (Collins and Tzima 2011). It is likely that a number of factors influence the development of atherosclerosis including biomechanical changes to the cell after exposure to uniform and laminar shear, fluid flow types, and the interaction of local hemodynamics and risk factors.

While there have been several comprehensive reviews on the effect of disturbed flow *in vitro* and *in vivo* on the endothelium (Chiu and Chien 2011, Harrison et al. 2006, Nigro et al. 2011), we will examine in this chapter how the environment surrounding the ECs affects their response to flow. We will describe the effect of biomechanical factors, smooth muscle cells, the properties of the subendothelium, endothelial cell age and cytokines upon the flow response of ECs. We will also discuss the interaction between hemodynamics and risk factors upon endothelial cell function. Finally, we will present future directions for the field.

The Biomechanical Environment Acting on Arterial Endothelium

In vivo, arterial endothelial cells are exposed to a complex biomechanical environment. Blood flow arises from the pressure waveform produced by the contraction of the heart (Table 1). The pressure acts normal to the endothelial cell surface with typical values ranging from 10,600 to 16,000 Pa, although pressures as high as 24,000 Pa can occur during hypertension. The pressure P generates a hoop stress $\sigma_{\theta\theta}$ in the elastic arteries, $\sigma_{\theta\theta} = PR/t$ where R is the vessel radius and t is the vessel thickness. The average hoop stress is approximately 10^5 Pa (Ku 1997) which acts on all of the vessel wall components including the endothelium. These pressures typically stretch the vessel diameter during systole by 6%–10% (Chamiot-Clerc et al. 1998, Dobrin 1978, L'Heureux et al. 2006). In contrast, the vessel length typically distends less than 1% during the cardiac cycle (Dobrin 1978).

Blood flow is laminar throughout most of the human cardiovascular system, although turbulence does arise around the aortic valve and ascending aorta during systole (Ku 1997). The unsteady nature of flow is characterized by the Womersley number $\alpha=R\,(\omega/v)^{1/2}$ where ω is the frequency of the arterial waveform in radians/s ($\omega=2\pi f$, where f is heartbeats

Table 1. Key Concepts in Arterial Hemodynamics.

Term	Meaning
Reynolds number, Re	Ratio of inertial (ρv^2) to viscous ($\mu v/D$) forces for which ρ is the fluid density, v is the average fluid velocity, μ is the fluid viscosity and D is the vessel diameter.
Stress	Force per unit area. Stresses arise in arteries due to pressure and shear stresses arising from fluid motion that act on endothelial cells and stretching of vessel due to pulsatile pressure that act on the entire vessel wall.
Shear stress, τ	Stress that acts tangent to a surface due to action of viscosity. Shear stress acting on endothelial cells is the key quantity affecting their response to fluid flow.
Pressure, P	Stress acting normal to endothelial surface arising from the pumping action of the heart.
Laminar flow	Flow when Re< 2100 for which fluid elements travel along streamlines under steady and unsteady conditions.
Flow reversal	Condition arising from an adverse pressure gradient due to increase in cross-sectional area in which direction of a flow in a region is in a direction opposite to the mean flow direction. Due to the pulsatile nature of blood flow such regions are transient for blood flow through arteries.
"Disturbed" flow	General term to describe complex flow around vessel branches or curved vessels which exhibit transient flow reversal with low time-averaged shear stresses.

per second) and v is the kinematic viscosity (Ku 1997, Truskey et al. 2009). The shape of the unsteady velocity profile and the shear stress acting on endothelium depend upon the magnitude of α which is affected by the relative contributions of inertia and the heart rate. When $\alpha < 1$, inertial forces are small and the flow waveform is parabolic at all times and the flow and pressure waveforms are in phase. As α increases, inertial forces become more important and the waveform becomes blunter. For $\alpha > 10$, which occurs in larger blood vessels, the profile is blunt and the phase difference with the pressure waveform is 90°. For coronary arteries, α is around 6, indicating inertial forces are important and the time varying shear stress acting on the endothelial cell surface is a complex function of time. The value of α is larger in the large arteries and approaches 1 in the arterioles and capillaries.

Due to the unsteady flow and curvature of arteries, the flow field is three-dimensional with spatial and temporal variations in the shear stress acting on the endothelium. In curved vessels, flow depends upon the ratio of the vessel radius R to the radius of curvature, R_c, and the Dean number, $De = (R/R_c)^{1/2} Re$, where the Reynolds number represents the ratio of inertial to viscous forces (Truskey et al. 2009). Inertial forces cause the velocity to be skewed towards the outer wall of the vessel. As a result, the outer wall is exposed to higher shear stresses whereas the inner wall is exposed to lower shear stresses. Values of the Dean number (20–700) and the ratio R/R_c (0.02–0.22) indicate significant skewing of the velocity profile in the aortic arch and coronary arteries. At branches, curvature induces inertial effects producing a region of high shear stress near the flow divider and lower shear stresses near the outer wall. If the cross-sectional area of both daughter vessels exceeds that of the parent vessel, then the flow may separate along the opposite surface during the diastolic portion of the waveform, producing a region with transient flow reversal (Fig. 1). The local flow reversal can be characterized by the oscillatory shear index, OSI, which represents the portion of the cycle during which the shear stress direction is opposite the flow direction (Ku 1997). The OSI is elevated at sites of intimal thickening in the carotid artery (Ku et al. 1985) and other sites at which atherosclerosis develops (Huo et al. 2007). Key fluid dynamic features influencing lesion development include a combination of low wall shear stresses, high OSI, wall shear stress gradients and a high probability of monocyte adhesion (Buchanan et al. 2003, Huo et al. 2007, Kleinstreuer et al. 2001).

The aorta and other large elastic arteries expand during systole, storing some of the energy from the accelerating blood and reducing the rise in the blood pressure. During diastole the vessels relax, pushing fluid down the arterial tree (Safar et al. 2003). Coupled with the branching nature of the arterial tree, a series of wave reflections alter the pressure and flow

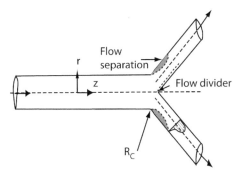

Figure 1. Schematic of flow at vessel branches leading to transient flow separation and reversal. Inertia produces high shear stress gradients near the flow divider and slower fluid velocity near the opposite wall. If the cross-sectional area in the daughter vessels exceeds the cross-sectional area in the parent vessel, then flow can separate leading to transient reversal during part of the cardiac cycle. Such regions of low and oscillating shear stresses are prone to develop atherosclerosis while the region near the flow divider is spared atherosclerosis. Rc is the radius of curvature. Adapted from Truskey et al. 2009.

waveforms leaving the heart. The speed at which the pressure pulse travels down arteries is known as the pulse wave velocity. Coupling between the blood flow dynamics and the elastic behavior of the blood vessels is described by the Moens-Korteweg equation which relates the pulse wave velocity, c, to the incremental elastic modulus of the vessel wall, E_{inc},[1] the vessel diameter D and the vessel wall thickness, t.

$$c = \sqrt{\frac{E_{inc}t}{\rho D}} \tag{1}$$

The pulse pressure, pulse wave velocity and systolic pressure increase during hypertension and are predictors of the severity of hypertension (Safar et al. 2003).

Although the arterial pressure is much higher than the shear stress, the endothelium responds specifically to the fluid shear stress. The *in vitro* responses to the endothelium have been ascribed to shear stress since equivalent responses have been observed in parallel plate flow channels in which flow is driven by a pressure gradient (Levesque and Nerem 1985) and in cone and plate viscometers in which the pressure is uniform and close to atmospheric pressure and flow is driven by the rotation of the cone (Dewey et al. 1981, Franke et al. 1984). By varying the shear rate and shear stress independently, *in vitro* responses to flow depend upon shear stress and *in vivo* dilation of vessels depends on the shear stress and the presence of the vascular endothelium (Melkumyants and Balashov 1990).

---

[1]The incremental elastic modulus represents the local value of the slope of the stress-strain curve.

Response of Endothelium to Steady and Pulsatile Unidirectional Flow

Unidirectional flow and shear stress promote an anti-thrombotic and anti-inflammatory phenotype

In humans the time averaged shear stress in arteries is between 0.2–1.6 Pa (Cheng et al. 2007), although values as high as 7.0 Pa occur in curved and branched vessels during systole. In mice, which are commonly used as models for human atherosclerosis due to the ready availability of specific gene mutations, the mean shear stress is 10–20 times higher (Weinberg and Ethier 2007). However, the Reynolds and Womersley numbers are much lower in mice suggesting that viscous forces are more significant and secondary flows are less significant (Suo et al. 2007). In spite of the difference in hemodynamics, atherosclerosis in mice in which the apolipoprotein E gene has been deleted (known as the ApoE–/– mouse) has a similar distribution as in humans, suggesting that the shear stress magnitude is not as important as the dynamics of shear stress.

In vitro, exposure of confluent endothelial cells to steady or unidirectional shear stress induces a change in shape and orientation such that the cells are elongated and aligned in the direction of flow (Davies 1995). Such EC organization is observed *in vivo* away from vessel branches whereas around vessel branches, ECs exhibit a rounder shape without any preferred orientation (Davies 1995). The endothelial cell actin cytoskeleton reorganizes in response to shear stress, aligning in the direction of flow. Prominent actin stress fibers appear (Franke et al. 1984) and cell proliferation is inhibited. *In vitro* the endothelial cells express a range of vasoactive and anti-thrombotic molecules, such as nitric oxide synthase, the vasodilator prostacyclin (PGI_2), and the anti-thrombotic molecules thrombomodulin and tissue plasminogen activator (Davies 1995).

A key physiological response to flow *in vivo* is the release of NO, which causes vessel dilation. NO is produced in ECs by endothelial nitric oxide synthase (eNOS or NOS III) (Cai and Harrison 2000). NO diffuses to smooth muscle where it acts upon soluble guanylatecyclase to alter production of cyclic GMP that decreases the activity of myosin light chain kinase, causing relaxation of the muscle. When ECs are exposed to steady or pulsatile unidirectional shear stress, NO production increases due to a change in the activity of NOS III followed by a sustained increase in NOS III levels (Harrison et al. 2006). Changes in brachial artery dilation are used to assess endothelial dysfunction in early atherosclerosis (Halcox et al. 2002). In addition to serving as a potent vasodilator, NO inhibits expression of monocyte chemoattractant protein-1 (MCP-1) and vascular cell adhesion molecule-1 (VCAM-1), prevents SMC proliferation, and decreases platelet aggregation (Palmer et al. 1987).

Application of unidirectional shear stress promotes other anti-inflammatory and anti-oxidative events including increases in intracellular glutathione, and sustained increases in the amount of superoxide dismutase, NOS III and cylcooxygenase-2 in ECs (Topper et al. 1996) as well as production of antioxidant genes such as hemeoxygenase-1 (HO-1), ferritin, and glutathione *S*-transferase (Chen et al. 2003).

The zinc-finger transcription factor Kruppel-like factor 2 (KLF-2) is a key regulator of oxidative and inflammatory responses to shear stress (Dekker et al. 2002, Parmar et al. 2006). This transcription factor is expressed by endothelial cells but not expressed by SMC in mature blood vessels (Dekker et al. 2002). KLF-2 exhibits continued high levels of expression after 7 days exposure to a pulsatile shear stress of 1.2 Pa *in vitro* (Dekker et al. 2002) due to stabilization of mRNA levels by shear stress (van Thienen et al. 2006) and down-regulation of microRNA-92 which inhibits KLF-2 protein levels (Wu et al. 2011). KLF-2 induces NOS III and the anti-thrombotic molecule thrombomodulin, and inhibits cytokine-induced expression of VCAM-1 and E-selectin, but does not affect cyclooxygenase 2, which is involved in PGI_2 production (SenBanerjee et al. 2004). KLF-2 overexpression blocks VCAM-1 and E-selectin expression by blocking activation of NFκB.

Antioxidant response elements are transcription factor binding sequences present on a number of anti-oxidant genes (e.g., HO-1) and are responsive to shear stress in endothelial cells (Chen et al. 2003). The sequences bind to Nrf2 (Nuclear factor-erythroid2-like2), which is a major transcriptional factor that regulates several antioxidant genes and redox balance in endothelial cells (Dai et al. 2007). Nrf2 is activated by unidirectional shear stress and moves from the cytoplasm to the nucleus where it binds to the antioxidant response elements (Chen et al. 2003). Shear stress acts upon Nrf2, in part through KLF2 expression which enhances the movement of Nrf2 from the cytoplasm to the nucleus, thereby increasing the antioxidant activity of Nrf2 (Fledderus et al. 2008). Together, KLF2 and Nrf2 control about 70% of all shear stress sensitive genes (Boon and Horrevoets 2009).

Another key regulator of metabolic activity sensitive to shear stress is SIRTUIN1 (SIRT1) that deacetylates proteins that regulate antioxidant activity and energy balance, promotes proliferation, and prevents senescence. Steady and pulsatile shear stress of 1.2 Pa increase SIRT1 levels and mitochondrial biosynthesis (Chen et al. 2010). Shear stress also elevates adenosine monophosphate (AMP)-activated protein kinase (AMPK), which regulates energy metabolism and facilitates the ability of SIRT1 to activate NOS III (Chen et al. 2010). Interestingly, SIRT1 elevates KLF2 levels providing an important regulatory loop that promotes an anti-oxidative environment and limits cell aging when ECs are exposed to unidirectional shear stress (Gracia-Sancho et al. 2010).

The application of shear stress causes a number of very rapid changes including ion channel opening, production of reactive oxygen species (ROS), activation of G proteins and focal adhesion movement (Davies 1995). Potential mechanisms initiating these processes likely involve the glycocalyx, a soft layer of proteoglycan on the cell surface; primary cilia, which extend from the endothelial cell surface; cytoskeleton deformation and focal adhesions (Davies 2009). However, the manner by which the applied stress is translated into a chemical signal has yet to be established.

Dynamically, cells respond to unidirectional shear stress with a short-term transient increase in NFκB activation and nuclear translocation leading to expression of leukocyte adhesion molecules VCAM-1, ICAM-1 and E-selectin and a longer-term anti-inflammatory and anti-thrombotic response initiated by expression of KLF2 and activation of Nrf2 (Boon and Horrevoets 2009).

The dynamic response to shear stress reflects the different signaling pathways activated by fluid shear stress (Fig. 2). Within 15 s of application of shear stress, a complex is formed in endothelial cell junctions involving VE-cadherin, platelet-endothelial cell adhesion molecule (PECAM) and vascular endothelial growth factor receptor 2 (VEGFR2) (Tzima et al. 2005).

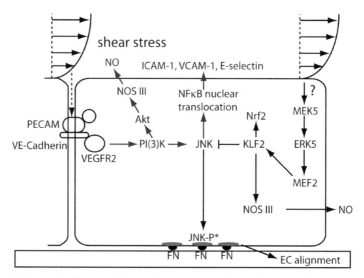

Figure 2. Schematic of key signaling pathways activated by unidirectional steady or pulsatile shear stress. Within the first few hours following exposure to shear stress, activation of PI(3) K by the complex of PECAM, VE-Cadherin and VEGFR2 leads to activation of JNK, NFκB nuclear translocation and expression of leukocyte adhesion molecules. As KLF2 and Nrf2 are expressed, JNK is inhibited and the long-term response to unidirectional shear stress is an anti-thrombotic and anti-inflammatory state. However, some JNK is phosphorylated and binds to activated integrins where it regulates EC alignment under flow. PI(3)K activates Akt which separately activates NOS III, increasing NO levels.

After binding VEGF, VEGFR2, as part of the mechanotransduction complex, binds to phosphatidylinositol-3-OH kinase (PI(3)K) which activates integrins initiating responses such as Jun NH2-terminal kinase (JNK) activation, nuclear factorκB (NFκB) activation and nuclear translocation (Hahn et al. 2009) which then leads to expression of adhesion molecules for leukocytes and secretion of MCP-1. PI(3)K also activates the protein kinase B or Akt which leads to activation of NOS III and other early anti-inflammatory responses. In addition to transient pro-inflammatory activity, JNK is phosphorylated and localizes to focal adhesions. There the JNK focal adhesion complex promotes EC alignment under unidirectional shear stress (Hahn et al. 2011).

The longer-term responses take several hours or days and involve activation of the mitogen activated protein kinase 5 (MEK5) by transient increases in reactive oxygen species. In turn, MEK5 phosphorylates extracellular signal-regulated kinase 5 (ERK5) which then activates the transcription factor myocyteenhancement factor 2 (MEF2) (Parmar et al. 2006). MEF2 promotes KLF2 expression. Further, KLF2 expression and Nrf2 produce antioxidants that control the intracellular levels of ROS, limiting activation of NFκB (Nigro et al. 2011). The PECAM/VE-Cadherin/PI(3)K and MEK/ERK5/MEF2 pathways activated by shear stress intersect in several ways to modulate the response to flow. Shear mediated activation of PI(3)K and Akt leads to the activation of Nrf2 (Dai et al. 2007)

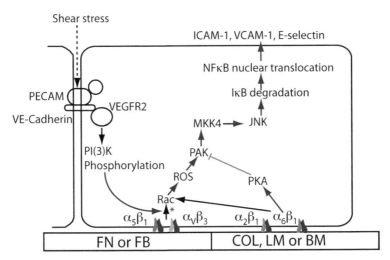

Figure 3. Schematic of signaling pathways by which shear stress activates integrins leading either to promotion or inhibition of NFκB nuclear transmigration and expression of leukocyte adhesion molecules ICAM-1, VCAM-1 and E-selectin in ECs. Adapted from (Hahn et al. 2009).

Color image of this figure appears in the color plate section at the end of the book.

which influences KLF2 activation (Huddleson et al. 2005). KLF2 regulates actin cytoskeleton organization which regulate the level of JNK and its downstream mediators, thereby limiting the pro-inflammatory response of steady and pulsatile shear stress (Boon et al. 2010).

Effect of shear stress on endothelial cell biomechanics

As a consequence of remodeling of the cytoskeleton, fluid shear stresses alter the mechanical properties of endothelial cells. The endothelial cell exhibits viscoelastic behavior (Cao et al. 2011, Theret et al. 1988), although the overall rheology is much more complex (del Alamo et al. 2008). After 24 hours exposure to flow, endothelial cells become stiffer with the elastic modulus increasing two to ten times, depending upon the measurement technique (Cao et al. 2011, Sato et al. 1987) and the magnitude of the applied shear stress (Theret et al. 1988). The elastic modulus is greater over the cell nucleus than over the cell body, and the actin stress fibers are considerably stiffer than the cell body or nucleus (Mathur et al. 2007). Exposure to fluid shear stress increases the elastic modulus of the nucleus and cell body with smaller increases in the stiffness of the actin stress fibers (Mathur et al. 2007).

The local shear stress exerted by steady flow of a Newtonian fluid over a confluent layer of endothelial cells is proportional to the local cell height and can be represented as (Barbee et al. 1995):

$$\tau / \tau_w = 1 + \kappa(h - h_{avg}) \tag{2}$$

where τ_w is the wall shear stress in the absence of the cell and $(h-h_{avg})$ is the variation in surface height about the mean height of a single cell. The slope in equation (2), κ, is sensitive to flow conditions and is 0.110 ± 0.005 μm^{-1} for unaligned cells not exposed to flow and, due to remodeling of the endothelial cell and nucleus, decreases to 0.085 ± 0.005 μm^{-1} after a 24-hour exposure of cells to an undisturbed shear stress of 1.2 Pa (Barbee et al. 1995). Consequently, the shear stresses vary as much as threefold over the surface of an endothelial cell. The nonuniform distributions of shear stresses may provide cues for the cells to respond to flow (Davies 1995).

Fluid shear stress produces complex rearrangement of the cytoskeleton (Helmke et al. 2001) and focal adhesions (Davies et al. 1994, Olivier et al. 1999). Displacements of the cytoskeleton are greater in the nuclear region (Helmke et al. 2001). These dynamic changes in the cytoskeleton are matched by rapid reductions in the cell compliance and elastic modulus (Dangaria and Butler 2007). While the shear stresses acting on endothelium are predicted to produce very small deformations on the order of 1–5 nm, the nonuniform shape of the endothelium and adhesion via specific focal adhesions leads to 10–100 fold stress concentration around the cell nucleus and strains around focal adhesions were on the order of 2–4% (Ferko et al. 2007). Such theoretical calculations are consistent with the greater displacement of the cytoskeleton near the nucleus and provide a mechanism by which focal adhesions act in mechanotransduction.

Endothelial cells exert stress on the substrate and on neighboring cells through the linkages formed by the cytoskeleton connections to integrins bound to their extracellular matrix protein ligands. Stresses between individual cells and substrate depend upon the type and amount of extracellular matrix ligand and the elasticity of the substrate. In spite of the nonuniform distribution of traction stresses, confluent layers of ECs exhibit a uniform distribution of intracellular tension (Hur et al. 2012). After exposure to flow for 16 hours, the EC traction force was slightly greater for cells exposed to steady shear stress than for cells under static conditions or exposed to a recirculating flow (Ting et al. 2012). Steady flow also caused the surface traction force vectors to align in the direction of flow, whereas the surface traction force vectors exhibited no preferred direction for cells under static conditions or exposed to the recirculating flow. These differences in the traction forces in response to unidirectional shear stress led to greater intercellular forces that were associated with increased levels of endothelial junction proteins. After 24 hours exposure to unidirectional flow, significant anisotropy in intracellular tension arose, with higher values normal to the direction of flow, possibly reflecting the reorganization of the cytoskeleton (Hur et al. 2012). Interestingly, the tension induced by shear stress is much less than the intracellular tension, suggesting that while changes to the intracellular tension are initiated by shear stress, the magnitude of the intracellular tension reflects biochemical changes within the cell (Hur et al. 2012) as well as cell-cell and cell-substrate interactions.

Risk factors for hypertension affect EC mechanical properties, influencing the changes to vessel wall. The hormone aldosterone is involved in the regulation of blood pressure and blood volume by enhancing NaCl reabsorption by the renal collecting duct epithelium. Quickly following addition of physiological levels of aldosterone, plasma sodium in the high physiological range increases EC stiffness 20% to 40% and the EC stiffening counteracts the endothelial formation of NO and reduces the ability ECs to release NO and trigger vasodilation (Oberleithner et al. 2007), while an acute increase of extracellular potassium in the physiological range does the opposite (Oberleithner et al. 2009). C-reactive protein, an inflammatory marker of cardiovascular diseases, enhances and sustains the effect of aldosterone on the EC stiffness (Kusche-Vihrog et al. 2011). C-reactive protein promotes the expression of the epithelial sodium channel and, in turn, the change in EC stiffness induced by C-reactive protein is dependent upon the epithelial sodium channel (Kusche-Vihrog et al. 2011).

Response of Endothelial Cells to Complex Flow Fields Found Around Vessel Branches

Flow around vessel branches is complex and often exhibits low mean shear stress that oscillate in with and against the main flow direction. Key features of flow around branches are large spatial and temporal gradients in shear stress, and low wall shear stress (Chien 2008). These regions tend be sites where atherosclerosis first develops. Regions with activated ECs also have high transcellular permeability to macromolecules, which makes them more prone to lipid deposition, endothelial activation and monocyte adhesion and migration into the subendothelium. These conclusions have also been made in a rabbit model (Okano and Yoshida 1993).

The reversing oscillatory flow observed *in vivo* has been modeled *in vitro* by reproducing the waveform at regions of the vessel wall that are and are not susceptible to atherosclerosis (Dai et al. 2004) or using idealized waveforms that represent key features of the flow (De Keulenaer et al. 1998), (Himburg et al. 2007). Endothelial cells exposed to this type of flow do not align or elongate. Other changes induced by reversing oscillatory shear stress that differ from unidirectional shear stress include an increase in endothelial cell proliferation, no change in NO and PGI_2 production above levels found with ECs under static conditions, and increased production of the vasoconstrictor endothelin-1 and pro-thrombotic molecules (Chiu and Chien 2011). In reversing oscillatory flow,the generation of ROS is sustained at much higher levels (Wasserman and Topper 2004) than in ECs exposed to unidirectional flow. Elevated ROS levels in ECs exposed to oscillatory shear stress are due to elevated NADH oxidase (De Keulenaer et al. 1998) and xanthine oxidase and decreased activity of the antioxidant enzyme xanthine dehydrogenase (McNally et al. 2003) as well as decreased intracellular glutathione levels (Harrison et al. 2006). Consequently, reversing oscillatory shear stresses produce sustained activation of NFκB and increases in mRNA and protein levels of pro-inflammatory molecules VCAM-1 and ICAM-1 (Chappell et al. 1998). Oscillatory shear stress elevates microRNA-92a levels repressing KLF2 and SIRT1 expression (Wu et al. 2011). Additionally, treatment of ECs with the antioxidant N-acetylcysteine reduces the effect of oscillatory shear on VCAM-1 and ICAM-1 expression, which further supports a role of ROS in the inflammatory response (Davis et al. 2001). Based on the findings from these studies, it is likely that oscillatory shear stress, via redox sensitive mechanisms, in the presence of risk factors for cardiovascular disease, promotes the earliest events in the development of atherosclerosis.

Spatial gradients in shear stress, which are also present at locations prone to develop atherosclerosis, affect endothelial cell function (Chiu and Chien 2011). Endothelial cells subjected to large spatial gradients in shear stress were round in shape at the reattachment point[2] (Chiu et al. 1998) as they are *in vivo* around branches. ECs exposed to spatial gradients exhibited rapid cell division and migration compared to cells exposed to uniform shear stress (DePaola et al. 1992) and increased nuclear localization of NFκB (Nagel et al. 1999) as well as the adhesion of monocytes (Barber et al. 1998), (McKinney et al. 2006). Shear stress gradients in regions of disturbed flow regulate the spatial distribution of an endothelial cell gap junction protein connexin 43 (DePaola et al. 1999). The spatial distribution of ICAM-1 expression was dependent upon both the shear stress and the component of normal stress caused by a sudden expansion in flow channel height (McKinney et al. 2006). Genes for NOS III, MCP-1, VCAM-1, ICAM-1, the transcription factor c-jun all responded differently to shear stress magnitude and gradient (LaMack and Friedman 2007). In mouse models of atherosclerosis, the transcription factor NFκB is more likely to be activated around branches, promoting a pro-inflammatory state (Hajra et al. 2000). In the pig aorta, regions of low and oscillating shear stress *in vivo* prime the endothelium, such that pro-oxidant and pro-inflammatory genes are elevated (Passerini et al. 2004). Such priming can facilitate the onset and progression of atherosclerosis in the presence of atherosclerosis. Thus, there is a fair degree of correspondence between the *in vitro* and *in vivo* results.

While the P(I)3K-Akt pathway promotes EC differentiation and growth (Morello et al. 2009) and is activated by flow, steady laminar flow also activates the AMPK pathway which blocks progression of ECs through the cell cycle (Guo et al. 2007). As a result, unidirectional laminar shear stress causes growth arrest of ECs. In contrast, reversing oscillatory shear stress timulates only the P(I)3K-Akt pathway and the AMPK pathway is not activated resulting in EC division and growth (Guo et al. 2007). Unlike unidirectional steady or pulsatile flow, reversing oscillatory flow and spatial gradients in shear stress produce prolonged activation of JNK, possibly due to increased levels of ROS (Chiu and Chien 2011) and inhibition of SIRT1 and AMPK (Chen et al. 2010).

Influence of the frequency of the shear stress waveform on endothelial cells

The complex blood flow patterns at vessel branches and bends produce shear stress fluctuations that include harmonic frequencies higher than the heart rate. The frequency of shear stress affects the inflammatory response of endothelial cells exposed to reversing and nonreversing oscillatory

---

[2]The reattachment point represents the location at which the flow separation ends and all of the fluid moves in the same direction.

shear stress. When porcine aortic endothelial cells (PAECs) were exposed to oscillatory shear stress of 1 Hz and different values of the time-averaged shear stress, transcriptional levels of inflammatory molecules VCAM-1 and IL-8 were repressed. At the same time, levels of atheroprotective genes increased. This gene expression profile though was reversed when the frequency was increased to 2 Hz. Interestingly, the pro-inflammatory response was most apparent under reversing oscillatory flow conditions, suggesting that both flow profiles and frequencies impact the development of atherosclerotic lesions. The results of this study also suggests that individuals whose arteries are chronically subjected to shear stress at frequencies of 2–3 Hz, such as those with high resting heart rates, may be at greater risk for atherosclerosis development (Himburg et al. 2007).

Interaction between Shear Stress and Cyclic Stretch on Endothelial Cell Function

Arterial pressure exerts normal stresses on the endothelium and vessel wall that, given the elastic behavior of arteries, causes a cyclic distention of arteries by 6–10%. Several studies have shown some weak effects due to the direct action of pressure on endothelium. When ECs grown on rigid substrates were exposed to physiological pressures, static or cyclic pressure alone can cause some modest increases in endothelial cell proliferation (Acevedo et al. 1993, Shin et al. 2002a, Sumpio et al. 1994) reorganization of the actin cytoskeleton (Acevedo et al. 1993, Salwen et al. 1998), alterations in integrin-α_v levels and focal adhesion organization (Schwartz et al. 1999) and changes in expression of genes in ECs involved in proliferation, inflammation and angiogenesis (Shin et al. 2002b). Since the ECs are grown on rigid surfaces for these *in vitro* studies, the compressive stresses induced by pressure may activate volume changes by unknown mechanisms.

A temporal phase shift between pressure and flow waveforms arises due to the inertia of blood and the reflection of the pressure wave by curved and branching arteries. This phase shift between flow and pressure waveforms produces a corresponding phase shift occurs between shear stress and the stretch of the vessel wall. The phase shift ranges from 0° in straight segments of the aorta to over 200° at the inner wall of the coronary arteries (Qiu and Tarbell 2000). In the carotid, the phase shift is very large at the sites of the internal carotid artery susceptible to atherosclerosis and which exhibit large values of OSI (Tada and Tarbell 2005). In curved vessels, the flow near the inner wall exhibits flow reversal when the phase shift between the wall shear stress and cyclic vessel wall distension exceed 60° (Klanchar et al. 1990).

The effect of the phase shift was examined *in vitro* with ECs grown on deformable cylindrical tubes for a circumferential strain of 5±4% at 1 Hz and shear stresses of 1.0±1 Pa (Dancu et al. 2004). In the absence of a phase shift between pressure and flow, the expression of NOS III and cyclooxygenase 2 mRNA and corresponding production of NO and PGI_2 are similar to values obtained with pulsatile unidirectional shear stress alone (Dancu et al. 2004). However, when the pressure and flow waveforms were out of phase (phase shift of 180°), NOS III and cyclooxygenase mRNA as well as NO and PGI_2 production changed more slowly following application of flow and were elevated relative to static conditions but not to the degree observed with shear stress alone (Dancu et al. 2004). The vasoconstrictor endothelin-1 was elevated after 12 hours exposure to flow that was out of phase with the pressure waveform whereas unidirectional shear stress alone suppressed release of endothelin-1. These results may explain how hypertension and curvature in coronary arteries influence the initiation and progression of atherosclerosis.

The Effect of Substrate on the Response to Flow

The subendothelium consists of collagen I as well as the basement membrane proteins collagen III, IV and laminin (Mayne 1986, Stenman and Vaheri 1978). During aging, the composition changes as the vessel wall stiffens. Collagen content can increase and collagen may fragment. During hypertension, the vessel stiffens in both the longitudinal and transverse directions due to changes in the extracellular matrix composition.

The pro- or anti-inflammatory response of vascular endothelium is dependent upon the interplay between the binding of specific integrins with extracellular matrix proteins in the subendothelium and the temporal and spatial dynamics of fluid shear stress (Orr et al. 2006). When cultured ECs adhere to fibronectin or fibrinogen, proteins that are deposited on the sub-endothelium after vascular injury, steady fluid shear stress activates the integrins $\alpha_5\beta_1$ and $\alpha_v\beta_3$ which subsequently promote transient NFκB activation (Orr et al. 2006, Tzima et al. 2005). When ECs are exposed to reversing oscillatory shear stresses for 18 hours or unidirectional shear stress for less than one hour, NFκB activation is sustained. Binding of $\alpha_5\beta_1$ and $\alpha_v\beta_3$ to fibronectin and subsequent flow-mediated integrin activation leads to activation of p21-activated kinase which then activates JNK and stimulates NFκB nuclear translocation (Hahn et al. 2009). There is a positive feedback loop in regions of complex flow around branches in which fibronectin synthesis by endothelial cells is increased by reversing oscillatory flow, further enhancing the likelihood of activation of the endothelium in these regions (Feaver et al. 2010). ECs attached to collagen or the basement membrane protein mimetic Matrigel bind to $\alpha_2\beta_1$. In contrast, when ECs attached to collagen via the integrin $\alpha_2\beta_1$ and were exposed to reversing oscillatory shear stress, they do not exhibit increased JNK activity (Hahn et al. 2009) and suppress activation of NFκB by elevating protein kinase A levels (Funk et al. 2010). In the ApoE–/– mouse, JNK activation and increased levels of subendothelial fibronectin is observed around curved and branching vessels where lesions develop, suggesting that the altered hemodynamics promote a pro-inflammatory state through sustained activation of JNK (Hahn et al. 2009). Thus, a normal subendothelium can limit the pro-inflammatory and pro-thrombotic effects of the complex flows around branches until additional risk factors appear.

In vitro, the dynamics of EC shape reorganization and focal adhesion dynamics are sensitive to the extracellular matrix to which the ECs are attached. After exposure to unidirectional flow, focal adhesions remodeling and EC alignment under flow is slower when ECs are attached to collagen than when attached to fibronectin (Davies et al. 1994, Hahn et al. 2011) and the dynamic response of ECs to flow is affected by JNK phosphorylation and localization to focal adhesions. Interestingly, the short-term pro-inflammatory response to flow can be eliminated by aligning ECs on micropatterned surfaces prior to exposure to flow (Vartanian et al. 2010), suggesting that promoting JNK interaction with focal adhesions by the extracellular matrix can influence the function of ECs.

The mechanical and functional behavior of endothelium is affected by the mechanical properties of the subendothelial surface. Cell elastic modulus often matches the soft substrate modulus within a range of values spanning that of soft tissues (Tee et al. 2011) and the elastic modulus of the bovine carotid endothelium is similar to the modulus of the subendothelium (Peloquin et al. 2011). Endothelial permeability increases when the substrate stiffness *in vitro* increases to levels corresponding to change in subendothelial stiffness observed in older mice and this change in stiffness and permeability is mediated by Rho kinase (Huynh et al. 2011). Increased EC stiffness facilitates neutrophil transmigration (Stroka and Aranda-Espinoza 2011). Senescence is promoted when ECs adhere to glycated collagen (Chen et al. 2002). These results suggest that changes to the mechanical properties of thesubendothelium affect transendothelial transport and the progression of atherosclerosis. An important issue to address is how exposure of ECs to stiffer subendothelium in the presence of flow affects the mechanical behavior and contractility of the endothelium.

Effect of Shear Stress on Response of ECs to Cytokines

Risk factors, including high cholesterol, smoking, aging, and hypertension play an important role in disease progression. These risk factors induce an inflammatory response in ECs and SMCs, which is key to disease progression (Hansson and Libby 2006). During atherosclerosis, cytokines are released by activated monocytes and macrophages that are recruited to regions of the vessel wall containing elevated or modified levels of lipoproteins. One key cytokine is tumor necrosis factor alpha (TNFα) which induces expression of leukocyte adhesion molecules on endothelial cells and smooth muscle cells and induces endothelial cell apoptosis (Sato et al. 1986) by activation of caspase 8. Prolonged exposure of ECs *in vitro* to steady laminar unidirectional shear stress stabilizes KLF2 levels (van Thienen et al. 2006) and inhibits NF-κB transcriptional activity (Chiu et al. 2004) thereby limiting the pro-inflammatory response to TNFα and other cytokines.

Like KLF2, the zinc finger transcription factor KLF4 is induced by shear stress. The expression of KLF4 is greater following exposure to arterial mean shear stress values greater than the mean venous value of shear stress (Hamik et al. 2007). Unlike KLF2, exposure of human umbilical vein ECs to TNFα, IL-1β, or interferon γ induces KLF4 expression. KLF4 inhibits the early pro-thrombotic and pro-inflammatory response induced by flow by blocking NFκB activation and promotes expression of NOS III and thrombomodulin (Hamik et al. 2007). Interestingly, KLF2 and KLF4 are regulated by microRNA 92a and are both elevated in regions of the aorta prone to atherosclerosis (Fang and Davies 2012). Under static conditions, the response to TNFα is reduced when microRNA 92a is blocked and this is due to the KLF4 and not KLF2. The relative contributions of KLF2 and KLF4 to the flow-mediated repression of TNFα or other cytokines has not yet been studied (Fang and Davies 2012).

The Influence of EC-SMC Interactions on EC Function

Proper blood vessel function depends on interactions between ECs and SMCs. ECs release vasodilators (PGI$_2$, NO), vasoconstrictors (endothelin-1) and endothelial derived relaxing factor that affect SMC contractile properties. NO and proteoglycans released by ECs inhibit SMC growth (Cornwell et al. 1994, Moses et al. 1995), and PGI$_2$ promotes a contractile phenotype (Fetalvero et al. 2006, Tsai et al. 2009). SMCs produce an inactive form of transforming growth factor β (TGF-β) (Sato et al. 1993). Urokinase on the EC surface activates TGF-β (Gleizes et al. 1997) which then inhibits SMC proliferation and induces contractile protein expression (Deaton et al. 2005). Conversely, SMCs influence EC function; cytokines or low flow cause SMCs to release cyclophilin A which has pro-inflammatory effects on ECs and SMCs (Satoh et al. 2008). Further, the response in coculture may differ from the response of either cell type cultured alone. For example, flow-mediated activation of the mammalian target of rapamycin (mTOR), which is critical for EC, SMC and mesenchymal stem cell proliferation and differentiation, is different when ECs are in monoculture than when they are cocultured with SMCs (Balcells et al. 2010); likewise, the sensitivity of the mTOR pathway in SMCs to growth factors was blocked when SMCs and ECs were cultured together (Balcells et al. 2010).

In the absence of disease, both ECs and SMCs are in a quiescent and nonproliferative state. During early atherogenesis, the endothelium becomes activated, and SMCs undergo phenotypic switching in which SMC exhibit decreased expression of smooth muscle contractile proteins and increased proliferation and migration (Gomez and Owens 2012). *In vitro*, ECs cocultured with proliferative SMCs exhibited greater adhesion of flowing leukocytes compared to ECs cocultured with quiescent SMCs (Rainger and Nash 2001). The ECs cocultured with the activated SMCs also became more sensitive and responded to TNFα concentrations that were 10,000 times lower compared to ECs cultured alone (Rainger and Nash 2001). In addition, coculture altered the endothelial selectin adhesion molecules used for leukocyte attachment. During co-culture pro-inflammatory genes are upregulated in ECs but not in SMCs (Chiu et al. 2005). Levels of eNOS are reduced in ECs cocultured with proliferating SMCs (Chiu et al. 2003), facilitating an inflammatory response in ECs and SMCs. The EC priming in response to the activated SMCs was attributed to TGF-β which was proteolytically activated by the serine protease plasmin (Rainger and Nash 2001). Application of ECs shear stress largely inhibited this inflammatory effect of proliferative SMCs upon ECs (Chiu et al. 2005). Thus, when SMCs are in a proliferative phenotype present in atherosclerosis, they promote a pro-inflammatory response by ECs. This response of SMCs may be exaggerated in regions of reversing oscillatory flow.

Using a direct coculture of ECs on SMCs in which ECs form a confluent monolayer on extracellular matrix produced by quiescent SMCs (Brown et al. 2009, Lavender et al. 2005, Wallace et al. 2007b), ECs were not pro-inflammatory. When cultured on quiescent SMCs, human ECs produce less tissue factor (Wallace et al. 2007b) and had similar levels of MCP-1, ICAM-1, VCAM-1 and E-selectin (Brown et al. 2009) relative to ECs on plastic surfaces. Under static and flow conditions, EC interactions with quiescent SMCs in coculture partially inhibit TNFα mediated EC activation (Wallace and Truskey 2010). This inhibition of TNFα responsiveness by ECs required contact between the ECs and SMCs and the extracellular matrix that they synthesized and involved an elevation of KLF2 levels in ECs.

When cultured on quiescent SMCs, ECs formed fibrillar adhesions and not focal adhesions, probably due to production of fibrillar fibronectin by SMCs (Wallace et al. 2007a). Due to the absence of focal adhesions, ECs in coculture had less RhoA mRNA expression than ECs in monoculture (Wallace et al. 2007a). RhoA activation mediates NF-κB nuclear translocation induced by TNFα (Hippenstiel et al. 2002). Further, EC interactions with fibrillar fibronectin lead to increased p38 activation (Katz et al. 2000), which can inhibit NF-κB nuclear activation translocation (Bowie and O'Neill 2000). Therefore, EC interactions with the extracellular matrix on the surface of the SMCs could inhibit the TNFα-induced inflammatory response by a decreased amount of active RhoA and/or increased p38 activation.

Hastings et al. showed that physiological flow waveforms that had unidirectional and reversing pulsatile patterns differentially regulate both EC and SMC phenotypes (Hastings et al. 2007). They used the direct contact coculture model and showed that hemodynamic forces applied directly and only to the endothelium modulate the SMC activated and quiescent phenotypes and influence SMC remodeling. The results reveal that both ECs and SMCs undergo differential phenotypic alterations in response to atheroprone compared with atheroprotective flow, indicating that vascular ECs and SMCs in an atheroprone environment are both susceptible to an early inflammatory response (Hastings et al. 2007).

In coculture of ECs with quiescent human aortic SMCs the elastic moduli increased 160–180% compared to those values measured for the ECs in monoculture (Cao et al. 2011). While the moduli of ECs almost doubled in monoculture and flow conditions, similar to previous reports (Theret et al. 1988), their corresponding values in coculture declined after exposure to flow. Both the number and diameter of cortical stress fiber per cell width increased in coculture and/or flow conditions, whereas the subcortical stress fiber density throughout the cell interior increased by a smaller amount. For ECs, fluid shear stress appeared to have greater effect on the elastic modulus than the presence of SMCs and changes to the elastic modulus in coculture may be due to EC-SMC communication. ECs became stiffer when cocultured on the SMC layer. Although ECs in coculture lacked focal adhesions and exhibited fibrillar adhesions, the difference in type of adhesion did not reduce actin stress fibers, possibly because cytoskeletal interactions in EC monolayers are dominated by connections at cell-cell junctions (Goeckeler et al. 2008). Although ECs in monoculture do secrete and remodel an extracellular matrix, their mechanical properties are largely influenced by the substrate (Byfield et al. 2009), suggesting the shear stress adaptation assists in the process of matching the modulus of ECs and the subendothelial surface.

Response of Aged ECs to Fluid Shear Stress

Aging at the cellular level is associated with replicative senescence or stress-induced senescence. For replicative senescence, the cell doubling time increases and ultimately ceases with increasing cell age due to telomere shortening which protects against genetic mutation (Erusalimsky and Skene 2009). Stress-induced senescence is primarily caused by environmental stresses such as intracellular oxidative stress, DNA damage by radiation, or chromatin decondensation. Endothelium at lesion-prone sites and on advanced lesions show signs of advanced aging, including elevated levels of β-galactosidase, a histochemical marker for senescence (Minamino et al. 2002), and telomere shortening (Ogami et al. 2004). ECs near atherosclerotic lesions often experience high cell turnover and high levels of oxidative stress, induced by a combination of the reversing oscillatory flow and presence of risk factors for cardiovascular disease.

Cell energy metabolism is altered during senescence. In particular, levels of SIRT1 are reduced which also reduces a number of anti-thrombotic and anti-inflammatory genes (Erusalimsky and Skene 2009). NO release is impaired as ECs age and NOS III production induced by shear stress is reduced in senescent human ECs (Matsushita et al. 2001). Following exposure to TNFα or short periods of shear stress, senescent ECs exhibit greater monocyte adhesion than do younger ECs (Matsushita et al. 2001). When levels of human telomerase, a reverse transcriptase, are elevated in late passage human ECs, NOS III levels are increased and monocyte adhesion is reduced.

Elevating SIRT1 levels reverses EC senescence. Resveratrol has been used to reduce age-associated metabolic phenotypes (Park et al. 2012) and permeability (Cheung et al. 2012) and it may mitigate symptoms associated with atherogenesis. Resveratrol acts upon SIRT1 via competitive inhibition of phosphodiesterase 4 which causes elevated cAMP levels (Park et al. 2012). The resulting elevation of cAMP activates effector protein Epac1, and sequentially increases Rap1, intracellular Ca^{2+} and activates the CamKKβ-AMPK pathway to activate SIRT1 (Park et al. 2012). Elevated cAMP levels cause a decrease in endothelial permeability by activation of protein kinase A which then inactivates myosin light chain kinase (Mehta and Malik 2006) or activates Epac-mediated Rap1 (Fukuhara et al. 2005). Further, membrane associated cAMP bound to Epac1 interacts with phosphodiesterase 4, to bind to VE-cadherin (Rampersad et al. 2010) and regulates endothelial permeability by redistributing tight junction molecules to cell junctions (Cullere et al. 2005). Given the shear stress sensitivity of SIRT1 and VE-cadherin, shear stress should influence EC senescence.

Older endothelial cells respond to shear stress in a manner different from the response of younger endothelial cells. Cyclin-dependent kinase inhibitor 2A and 2B which reduce cell proliferation are elevated by both aging and exposure to 1.2 Pa shear stress for 24 hours (Mun et al. 2009). Expression of NADPH oxidase 4 is reduced by unidirectional shear stress and aging. NADPH oxidase 4 promotes production of oxygen radicals, and a reduction in the level of this enzyme protects the aging ECs from oxidative damage. This is offset by a reduction in aging cells of NOS III and argininosuccinate synthetase 1 although the levels of these enzymes are elevated by shear stress in older ECs, although to a less extent than younger cells (Mun et al. 2009). Argininosuccinate synthetase 1 catalyzes the rate-limiting step of arginine biosynthesis and can regulate NO release and monocyte adhesion independent of NOS III levels (Mun et al. 2011). Increased monocyte adhesion in older ECs exposed to TNFα is due to a reduction in NO release and an increase in the expression of CD44, which can bind to leukocytes, but not to changes in the levels of leukocyte adhesion molecules VCAM-1, ICAM-1 and E-selectin (Mun and Boo 2010). Interestingly, CD44 levels are not affected by exposure of older ECs to shear stress (Mun and Boo 2010).

The results of these studies suggest that cell age may be a key biological mechanism in atherogenesis. This is of particular interest because senescent cells are known to be present in areas with atherosclerosis. ECs in these regions also undergo more cycles of replication and, due to the local hemodynamics, are more sensitive to extracellular stresses. The body has several intrinsic mechanisms for repairing damaged endothelium. Although circulating EPCs do not appear to contribute significantly to the endothelium overlying atherosclerotic lesions (Hagensen et al. 2010), progenitor cells from young animals preferentially localize to sites of endothelial dysfunction and can reverse atherosclerosis in older animals (Rauscher et al. 2003), suggesting that a decline in progenitor cell adhesion and function with age may influence the onset and progression of atherosclerosis. Interestingly, the localization of progenitor cells exhibits the same distribution of monocytes attaching to arterial endothelium (Buchanan et al. 2003), suggesting that that localization is influenced by the local flow field.

Conclusion

In the thirty years since the first demonstration that vascular endothelial cells responded to steady laminar flow, a considerable body of data has developed to show that the local flow field influences the functional state of the endothelium. While many details of the signaling pathways have been elucidated, the initial steps involved in the transduction of the fluid shear stress to a biochemical event are not known. The interaction of the signaling pathways is likely to influence how ECs respond to different flow fields and key details still need to be identified. Many studies of the effects of flow on the endothelium have been done *in vitro*. The direct contact co-culture model has become the most popular for modeling the blood vessel. Whether this model can account for the complexities of vessel structure and function still remains to be proven.

The interaction of hemodynamics and the various risk factors influences whether or not atherosclerosis develops. This review highlighted our current understanding of several key risk factors. While cell-cell inflammatory reactions, cell-substrate interactions and cell age appear to accelerate disease progression, future studies should focus upon how these disease-associated risk factor interact with different shear stress patterns to influence the onset and progression of atherosclerosis. From such studies new therapeutics may be identified to treat those most prone to develop cardiovascular disease before clinical symptoms appear.

Acknowledgements

This work was supported by NIH grants HL44972, HL88825 and an NSF graduate research fellowship to Tracy Cheung.

References

Acevedo, A.D., S.S. Bowser, M.E. Gerritsen and R. Bizios. 1993. Morphological and proliferative responses of endothelial-cells to hydrostatic-pressure-role of fibroblast growth-factor. *J. Cell Physiol.* 157: 603–614.

Balcells, M., J. Martorell, C. Olive, M. Santacana, V. Chitalia, A.A. Cardoso and E.R. Edelman. 2010. Smooth muscle cells orchestrate the endothelial cell response to flow and injury. *Circulation* 121: 2192–2199.

Barbee, K.A., T. Mundel, R. Lal and P.F. Davies. 1995. Subcellular distribution of shear stress at the surface of flow-aligned and nonaligned endothelial monolayers. *Am. J. Physiol.* 268: HI765–H1772.

Barber, K.M., A. Pinero and G.A. Truskey. 1998. Effects of recirculating flow on U-937 cell adhesion to human umbilical vein endothelial cells. *Am. J. Physiol. Heart Circ. Physiol.* 275: H591–599.

Boon, R. and A. Horrevoets. 2009. Key transcriptional regulators of the vasoprotective effects of shear stress. *Hamostaseologie* 29: 39–43.

Boon, R., T. Leyen, R. Fontijn, J. Fledderus, J. Baggen, O. Volger, G. van Nieuw Amerongen and A. Horrevoets. 2010. KLF2-induced actin shear fibers control both alignment to flow and JNK signaling in vascular endothelium. *Blood* 115: 2533–2542.

Bowie, A.G. and L.A. O'Neill. 2000. Vitamin C inhibits NF-kappa B activation by TNF via the activation of p38 mitogen-activated protein kinase. *J. Immunol.* 165: 7180–7188.

Brown, M.A., C.S. Wallace, M. Angelos and G.A. Truskey. 2009. Characterization of Umbilical Cord Blood Derived Late Outgrowth Endothelial Progenitor Cells Exposed to Laminar Shear Stress. *Tissue Eng. Part A* 15: 3575–3587.

Buchanan, J., C. Kleinstreuer, S. Hyun and G. Truskey. 2003. Hemodynamics simulation and identification of susceptible sites of atherosclerotic lesion formation in a model abdominal aorta. *J. Biomech.* 36: 1185–1196.

Byfield, F.J., R.K. Reen, T.P. Shentu, I. Levitan and K.J. Gooch. 2009. Endothelial actin and cell stiffness is modulated by substrate stiffness in 2D and 3D. *J. Biomech.* 42: 1114–1119.

Cai, H. and D.G. Harrison. 2000. Endothelial Dysfunction in Cardiovascular Diseases: The Role of Oxidant Stress. *Circ. Res.* 87: 840–844.

Cao, L., A. Wu and G. Truskey. 2011. Biomechanical effects of flow and coculture on human aortic and cord blood-derived endothelial cells. *J. Biomech.* 44: 2150–2157.

Chamiot-Clerc, P., X. Copie, J.-F. Renaud, M. Safar and X. Girerd. 1998. Comparative reactivity and mechanical properties of human isolated internal mammary and radial arteries. *Cariovasc. Res.* 37: 811–819.

Chappell, D.C., S.E. Varner, R.M. Nerem, R.M. Medford and R.W. Alexander. 1998. Oscillatory Shear Stress Stimulates Adhesion Molecule Expression in Cultured Human Endothelium. *Circ. Res.* 82: 532–539.

Cheung, T.M., M.P. Ganatra, E.B. Peters and G.A. Truskey. 2012. The Effect of Cellular Senescence on the Albumin Permeability of Blood-Derived Endothelial Cells. *Am. J. Physiol.* 303: H1374–H1383.

Chen, J., S.V. Brodsky, D.M. Goligorsky, D.J. Hampel, H. Li, S.S. Gross and M.S. Goligorsky. 2002. Glycated Collagen I Induces Premature Senescence-Like Phenotypic Changes in Endothelial Cells. *Circ. Res.* 90: 1290–1298.

Chen, X., S. Varner, A. Rao, J. Grey, S. Thomas, C. Cook, M. Wasserman, R. Medford, A. Jaiswal and C. Kunsch. 2003. Laminar flow induction of antioxidant response element-mediated genes in endothelial cells. *J. Biol. Chem.* 278: 703–711.

Chen, Z., I.-C. Peng, X. Cui, Y.-S. Li, S. Chien and J.-J. Shyy. 2010. Shear stress, SIRT1, and vascular homeostasis. *Proc. Natl. Acad. Sci.* 107: 10268–10273.

Cheng, C., F. Helderman, D. Tempel, D. Segers, B. Hierck, R. Poelman, A. van Tol, D.J. Duncker, D. Robbers-Visser, N.-T. Ursem, R. van Haperen, J.-J. Wentzel, F. Gijsen, A.-F. van der Steen, R. de Crom and R. Krams. 2007. Large variations in absolute wall shear stress levels within one species and between species. *Atherosclerosis* 195: 225–235.

Chien, S. 2008. Effects of Disturbed Flow on Endothelial Cells. *Ann. Biomed. Eng.* 36: 554–562.

Chiu, J.-J., L.-J. Chen, S.-F. Chang, P.-L. Lee, C.-I. Lee, M.-C. Tsai, D.-Y. Lee, H.-P. Hsieh, S. Usami and S. Chien. 2005. Shear Stress Inhibits Smooth Muscle Cell-Induced Inflammatory Gene Expression in Endothelial Cells: Role of NF-{kappa}B. *Arterioscler. Thromb. Vasc. Biol.* 25: 963–969.

Chiu, J.J. and S. Chien. 2011. Effects of Disturbed Flow on Vascular Endothelium: Pathophysiological Basis and Clinical Perspectives. *Physiol. Rev.* 91: 327–387.

Chiu, J.J., D.L. Wang, S. Chien, R. Skalak and S. Usami. 1998. Effects of disturbed flow on endothelial cells. *J. Biomech. Eng.* 120: 2–8.

Chiu, J.J., L.J. Chen, P.L. Lee, C.I. Lee, L.W. Lo, S. Usami and S. Chien. 2003. Shear stress inhibits adhesion molecule expression in vascular endothelial cells induced by coculture with smooth muscle cells. *Blood* 101: 2667–2674.

Chiu, J.J., P.L. Lee, C.N. Chen, C.I. Lee, S.F. Chang, L.J. Chen, S.C. Lien, Y.C. Ko, S. Usami and S. Chien. 2004. Shear Stress Increases ICAM-1 and Decreases VCAM-1 and E-selectin Expressions Induced by Tumor Necrosis Factor-α in Endothelial Cells. *Arterioscler. Thromb. Vasc. Biol.* 24: 73–79.

Chu, T. and D. Peters. 2008. Serial analysis of the vascular endothelial transcriptome under static and shear stress conditions. *Physiol. Genom.* 34: 185–192.

Cines, D.B., E.S. Pollak, C.A. Buck, J. Loscalzo, G.A. Zimmerman, R.P. McEver, J.S. Pober, T.M. Wick, B.A. Konkle, B.S. Schwartz, E.S. Barnathan, K.R. McCrae, B.A. Hug, A.M. Schmidt and D.M. Stern. 1998. Endothelial cells in physiology and in the pathophysiology of vascular disorders. *Blood* 91: 3527–3561.

Collins, C. and E. Tzima. 2011. Hemodynamic forces in endothelial dysfunction and vascular aging. *Experimental Gerontology* 46: 185–188.

Cornwell, T.L., E. Arnold, N.J. Boerth and T.M. Lincoln. 1994. Inhibition of smooth muscle cell growth by nitric oxide and activation of cAMP-dependent protein kinase by cGMP. *Am. J. Physiol. - Cell Physiol.* 267: C1405–C1413.

Cullere, X., S.K. Shaw, L. Andersson, J. Hirahashi, F.W. Luscinskas and T.N. Mayadas. 2005. Regulation of vascular endothelial barrier function by Epac, a cAMP-activated exchange factor for Rap GTPase. *Blood* 105: 1950–1955.

Dai, G., S. Vaughn, Y. Zhang, E.T. Wang, G. Garcia-Cardena and M.A. Gimbrone. 2007. Biomechanical Forces in Atherosclerosis-Resistant Vascular Regions Regulate Endothelial Redox Balance via Phosphoinositol 3-Kinase/Akt-Dependent Activation of Nrf2. *Circulation Research* 101: 723–733.

Dai, G., M. Kaazempur-Mofrad, S. Natarajan, Y. Zhang, S. Vaughn, B. Blackman, R. Kamm, G. Carcia-Cardena and M.A. Gimbrone. 2004. Distinct endothelial phenotypes evoked by arterial waveforms derived from atherosclerosis-susceptible and -resistant regions of human vasculature. *Proc. Natl. Acad. Sci.* 101: 14871–14876.

Dancu, M.B., D.E. Berardi, J.P. Vanden Heuvel and J.M. Tarbell. 2004. Asynchronous shear stress and circumferential strain reduces endothelial NO synthase and cyclooxygenase-2 but induces endothelin-1 gene expression in endothelial cells. *Arterioscler. Thromb. Vasc. Biol.* 24: 2088–2094.

Dangaria, J.H. and P.J. Butler. 2007. Macrorheology and adaptive microrheology of endothelial cells subjected to fluid shear stress. *Am. J. Physiol.* 293: C1568–C1575.

Davies, P.F. 1995. Flow-mediated endothelial mechanotransduction. *Physiol. Rev.* 75: 519–560.

Davies, P.F. 2009. Hemodynamic shear stress and the endothelium in cardiovascular pathophysiology. *Nature Clin. Pract. Cardiovasc. Med.* 6: 16–26.

Davies, P.F., A. Robotewskyj and M.L. Griem. 1994. Quantitative studies of endothelial cell adhesion: Directional remodeling of focal adhesion sites in response to flow forces. *J. Clin. Invest.* 93: 2031–2038.

Davis, M.E., H. Cai, G.R. Drummond and D.G. Harrison. 2001. Shear Stress Regulates Endothelial Nitric Oxide Synthase Expression Through c-Src by Divergent Signaling Pathways. *Circ. Res.* 89: 1073–1080.

De Keulenaer, G.W., D.C. Chappell, N. Ishizaka, R.M. Nerem, R.W. Alexander and K.K. Griendling. 1998. Oscillatory and steady laminar shear stress differentially affect human endothelial redox state : role of a superoxide-producing NADH oxidase. *Circ. Res.* 82: 1094–1101.

Deaton, R.A., C. Su, T.G. Valencia and S.R. Grant. 2005. Transforming Growth Factor-Beta-induced Expression of Smooth Muscle Marker Genes Involves Activation of PKN and p38 MAPK. *J. Biol. Chem.* 280: 31172–31181.

Dekker, R., S. van Soest, R. Fontijn, S. Salamanca, P. de Groot, E. VanBavel, H. Pannekoek and A. Horrevoets. 2002. Prolonged fluid shear stress induces a distinct set of endothelial cell genes, most specifically lung Kruppel-like factor (KLF2). *Blood* 100: 1689–1698.

del Alamo, J.C., G.N. Norwich, Y.J. Li, J.C. Lasheras and S. Chien. 2008. Anisotropic rheology and directional mechanotransduction in vascular endothelial cells. *Proc. Natl. Acad. Sci.* 105.

DePaola, N., M.A. Gimbrone, P.F. Davies and C.F. Dewey. 1992. Vascular endothelium responds to fluid shear stress gradients. *Arterioscler. Thromb.* 1254–1257.

DePaola, N., P.F. Davies, W.F.J. Pritchard, L. Florez, N. Harbeck and D.C. Polacek. 1999. Spatial and temporal regulation of gap junction connexin43 in vascular endothelial cells exposed to controlled disturbed flows *in vitro*. *Proc. Natl. Acad. Sc.* 96: 3154–3159.

Dewey, C., S. Bussolari, M. Gimbrone and P.F. Davies. 1981. The dynamic response of vascular endothelial cells to fluid shear stress. *J. Biomech. Eng.* 103: 177–185.

Dobrin, P.B. 1978. Mechanical properties of arteries. *Physiol. Rev.* 58: 397–460.

Erusalimsky, J.D. and C. Skene. 2009. Mechanisms of endothelial senescence. *Exptl. Physiol.* 94: 299–304.

Fang, Y. and P.F. Davies. 2012. Site-Specific MicroRNA-92a Regulation of Kruppel-Like Factors 4 and 2 in Atherosusceptible Endothelium. *Arterioscler. Thromb. Vasc. Biol.* 32: 979–987.

Feaver, R.E., B.D. Gelfand, C. Wang, M.A. Schwartz and B.R. Blackman. 2010. Atheroprone Hemodynamics Regulate Fibronectin Deposition to Create Positive Feedback That Sustains Endothelial Inflammation. *Circ. Res.* 106: 1703–1711.

Ferko, M.C., A. Bhatnagar, M.B. Garcia and P.J. Butler. 2007. Finite-element stress analysis of a multicomponent model of sheared and focally-adhered endothelial cells. *Ann. Biomed. Eng.* 35: 208–223.

Fetalvero, K.M., M. Shyu, A.P. Nomikos, Y.F. Chiu, R.J. Wagner, R.J. Powell, J. Hwa and K.A. Martin. 2006. The prostacyclin receptor induces human vascular smooth muscle cell differentiation via the protein kinase A pathway. *Am. J. Physiol. Heart Circ. Physiol.* 290: H1337–1346.

Fledderus, J.O., R.A. Boon, O.L. Volger, H. Hurttila, S. Ylä-Herttuala, H. Pannekoek, A.L. Levonen and A.J. Horrevoets. 2008. KLF2 primes the antioxidant transcription factor Nrf2 for activation in endothelial cells. *Arterioscler. Thromb. Vasc. Biol.* 28: 1339–1346.

Franke, R.-P., M. Gräfe, H. Schnittler, D. Seiffge, C. Mittermayer and D. Drenckhahn. 1984. Induction of human vascular endothelial stress fibres by fluid shear stress. *Nature* 307: 648–649.

Fukuhara, S., A. Sakurai, H. Sano, A. Yamagishi, S. Somekawa, N. Takakura, Y. Saito, K. Kangawa and N. Mochizuki. 2005. Cyclic AMP potentiates vascular endothelial cadherin-mediated cell-cell contact to enhance endothelial barrier function through an Epac-Rap1 signaling pathway. *Mol. Cell Biol.* 25: 136–146.

Funk, S.D., A. Yurdagul, J.M. Green, K.A. Jhaveri, M.A. Schwartz and A.W. Orr. 2010. Matrix-Specific Protein Kinase A Signaling Regulates p21-Activated Kinase Activation by Flow in Endothelial Cells. *Circ. Res.* 106: 1394–1403.

Gleizes, P.E., J.S. Munger, I. Nunes, J.G. Harpel, R. Mazzieri, I. Noguera and D.B. Rifkin. 1997. TGF-beta latency: biological significance and mechanisms of activation. *Stem Cells* 15: 190–197.

Goeckeler, Z.M., P.C. Bridgman and R.B. Wysolmerski. 2008. Nonmuscle myosin II is responsible for maintaining endothelial cell basal tone and stress fiber integrity. *Am. J. Physiol.* 295: C994–C1006.

Gomez, D. and G.K. Owens. 2012. Smooth muscle cell phenotypic switching in atherosclerosis. *Cardiovasc. Res.* 95: 156–164.

Gracia-Sancho, J., G. Villarreal Jr., Y. Zhang and G. García-Cardena. 2010. Activation of SIRT1 by resveratrol induces KLF2 expression conferring an endothelial vasoprotective phenotype. *Cardiovasc. Res.* 85: 514–519.

Guo, D., S. Chien and J. Shyy. 2007. Regulation of Endothelial Cell Cycle by Laminar Versus Oscillatory Flow: Distinct Modes of Interactions of AMP-Activated Protein Kinase and Akt Pathways. *Circ. Res.* 100: 564–571.

Hagensen, M.K., J. Shim, T. Thim, E. Falk and J.F. Bentzon. 2010. Circulating endothelial progenitor cells do not contribute to plaque endothelium in murine atherosclerosis. *Circulation* 121: 898–905.

Hahn, C., A.W. Orr, J.M. Sanders, K.A. Jhaveri and M.A. Schwartz. 2009. The Subendothelial Extracellular Matrix Modulates JNK Activation by Flow. *Circ. Res.* 104: 995–1003.

Hahn, C., C. Wang, A.W. Orr, B.G. Coon and M.A. Schwartz. 2011. JNK2 Promotes Endothelial Cell Alignment under Flow. *PLoS ONE* 6: e24338.

Hajra, L., A.I. Evans, M. Chen, S.J. Hyduk, T. Collins and M.I. Cybulsky. 2000. The NF-kappa B signal transduction pathway in aortic endothelial cells is primed for activation in regions predisposed to atherosclerotic lesion formation. *PNAS* 97: 9052–9057.

Halcox, J., W. Schenke, G. Zalos, R. Mincemoyer, A. Prasad, M. Waclawiw, K. Nour and A. Quyyumi. 2002. Prognostic value of coronary vascular endothelial dysfunction. *Circulation* 106: 653–658.

Hamik, A., Z. Lin, A. Kumar, M. Balcells, S. Sinha, J. Katz, M.W. Feinberg, R.E. Gerszten, E.R. Edelman and M.K. Jain. 2007. Kruppel-like Factor 4 Regulates Endothelial Inflammation. *J. Biol. Chem.* 282: 13769–13779.

Hansson, G.K. and P. Libby. 2006. The immune response in atherosclerosis: a double-edged sword. *Nat. Rev. Immunol.* 6: 508–519.

Harrison, D.G., J. Widder, I. Grumbach, W. Chen, M. Weber and C. Searles. 2006. Endothelial mechanotransduction, nitric oxide and vascular inflammation. *J. Int. Med.* 259: 351–363.

Hastings, N.E., M.B. Simmers, O.G. McDonald, B.R. Wamhoff and B.R. Blackman. 2007. Atherosclerosis-prone hemodynamics differentially regulates endothelial and smooth muscle cell phenotypes and promotes pro-inflammatory priming. *Am. J. Physiol. - Cell Physiol.* 293: C1824–C1833.

Helmke, B.P., D.B. Thakker, R.D. Goldman and P.F. Davies. 2001. Spatiotemporal analysis of flow-induced intermediate filament displacement in living endothelial cells. *Biophys. J.* 80: 184–194.

Himburg, H.A., S.E. Dowd and M.H. Friedman. 2007. Frequency-dependent response of the vascular endothelium to pulsatile shear stress. *Am. J. Physiol. - Heart Circ. Physiol.* 293: H645–H653.

Hippenstiel, S., B. Schmeck, J. Seybold, M. Krull, C. Eichel-Streiber and N. Suttorp. 2002. Reduction of tumor necrosis factor-alpha (TNF-alpha) related nuclear factor-kappaB (NF-kappaB) translocation but not inhibitor kappa-B (Ikappa-B)-degradation by Rho protein inhibition in human endothelial cells. *Biochem. Pharmacol.* 64: 971–977.

Huddleson, J.P., N. Ahmad, S. Srinivasan and J.B. Lingrel. 2005. Induction of KLF2 by Fluid Shear Stress Requires a Novel Promoter Element Activated by a Phosphatidylinositol 3-Kinase-dependent Chromatin-remodeling Pathway. *Journal of Biological Chemistry* 280: 23371–23379.

Huo, Y., T. Wischgoll and G.S. Kassab. 2007. Flow patterns in three-dimensional porcine epicardial coronary arterial tree. *Am. J. Physiol. - Heart Circ. Physiol.* 293: H2959–H2970.

Hur, S.S., C. del Álamo, J.S. Park, Y.S. Li, H.A. Nguyen, D. Teng, K.C. Wang, L. Flores, B. Alonso-Latorre, J.C. Lasheras and S. Chien. 2012. Roles of cell confluency and fluid shear in 3-dimensional intracellular forces in endothelial cells. *Proc. Natl. Acad. Sc.* 109: 11110–11115.

Huynh, J., N. Nishimura, K. Rana, J.M. Peloquin, J.P. Califano, C.R. Montague, M.R. King, C.B. Schaffer and C.A. Reinhart-King. 2011. Age-Related Intimal Stiffening Enhances Endothelial Permeability and Leukocyte Transmigration. *Sci. Trans. Med.* 3: 112ra122.

Karino, T., H.L. Goldsmith, M. Motomiya, S. Mabuchi and Y. Sohara. 1987. Flow Patterns in Vessels of Simple and Complex Geometriesa. *Ann. NY Acad. Sci.* 516: 422–441.

Katz, B.Z., M. Zohar, H. Teramoto, K. Matsumoto, J.S. Gutkind, D.C. Lin, S. Lin and K.M. Yamada. 2000. Tensin can induce JNK and p38 activation. *Biochem. Biophys. Res. Commun.* 272: 717–720.

Klanchar, M., J.M. Tarbell and D.M. Wang. 1990. *In vitro* study of the influence of radial wall motion on wall shear stress in an elastic tube model of the aorta. *Circ. Res.* 66: 1624–1635.

Kleinstreuer, C., S. Hyun, J. Buchanan Jr., P. Longest, J. Archie Jr. and G.A. Truskey. 2001. Hemodynamic Factors and Early Intimal Thickening in Branching Blood Vessels. *Critical Reviews in Biomedical Engineering* 29: 1–64.

Ku, D. 1997. Blood Flow in Arteries. *Ann. Rev. Fl. Mech.* 29: 399–434.

Ku, D., D. Giddens, C. Zarins and S. Glasgov. 1985. Pulsatile flow and atherosclerosis in the human carotid bifurcation: positive correlation between plaque location and low and oscillating shear stress. *Arteriosclerosis* 5: 293–302.

Kusche-Vihrog, K., K. Urbanova, A. Blanque, M. Wilhelmi, H. Schillers, K. Kliche, H. Pavenstadt, E. Brand and H. Oberleithner. 2011. C-reactive protein makes human endothelium stiff and tight. *Hypertension* 57: 231–237.

L'Heureux, N., N. Dusserre, G. Konig, B. Victor, P. Keire, T.N. Wight, N.A. Chronos, A.E. Kyles, C.R. Gregory, G. Hoyt, R.C. Robbins and T.N. McAllister. 2006. Human tissue-engineered blood vessels for adult arterial revascularization. *Nat. Med.* 12: 361–365.

LaMack, J.A. and M.H. Friedman. 2007. Individual and combined effects of shear stress magnitude and spatial gradient on endothelial cell gene expression. *Am. J. Physiol.* 293: H2853–H2859.

Lavender, M.D., Z. Pang, C.S. Wallace, L.E. Niklason and G.A. Truskey. 2005. A system for the direct co-culture of endothelium on smooth muscle cells. *Biomaterials* 26: 4642–4653.

Levesque, M.J. and R.M. Nerem. 1985. The elongation and orientation of cultured endothelial cells in response to shear-stress. *J. Biomech. Eng.* 107: 341–347.

Mathur, A., W. Reichert and G. Truskey. 2007. Flow and High Affinity Binding Affect the Elastic Modulus of the Nucleus, Cell Body and the Stress Fibers of Endothelial Cells. *Ann. Biomed. Eng.* 35: 1120–1130.

Matsushita, H., E. Chang, A.J. Glassford, J.P. Cooke, C.-P. Chiu and P.S. Tsao. 2001. eNOS Activity Is Reduced in Senescent Human Endothelial Cells. *Circ. Res.* 89: 793–798.

Mayne, R. 1986. Collagenous proteins of blood vessels. *Arteriosclerosis* 6: 585–593.

McKinney, V.Z., K.D. Rinker and G.A. Truskey. 2006. Spatial distribution of intracellular adhesion molecule-1 expression in human umbilical vein endothelial cells exposed to sudden expansion flow. *J. Biomech.* 39: 806–817.

McNally, J.S., M.E. Davis, D.P. Giddens, A. Saha, J. Hwang, S. Dikalov, H. Jo and D.G. Harrison. 2003. Role of xanthine oxidoreductase and NAD(P)H oxidase in endothelial superoxide production in response to oscillatory shear stress. *Am. J. Physiol.* 285: H2290–2297.

Mehta, D. and A.B. Malik. 2006. Signaling mechanisms regulating endothelial permeability. *Physiol. Rev.* 86: 279–367.

Melkumyants, A.M. and S.A. Balashov. 1990. Effect of blood viscocity on arterial flow induced dilator response. *Cardiovasc. Res.* 24: 165–168.

Minamino, T., H. Miyauchi, T. Yoshida, Y. Ishida, H. Yoshida and I. Komuro. 2002. Endothelial Cell Senescence in Human Atherosclerosis. *Circulation* 105: 1541–1544.

Morello, F., A. Perino and E. Hirsch. 2009. Phosphoinositide 3-kinase signalling in the vascular system. *Cardiovasc. Res.* 82: 261–271.

Moses, M.A., M. Klagsbrun and Y. Shing. 1995. The Role of Growth Factors in Vascular Cell Development and Differentiation. *Intl. Rev. Cytol.* 161: 1–48.

Mun, G.I. and Y.C. Boo. 2010. Identification of CD44 as a senescence-induced cell adhesion gene responsible for the enhanced monocyte recruitment to senescent endothelial cells. *Am. J. Physiol.* 298: H2102–H2111.

Mun, G.I., I.S. Kim, B.H. Lee and Y.C. Boo. 2011. Endothelial Argininosuccinate Synthetase 1 Regulates Nitric Oxide Production and Monocyte Adhesion under Static and Laminar Shear Stress Conditions. *J. Biol. Chem.* 286: 2536–2542.

Mun, G.I., S.J. Lee, S.M. An, I.K. Kim and Y.C. Boo. 2009. Differential gene expression in young and senescent endothelial cells under static and laminar shear stress conditions. *Free Rad. Biol. Med.* 47: 291–299.

Nagel, T., N. R, C.F. Dewey and M.A.J. Gimbrone. 1999. Vascular endothelial cells respond to spatial gradients in fluid shear stress by enhanced activation of transcription factors. *Arterioscler. Thromb. Vasc. Biol.* 19: 1925–1834.

Nerem, R.M. 1984. Atherogenesis: hemodynamics, vascular geometry, and the endothelium. *Biorheology* 21: 565–569.

Nigro, P., J. Abe and B. Berk. 2011. Flow Shear Stress and Atherosclerosis: A Matter of Site Specificity. *Antiox. Redox Sig.* 15: 1405–1414.

Oberleithner, H., C. Riethmuller, H. Schillers, G.A. MacGregor, H.E. de Wardener and M. Hausberg. 2007. Plasma sodium stiffens vascular endothelium and reduces nitric oxide release. *Proc. Natl. Acad. Sci.* 104: 16281–16286.

Oberleithner, H., C. Callies, K. Kusche-Vihrog, H. Schillers, V. Shahin, C. Riethmuller, G.A. Macgregor and H.E. de Wardener. 2009. Potassium softens vascular endothelium and increases nitric oxide release. *Proc. Natl. Acad. Sci. USA* 106: 2829–2834.

Ogami, M.,Y. Ikura, M. Ohsawa, T. Matsuo, S. Kayo, N. Yoshimi, E. Hai, N. Shirai, S. Ehara, R. Komatsu, T. Naruko and M. Ueda. 2004. Telomere shortening in human coronary artery diseases. *Arterioscler. Thromb. Vasc. Biol.* 24: 546–550.

Okano, M. and Y. Yoshida. 1993. Influence of shear stress on endothelial cell shapes and junction complexes of endothelial cells in atherosclerosis-prone and atherosclerosis-resistant regions on flow dividers of brachiocephalic bifurcations in the rabbit aorta. *Frontiers of Medical and Biological Engineering* 5: 95–120.

Olivier, L.A., J. Yen, W.M. Reichert and G.A. Truskey. 1999. Short-term cell/substrate contact dynamics of subconfluent endothelial cells following exposure to laminar flow. *Biotechnol. Prog.* 15: 33–42.

Orr, A.W., M.H. Ginsberg, S.J. Shattil, H. Deckmyn and M.A. Schwartz. 2006. Matrix-specific Suppression of Integrin Activation in Shear Stress Signaling. *Molecular Biology Cell* 17: 4686–4697.

Palmer. R.M.J., A.G. Ferrige and S. Moncada. 1987. Nitric oxide release accounts for the biological activity of endothelium-derived relaxing factor. *Nature* 327: 524–526.

Park, S.J., F. Ahmad, A. Philp, K. Baar, T. Williams, H. Luo , H. Ke, H. Rehmann, R. Taussig, A.L. Brown, M.K. Kim, M.A. Beaven, A.B. Burgin, V. Manganiello and J.H. Chung. 2012. Resveratrol ameliorates aging-related metabolic phenotypes by inhibiting cAMP phosphodiesterases. *Cell* 148: 421–433.

Parmar, K., H.B. Larman, G. Dai, Y. Zhang, E.T. Wang, S.N. Moorthy, J.R. Kratz, Z. Lin, M.K. Jain, M.A. Gimbrone, Jr. and G. García-Cardeña. 2006. Integration of flow-dependent endothelial phenotypes by Kruppel-like factor 2. *J. Clin. Invest.* 116: 49–58.

Passerini, A.G., D.C. Polacek, C. Shi, N.M. Francesco, E. Manduchi, G.R. Grant, W.F. Pritchard, S. Powell, G.Y. Chang, C.J. Stoeckert, Jr. and P.F. Davies. 2004. Coexisting proinflammatory and antioxidative endothelial transcription profiles in a disturbed flow region of the adult porcine aorta. *Proc. Natl. Acad. Sc.* 101: 2482–2487.

Peloquin, J., J. Huynh, R.M. Williams and C.A. Reinhart-King. 2011. Indentation measurements of the subendothelial matrix in bovine carotid arteries. *J. Biomech.* 44: 815–821.

Qiu, Y. and J.M. Tarbell. 2000. Numerical Simulation of Pulsatile Flow in a Compliant Curved Tube Model of a Coronary Artery. *J. Biomech. Eng.* 122: 77–85.

Rainger, G.E. and G.B. Nash. 2001. Cellular Pathology of Atherosclerosis: Smooth Muscle Cells Prime Cocultured Endothelial Cells for Enhanced Leukocyte Adhesion. *Circulation Research* 88: 615–622.

Rampersad, S.N., J.D. Ovens, E. Huston, M.B. Umana, L.S. Wilson, S.J. Netherton, M.J. Lynch, G.S. Baillie, M.D. Houslay and D.H. Maurice. 2010. Cyclic AMP phosphodiesterase 4D (PDE4D) Tethers EPAC1 in a vascular endothelial cadherin (VE-Cad)-based signaling complex and controls cAMP-mediated vascular permeability. *J. Biol. Chem.* 285: 33614–33622.

Rauscher, F.M., P.J. Goldschmidt-Clermont, B.H. Davis, T. Wang, D. Gregg, P. Ramaswami, A.M. Pippen, B.H. Annex, C. Dong and D.A. Taylor. 2003. Aging, progenitor cell exhaustion, and atherosclerosis. *Circulation* 108: 457–463.

Roger, V.L. et al. 2012. Executive Summary: Heart Disease and Stroke Statistics—2012 Update. *Circulation* 125: 188–197.

Safar, M., B. Bernard I. Levy and H. Struijker-Boudierm. 2003. Current Perspectives on Arterial Stiffness and Pulse Pressure in Hypertension and Cardiovascular Diseases. *Circulation* 107: 2864–2869.

Salwen, S.A., D.H. Szarowski, J.N. Turner and R. Bizios. 1998. Three dimensional changes of the cytoskeleton of vascular endothelial cells exposed to sustained hydrostatic pressure. *Cell Eng.* 36: 520–527.

Sato, M., M. Levesque and R. Nerem. 1987. Micropipette aspiration of cultured bovine aortic endothelial cells exposed to shear stress. *Arteriosclerosis* 7: 276–286.

Sato, N., T. Goto, K. Haranaka, N. Satomi, H. Nariuchi, Y. Mano-Hirano and Y. Sawasaki. 1986. Actions of tumor necrosis factor on cultured vascular endothelial cells: Morphologic modulation, growth inhibition, and cytotoxicity. *J. Natl. Cancer Inst.* 76: 1113–1121.

Sato, Y., F. Okada, M. Abe, T. Seguchi, M. Kuwano, S. Sato, A. Furuya, N. Hanai and T. Tamaoki. 1993. The mechanism for the activation of latent TGF-beta during co-culture of endothelial cells and smooth muscle cells: cell-type specific targeting of latent TGF-beta to smooth muscle cells. *J. Cell Biol.* 123: 1249–1254.

Satoh, K., T. Matoba, J. Suzuki, M.R. O'Dell, P. Nigro, Z. Cui, A. Mohan, S. Pan, L. Li, Z.G. Jin, C. Yan, J. Abe and B.C. Berk. 2008. Cyclophilin A mediates vascular remodeling by promoting inflammation and vascular smooth muscle cell proliferation. *Circulation* 117: 3088–3098.

Schwartz, E.A., R. Bizios, M.S. Medow and M.E. Gerritsen. 1999. Exposure of Human Vascular Endothelial Cells to Sustained Hydrostatic Pressure Stimulates Proliferation: Involvement of the αV Integrins. *Circ. Res.* 84: 315–322.

SenBanerjee, S., Z. Lin, G.B. Atkins, D.M. Greif, R.M. Rao, A. Kumar, M.W. Feinberg, Z. Chen, D.I. Simon, F.W. Luscinskas, T.M. Michel, M.A. Gimbrone, Jr., G. García-Cardeña and M.K. Jain. 2004. KLF2 Is a novel transcriptional regulator of endothelial proinflammatory activation. *J. Exp. Med.* 199: 1305–1315.

Shin, H.Y., M.E. Gerritsen and R. Bizios. 2002a. Regulation of Endothelial Cell Proliferation and Apoptosis by Cyclic Pressure. *Ann. Biomed. Eng.* 30: 297–304.

Shin, H.Y., M.L. Smith, K.J. Toy, P.M. Williams, R. Bizios and M.E. Gerritsen. 2002b. VEGF-C mediates cyclic pressure-induced endothelial cell proliferation. *Physiol. Genom.* 11: 245–251.

Stenman, S. and A. Vaheri. 1978. Distribution of a major connective tissue protein, fibronectin, in normal human tissues. *J. Exp. Med.* 147: 1054–1064.

Stroka, K.M. and H. Aranda-Espinoza. 2011. Endothelial cell substrate stiffness influences neutrophil transmigration via myosin light chain kinase-dependent cell contraction. *Blood* 118: 1632–1640.

Sumpio, B.E., M.D. Widmann, J. Ricotta, M.A. Awolesi and M. Watse. 1994. Increased ambient-pressure stimulates proliferation and morphologic changes in cultured endothelial cells. *J. Cell Physiol.* 158: 133–139.

Suo, J., D.E. Ferrara, D. Sorescu, R.E. Guldberg, W.R. Taylor and D.P. Giddens. 2007. Hemodynamic Shear Stresses in Mouse Aortas. *Arteriosclerosis, Thrombosis, and Vascular Biology* 27: 346–351.

Tada, S. and J.M. Tarbell. 2005. A computational study of flow in a compliant carotid bifurcation-stress phase angle correlation with shear stress. *Ann. Biomed. Eng.* 33: 1202–1212.

Tee, S.Y., J. Fu, C.S. Chen and P.A. Janmey. 2011. Cell shape and substrate rigidity both regulate cell stiffness. *Biophys. J.* 100: L25–27.

Theret, D., M.J. Levesque, M. Sato, R. Nerem and L. Wheeler. 1988. The application of a homogeneous half-space model in the analysis of endothelial cell micropipette measurements. *J. Biomech. Eng.* 110: 190–199.

Ting, L., J. Jessica R. Jahn, J. Jung, B. Shuman, S. Feghhi, S. Han, M. Rodriguez and N. Sniadecki. 2012. Flow mechanotransduction regulates traction forces, intercellular forces, and adherens junctions. *Am. J. Physiol.* 302: H2220–H2229.

Topper, J.N., J. Cai, D. Falb and M.A. Gimbrone. 1996. Identification of vascular endothelial genes differentially responsive to fluid mechanical stimuli: cyclooxygenase-2, manganese superoxide dismutase, and endothelial cell nitric oxide synthase are selectively up-regulated by steady laminar shear stress. *Proc. Natl. Acad. Sci. USA* 93: 10417–10422.

Truskey, G., K. Barber and K. Rinker. 2002. Factors influencing the nonuniform localization of monocytes in the arterial wall. *Biorheology* 39: 325–329.

Truskey, G., F. Yuan and D. Katz. 2009. Transport Phenomenon in Biological Systems. Upper Saddle River, NJ: Pearson/Prentice Hall.

Tsai, M.-C., L. Chen, J. Zhou, Z. Tang, T.F. Hsu, Y. Wang, Y.T. Shih, H.H. Peng, N. Wang, Y. Guan, S. Chien and J.J. Chiu. 2009. Shear Stress Induces Synthetic-to-Contractile Phenotypic Modulation in Smooth Muscle Cells via Peroxisome Proliferator-Activated Receptor alpha/delta Activations by Prostacyclin Released by Sheared Endothelial Cells. *Circ. Res.* 105: 471–480.

Tzima, E., M. Irani-Tehrani, W. Kiosses, E. Dejana, D. Schultz, B. Engelhardt, G. Cao, H. DeLisser and M. Schwartz. 2005. A mechanosensory complex that mediates the endothelial cell response to fluid shear stress. *Nature* 437: 426–431.

van Thienen, J.V., J.O. Fledderus, R.J. Dekker, J. Rohlena, G.A. van IJzendoorn, N.A. Kootstra, H. Pannekoek and A.J.G. Horrevoets. 2006. Shear stress sustains atheroprotective endothelial KLF2 expression more potently than statins through mRNA stabilization. *Circ. Res.* 72: 231–240.

Vartanian, K.B., M.A. Berny, O.J. McCarty, S.R. Hanson and M.T. Hinds. 2010. Cytoskeletal structure regulates endothelial cell immunogenicity independent of fluid shear stress. *Am. J. Physiol.* 298: C333–C341.

Wallace, C. and G. Truskey. 2010. Direct-Contact Co-culture between Smooth Muscle and Endothelial Cells Inhibit TNF-α Mediated Endothelial Cell Activation. *Am. J. Physiol. Heart Circ. Physiol.* 299: 338–346.

Wallace, C., S. Strike and G. Truskey. 2007a. Smooth muscle cell rigidity and extracellular matrix organization influence endothelial cell spreading and adhesion formation in coculture. *Am. J. Physiol. Heart Circ. Physiol.* 293: H1978–H1986.

Wallace, C., J. Champion and G. Truskey. 2007b. Adhesion and Function of Human Endothelial Cells Co-cultured on Smooth Muscle Cells. *Ann. Biomed. Eng.* 35: 375–386.

Wasserman, S.M. and J.N. Topper. 2004. Adaptation of the endothelium to fluid flow: *in vitro* analyses of gene expression and *in vivo* implications. *Vascular Medicine* 9: 35–45.

Weinberg, P.D. and C.R. Ethier. 2007. Twenty-fold difference in hemodynamic wall shear stress between murine and human aortas. *J. Biomech.* 40: 1594–1598.

Wu, W., H. Xiao, A. Laguna-Fernandez, G. Villarreal, Jr., K.C. Wang, G.G. Geary, Y. Zhang, W.C. Wang, H.D. Huang, J. Zhou, Y.S. Li, S. Chien, G. Garcia-Cardena and J.Y. Shyy. 2011. Flow-Dependent Regulation of Kruppel-Like Factor 2 Is Mediated by MicroRNA-92a. *Circulation* 124: 633–641.

6

Differential Response of the Endothelium to Simultaneous Chemical and Mechanical Stimulation in Inflammation Response

Ryan B. Huang[1] and Omolola Eniola-Adefeso[2,] *

Introduction

As discussed in previous chapters, vascular-targeting via endothelial cell (EC)-expressed molecules has potential for diagnostic and therapeutic applications in several human diseases, particularly in many cancers and cardiovascular diseases. However, the capacity to localize drug therapy to diseased vasculature while avoiding healthy vessels would depend on the uniqueness of the targeted receptors' expression to the disease state or tissue and a precise match between the targeting ligands on the delivery

[1]Department of Chemical and Biomolecular Engineering, University of Illinois, 600 S. Mathews, RAL 4, C3, Urbana, Illinois 61801.
Email: rbhuang@illinois.edu
[2]Department of Chemical Engineering, University of Michigan, 2300 Hayward Street, 3074 H.H. Dow, Ann Arbor, Michigan 48109.
*Corresponding author: lolaa@umich.edu

vehicle and their endothelial-based receptors. This is particularly the case for targeting endothelial cell surface-expressed molecules associated with chronic inflammation, which is known to be associated with a number of disease pathologies. Here, vascular-targeted carriers must be able to successfully differentiate protein expression patterns observed during chronic inflammation under disease conditions from expression patterns associated with acute inflammation of the healthy endothelium. With ethical practices and government regulations precluding the use of human *in vivo* models and the fact that endothelial response in animal models are known to not translate well to human physiology, researchers have focused on developing *in vitro* models of human inflammatory response. To this end, current *in vitro* models have attempted to mimic relevant endothelial cell inflammatory response in health and in disease; however, many fail to simulate the actual physiological conditions under which inflammation occurs. The bulk of the literature is represented by evaluation of endothelial cell protein expression in static culture, typically on glass or plastic substrates. While these works have been critical in elucidating the roles of different chemical cues in eliciting protein expression in many physiological and pathological processes, these studies often fail to encompass the true *in vivo* environment under which such endothelial activation occurs.

By nature of their location, endothelial cells are constantly exposed to shear forces imparted by blood flow. It is well documented that this mechanical stimulus in isolation has a significant impact on endothelial cell behavior at both cellular and molecular levels to influence their morphology, function, and gene expression. For example, studies have identified the importance of blood fluid shear stress in regulating angiogenesis and signal transduction through stress-sensing mechanotransductive receptors (Ando and Kamiya 1993, Ando and Yamamoto 2009, Reinhart-King 2008, Reinhart-King et al. 2008, Reinhart-King et al. 2008, Tarbell and Pahakis 2006). Therefore, disturbance in normal fluid shear patterns can result in abnormal cell behavior leading to disease pathogenesis (Secomb et al. 2001). Indeed, there is overwhelming evidence that prolonged exposure to *disturbed* flow profiles, typically encountered in branches, curvatures, and bifurcations where low shear zones and stagnation are prevalent, presents a diseased endothelial cell phenotype characterized by a cobblestone morphology that is associated with diseases such as atherosclerosis and aortic stenosis (Chiu and Chien 2011, Chiu et al. 2009, Du and Soon 2011, New and Aikawa 2011, New and Aikawa 2011). However, exposure to high laminar fluid shear—especially found in non-branched straight lengths of the vasculature—modulates vascular endothelium toward a healthy phenotype where endothelial cells align in the direction of flow (Fig. 1) adopting a more streamlined morphology in order to minimize the impact of the tangential fluid shear. Despite the understanding of the profound influence fluid shear

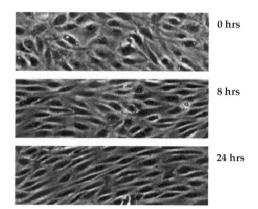

Figure 1. **Cell morphology under shear flow.** HUVEC monolayer perfused under 10 dyn cm^{-2} shear stress of fluid flow (left to right). Brightfield images (20x magnification) were taken at 0, 8, and 24 hr into perfusion, highlighting the shear-induced elongation and realignment of individual cells over time.

has on endothelial function, many *in vitro* assays of inflammation response presented in the literature still often overlook or downplay the contributions of fluid shear stress to chemical (i.e., cytokine)-induced activation. Only in the last decade have studies reported the differential response of endothelial cells to both combined chemical and mechanical cues. The present chapter discusses key differences in the expression of inflammatory molecules by human endothelial cells in response to simultaneous mechanical and chemical stimuli. Emphasis is placed on surface protein expressions, rather than at the mRNA levels, since these apical markers directly participate in cell and, presumably, drug carrier recruitment to the vascular wall. Additionally, previous reports have shown that mRNA expression does not correlate with surface protein expression levels for some adhesion molecules (Kluger et al. 1997, Kraiss et al. 2003).

Endothelial Cell Adhesion Molecule Expression in Inflammation

While several cell adhesion molecules (CAMs) are known to be involved in inflammation response and leukocyte recruitment, endothelial cell-expressed E-selectin, intercellular CAM (ICAM)-1, and vascular CAM (VCAM)-1 are the key inflammation-associated molecules that are often proposed for use as vascular targets due to their differential expression by inflamed relative to healthy endothelial cells. In quiescent, healthy vascular tissue, endothelial cells do not express E-selectin molecules on their surface (Bevilacqua et al. 1985, Cotran et al. 1986). However, upon activation with inflammatory cytokines, E-selectin is upregulated to a significantly elevated level promoting increased leukocyte recruitment to the vascular wall (Munro et al. 1991). VCAM-1 is similarly not expressed by healthy, non-inflamed endothelial cells in most vascular beds. While pro-inflammatory factors also induce higher ICAM-1 expression levels, a basal level of this molecule is constitutively expressed by non-inflamed ECs, possibly making it an unsuitable target for use in discriminating between healthy and diseased cells (Dustin et al. 1986, Pober et al. 1986, Swerlick et al. 1992). P-selectin, an additional CAM expressed by ECs in response to inflammatory cues, is not often associated with chronic inflammation.

Overview of Cytokines in Inflammation Response

Inflammation literature has primarily observed endothelial cell response to one particular cytokine, tumor necrosis factor alpha (TNFα) perhaps due to its role in a large number of cardiovascular diseases (CVDs) (Hallenbeck 2002, Javaid et al. 2003, Rahman et al. 2000) and its attractiveness as a potential anti-cancer therapy in the selective targeting of tumor vasculature (van Horssen et al. 2006). However, TNFα is only one of *three* primary pro-inflammatory factors: the cytokine interleukin-1 beta (IL-1β) has also been shown to be critical in the development of a number of pathologies including Alzheimer's disease (Griffin and Mrak 2002), chronic autoimmune disorders (Dinarello 2011), and metabolic diseases (Dinarello 2009, Maedler et al. 2011). A third factor, lipopolysaccharide (LPS), an endotoxin shed from bacteria during infection, is responsible for septic shock, a large number of innate immune responses (Cines et al. 1998, Raetz and Whitfield 2002), and the secretion of other major pro-inflammatory cytokines including IL-4, IL-6, and nitric oxide. However, in addition to acting in a pro-inflammatory capacity, especially in the development of diabetes, rheumatoid arthritis, and prostate cancer (Kristiansen and Mandrup-Poulsen 2005, Nishimoto 2006, Smith et al. 2001), IL-6 is also known for its anti-inflammatory roles in the inhibition of TNFα and IL-1β activity and induction of IL-10 production (an anti-inflammatory agent (Braat et al. 2006)). Table 1 describes some functions of a few of the more common factors involved in inflammation.

Differential Response of HUVECs Exposed to Simultaneous Mechanical and Chemical Cues

As highlighted in previous chapters, human umbilical vein endothelial cells (HUVECs) are the most common cell line used for many *in vitro* studies as they are readily obtained from commercial stocks or procured from human waste tissue ubiquitous with every childbirth. The differential expression

Table 1. A Partial List of Mediators of Inflammation.

Mediator	Role/Prominence
Lipopolysaccharides (LPS)	- Promotes secretion of IL-1, IL-6, TNFα, chemokines (Cines et al. 1998) - Leads to septic shock
Interleukin-1 (IL-1) and Tumor necrosis factor (TNF)	- Stimulates and activates lymphocytes (T, B, and NK cells) - Regulates immune cell response (Cines et al. 1998)
Interleukin-6 (IL-6)	- Anti-inflammatory mediator (produced during exercise) (Petersen and Pedersen 2005) - Implicated in diabetes, rheumatoid arthritis, prostate cancer, myocardial infarction (Ikeda 2003, Kristiansen and Mandrup-Poulsen 2005, Nishimoto 2006, Smith et al. 2001) - Inhibits TNFα and IL-1 activity (Abbas 1996) - Induces IL-10 production (Abbas 1996)
Interleukin-4 (IL-4)	- Activates macrophages in tissue repair (Abbas 1996) - Produced during allergic response (Hershey et al. 1997)
Histamine	- Increases vessel permeability leading to allergic reaction - Mediates chemotaxis of several immune cells
Nitric oxide	- Inhibits leukocyte adhesion, maintains homeostasis (Hill and Whitten 1997) - Overproduction contributes to reperfusion injury (Masullo et al. 2000) - Implicated in pulmonary arterial hypertension (Crosswhite and Sun 2010)
Interleukin-10 (IL-10)	- Inhibits synthesis of pro-inflammatory molecules (Abbas 1996)

of the three key CAMs, E-selectin, ICAM-1, and VCAM-1, in response to simultaneous exposure to fluid shear and cytokine stimulation have primarily been evaluated for the inflammatory cytokines TNFα and, to a lesser extent, IL-1β.

Cell adhesion molecule expression in shear-cytokine activated naïve HUVECs

In culture, HUVECs are typically maintained in a static environment with regular growth mediums and in the absence of inflammatory agonists. These cells thus acquire a "naïve" phenotype during culture. In this naïve state, cells in the monolayer maintain a cobblestone-like pattern, where individual cells are circular in shape. This static *in vitro* condition is hardly the normal physiological morphology observed *in vivo* due to the absence of the aforementioned shape-changing forces exerted by blood fluid flow. However, this morphology may exist in some pathophysiological conditions, such as in ischemia, where blood flow in a vessel is interrupted downstream of a blockage (e.g., due to restrictions such as clots or plaque buildup).

Non-stimulated, naïve HUVECs constituently express ICAM-1 at an appreciable level relative to the near-zero expression of E-selectin and VCAM-1 by these cells. During static inflammatory activation (*in absentia* shear flow) with either TNFα or IL-1β, naïve HUVECs show a considerable increase in ICAM-1 expression, doubling to tripling in magnitude over a 24 hr period. E-selectin densities similarly spike upward within a couple of hours in response to TNFα or IL-1β with peak expression occurring between 4–6 hr but quickly drop to near baseline values by 12 hr post activation irrespective of continuous exposure to activating agent. VCAM-1 expression induced by static activation linearly increases and plateaus at 20–24 hr (Huang and Eniola-Adefeso 2012). However, a differentiable expression pattern for these three adhesion molecules as well as leukocyte capture is reported when naïve HUVECs are chemically stimulated with cytokine under shear flow conditions (or *shear-cytokine activation*) when compared to activation under static conditions. For instance, a 4 hr exposure of naïve HUVEC to both a high shear (20 dyn/cm^2) laminar flow and TNFα cytokine is reported to induce significantly higher ICAM-1 but decreased VCAM-1 protein expression relative to expression by naïve cells stimulated with TNFα in static (Chiu et al. 2004). Similarly, instead of achieving peak expression density or functional leukocyte capture in the typical 4–6 hr activation time range, maximum E-selectin expression and leukocyte adhesion by naïve HUVECs shifts to the 8–12 hr range with shear magnitudes equal to or greater than 1 dyn/cm^2 for either TNFα or IL-1β (Urschel et al. 2010). Figure 2 shows the temporal E-selectin expression by naïve HUVECs exposed to laminar shear and IL-1β. From this image, it is apparent that the mere presence of shear with cytokine activation induces an enhanced and prolonged E-selectin expression indicative of chronic inflammation. However, slight differences exist in the pattern of E-selectin expression when comparing data between shear-TNFα and shear-IL-1β activation of naïve HUVECs. Specifically, while E-selectin expression decreases at a more gradual rate with both shear-TNFα and shear-IL-1β, expression remains significantly elevated relative to static expression after 24 hr only for shear-IL-1β activation, particularly for shear magnitudes greater than 1 dyn/cm^2 (Huang and Eniola-Adefeso 2012). A 24 hr laminar shear-TNFα activation of naïve HUVECs produces an E-selectin expression level that is not significantly different from expression by cells activated with TNFα in static whereas E-selectin expression after 4 hr of shear-TNFα activation is significantly lower than that of static expression. Laminar shear-TNFα

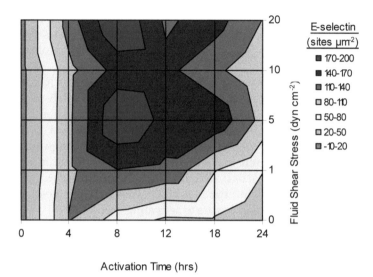

Activation Time (hrs)

Figure 2. E-selectin expression (sites μm⁻²) on 0–20 dyn cm⁻² shear-cytokine activated monolayers. HUVEC monolayers activated with 0.1 ng ml⁻¹ IL-1ß either under static conditions (no fluid shear) or simultaneously exposed to 1–20 dyn cm⁻² of laminar fluid shear at activation periods ranging from 0 to 24 hr.

Color image of this figure appears in the color plate section at the end of the book.

activation only induces elevated E-selectin density in naïve HUVECs relative to static activation in the 8–12 hr activation window.

While there are a few studies reporting CAM expression by HUVEC stimulated under static culture via multiple cytokines (Raab et al. 2002), there is currently scarce literature on HUVEC response to multiple cytokine challenge in the presence of shear flow. Based on unpublished observations by the authors, it appears that E-selectin expression in naïve HUVECs when simultaneously stimulated with high laminar shear (10 dyn/cm²) and both TNFα and IL-1β follow the same trend as that observed for shear-TNFα-only activation. Specifically, E-selectin expression with shear-dual cytokine stimulation is significantly lower for a 4 hr activation time but is significantly higher at 8 and 12 hr of activation relative to expression induced by static-dual cytokine stimulation (unpublished data). As for other adhesion molecules, e.g., ICAM-1 and VCAM-1, no differential response has been reported for naïve HUVECs subjected to shear-dual cytokine activation relative to static activation.

As previously mentioned, observations with naïve HUVECs are of non-ideal or diseased HUVEC phenotypes (i.e., naïve, cobblestone-like morphology). In order to effectively determine the behavioral expression patterns of healthy HUVECs in their natural environment *in vivo*, the cell monolayers need to have acquired the streamlined, elongated shape prior to exposure to cytokine challenge in the presence of shear. The next section summarizes CAM expression patterns of such preconditioned, shear-cytokine activated HUVECs.

Cell adhesion molecule expression in shear-cytokine activated shear preconditioned HUVECs

Cells preconditioned, or pre-exposed to laminar fluid shear stimulus-only (e.g., in the absence of any inflammatory cytokine) display different trends of CAM expression levels upon subsequent shear-cytokine activation. Numerous studies have investigated both the short (1–4 hr) and long (24+ hrs) term inflammatory insults of long-term (20+ hr) laminar shear preconditioned HUVECs and have found that almost all ICAM-1 and VCAM-1 expression is abolished relative to naïve monolayers activated in static, regardless of cytokine type (TNFα or IL-1β) (Glen et al. 2011, Honda et al. 2001, Ji et al. 2008, Kraiss et al. 2003, Luu et al. 2010, Matharu et al. 2008, Sheikh et al. 2005, Sheikh et al. 2003, Tsou et al. 2008, Urschel et al. 2011) or activation timeframe. These studies suggest that a preconditioning laminar fluid shear prescribes a quiescent HUVEC phenotype typically seen in healthy physiological ECs.

Fewer studies have identified the effect of shear on E-selectin expression by shear preconditioned HUVECs. The initial reports that compared shear-cytokine-induced ICAM-1 expression between long-term preconditioned and naïve HUVECs also investigated E-selectin densities, but found no significant differences relative to naïve shear-cytokine activation for the long-term activation period. However, when a broader range of preconditioning duration is explored, E-selectin expression pattern by shear-cytokine activation of preconditioned HUVECs can differ from shear-cytokine or static activation of naïve cells. Specifically, short-term, 2–4 hr, simultaneous shear-cytokine stimulation with TNFα or IL-1β of 20+ hr preconditioned HUVECs results in a decrease in E-selectin expression relative to the 2–4 hr static activation of naïve HUVECs. However, as shown in Fig. 3, a short-term shear preconditioning period between 8 and 16 hr followed by shear-cytokine activation with IL-1β is known to result in an early spike of E-selectin expression—nearly two-folds higher than that observed in shear-cytokine or static activation of naïve HUVECs. E-selectin expression levels return to densities observed in static-IL-1β activated naïve HUVECs by 12 hr of activation. Thus, HUVEC response to shear-IL-1β activation of preconditioned monolayers appears to be dependent on the duration of both the preconditioning and shear-cytokine activation. However, this may not be the case for shear-TNFα activation of preconditioned cells since HUVECs preconditioned with high laminar shear stress for 4 or 24 hr have been previously reported by Chiu and co-workers to exhibit lower E-selectin

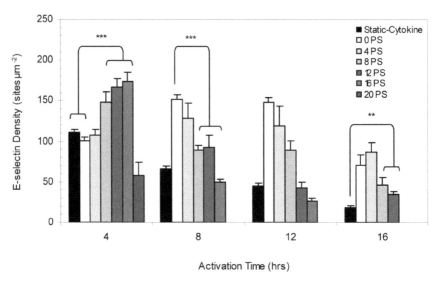

Figure 3. E-selectin expression (sites μm⁻²) on 10 dyn cm⁻² shear-cytokine activated preconditioned HUVEC monolayers. Monolayers were subjected to 4, 8, 12, 16, or 20 hrs of 10 dyn cm⁻² magnitude shear preconditioning (4 PS, 8 PS, 12 PS, 16 PS, and 20 PS, respectively) followed by up to 16 hrs of shear-cytokine activation with 0.1 ng mL⁻¹ of IL-1β. E-selectin density values were measured after each 4 hr time point. Non-preconditioned monolayers (0 PS) and static-activated monolayers (Static-Cytokine) were used as controls.

Color image of this figure appears in the color plate section at the end of the book.

mRNA and protein expression with 4 hr TNFα stimulation (Chiu et al. 2004). This study, however, activated preconditioned cells under static and not shear-cytokine conditions.

Shear-Cytokine Induced Cell Adhesion Molecule Expression in ECs from Other Vascular Beds

Endothelial cells from different vascular sources are exposed to various types of flow, which include the high magnitude shear, and pulsatile-like profile of the arteries to the low, creeping-like steady flow of capillaries. Cells procured from these different vascular beds have been shown to exhibit a variable range of CAM expression in response to shear-cytokine activation—especially when subjected to non-native flow profiles (Methe et al. 2007). For instance, cell adhesion molecule expression on naïve human aortic endothelial cells (HAECs) in response to short-term (4 hr) shear-cytokine challenge is dependent on the magnitude of laminar shear (Tsou et al. 2008). Lower laminar shear stress levels (2–4 dyn/cm²) induced higher E-selectin and VCAM-1 expression by shear-TNFα activation relative to static activation, while at higher shear stress levels (6–14 dyn/

cm^2) expression of E-selectin or VCAM-1 was either the same or lower than static activation. Expression of ICAM-1 under these shear-cytokine activation conditions was higher relative to static activation with shear stress from 0–14 dyn/cm^2.

On both human umbilical artery and coronary artery endothelial cells (HUAEC and HCAEC, respectively), long-term high laminar (20 dyn/cm^2) shear preconditioning followed by a 4 hr shear (20 dyn/cm^2)-cytokine activation with either TNFα or IL-1β resulted in a significant decrease in neutrophil capture relative to static, which would suggest a significant decrease in the expression levels of key CAMs in inflammation response. The expressions of E-selectin mRNA in these long-term preconditioned cells were also found to be significantly lower under the shear-cytokine activation condition explored. These observations indicate there may be a differentiable response to shear-IL-1β activation between arterial ECs and venous ECs (e.g., HUVECs) (Luu et al. 2010). As described in the previous section, under similar conditions (16 hr preconditioning followed by 4 hr shear-IL-1β activation), HUVECs displayed an increased E-selectin expression (relative to static); however, actual CAM expression levels in arterial ECs for these conditions have yet to be directly measured.

In a study by Methe et al., the differentiable behavior between preconditioned venous and arterial ECs in response to shear-TNFα activation under arterial and venous flow profiles relative to static activation are clearly shown (Methe et al. 2007). Human saphenous vein ECs (HSVECs), when exposed to venous-like (laminar shear at 2.2 dyn/cm^2) or arterial-like (a 1 Hz pulsatile, 17 dyn/cm^2 magnitude) flow patterns typically observed *in vivo*, expressed higher densities of E-selectin and ICAM-1 relative to HSVECs under non-flow, static-TNFα activation conditions. VCAM-1 expression induced by shear-TNFα stimulation of preconditioned HSVECs was only significantly higher than the static control for the arterial-like flow. However, in preconditioned HCAECs, exposure to such shear-cytokine activation conditions resulted in a slightly muted CAM expression density for both E-selectin and VCAM-1 relative to static activation. Interestingly, HCAEC ICAM-1 expression significantly increases for the arterial-like shear-TNFα activation but remains relatively unchanged under non-native venous-like flow when compared with densities measured during static-TNFα activation of HCAECs.

Conclusions and Perspective

In conclusion, endothelial cells differentially respond to chemical cues in the presence of fluid shear. The nature of this shear-induced differential response, attenuation or augmentation, depends on the vascular bed origin of the endothelial cells, properties of the fluid flow (i.e., shear magnitude and flow profile type), the shear history of the cells, the duration of the applied cytokine stimulation under shear, and the type of activating cytokine present. In general, a prolonged (>24 hr) exposure to laminar fluid shear prior to induction by activating cytokine as would be present *in vivo* in a healthy, well-perfused vessel results in a muted endothelial cell response to inflammatory cytokines in both acute and chronic time frames. In culture, however, naïve cells or short-term laminar shear preconditioned (<12 hr) endothelial cells are more sensitive to cytokine activation under shear, often displaying an enhanced expression of inflammatory molecules. This type of naïve or short-term conditioned EC phenotypes are likely relevant in cardiovascular diseases wherein endothelial cells downstream of a vessel occlusion acquire a naïve phenotype (Chiu et al. 2004). These naïve-like cells can have an enhanced response to shear-cytokine stimulation upon reperfusion of the occluded vessels. While there is ample evidence that disturbed and pulsatile flow independently induce pro-inflammatory phenotypes in endothelial cells from several vascular beds, there is currently very limited research on endothelial cell response to simultaneous cytokine and disturbed shear activation of either naïve or preconditioned ECs. As such, the general conclusion drawn here on the effect of laminar shear-cytokine stimulation cannot yet be extended to ECs exposed to disturbed flow in the absence of more detailed experimental works.

While the exact mechanism by which shear modulates EC response to cytokine stimulation remain unclear, protein complexes such as NF-κB that are known to independently regulate cell shear and cell cytokine response has been suggested to be involved in the differential EC response conferred by shear-cytokine stimulation (Chiu et al. 2004). However, it is possible that the shear-induced secretion of pro- or anti-inflammatory soluble factors during shearing also help modulate the endothelial cell response to specific cytokines, particularly in the case of shear-preconditioned monolayers. Furthermore, the bulk of the current literature on endothelial cell response to shear and cytokine do not explore the possible contribution of other cells of the vascular wall, e.g., smooth muscle cells and pericytes, in regulating endothelial cells in health and especially in disease. Future research direction must address these possibilities along with many others to begin to fully understand the *in vivo* endothelial cell protein expression patterns throughout the circulatory system in health and disease.

Acknowledgements

The authors acknowledge Mr. Peter Onyskiw and Mr. Alex J. Thompson for their technical assistance with literature research.

References

Abbas, A.K. 1996. Die and let live: eliminating dangerous lymphocytes. *Cell* 84(5): 655–657.

Ando, J. and A. Kamiya. 1993. Blood flow and vascular endothelial cell function. *Front Med. Biol. Eng.* 5(4): 245–264.

Ando, J. and K. Yamamoto. 2009. Vascular mechanobiology: endothelial cell responses to fluid shear stress. *Circ. J.* 73(11): 1983–1992.

Bevilacqua, M.P., J.S. Pober, M.E. Wheeler, R.S. Cotran and M.A. Gimbrone, Jr. 1985. Interleukin 1 acts on cultured human vascular endothelium to increase the adhesion of polymorphonuclear leukocytes, monocytes, and related leukocyte cell lines. *J. Clin. Invest.* 76(5): 2003–2011.

Braat, H., P. Rottiers, D.W. Hommes, N. Huyghebaert, E. Remaut, J.P. Remon, S.J. van Deventer, S. Neirynck, M.P. Peppelenbosch and L. Steidler. 2006. A phase I trial with transgenic bacteria expressing interleukin-10 in Crohn's disease. *Clin. Gastroenterol. Hepatol.* 4(6): 754–759.

Chiu, J.J. and S. Chien. 2011. Effects of disturbed flow on vascular endothelium: pathophysiological basis and clinical perspectives. *Physiol. Rev.* 91(1): 327–387.

Chiu, J.J., P.L. Lee, C.N. Chen, C.I. Lee, S.F. Chang, L.J. Chen, S.C. Lien, Y.C. Ko, S. Usami and S. Chien. 2004. Shear stress increases ICAM-1 and decreases VCAM-1 and E-selectin expressions induced by tumor necrosis factor-[alpha] in endothelial cells. *Arterioscler. Thromb. Vasc. Biol.* 24(1): 73–79.

Chiu, J.J., S. Usami and S. Chien. 2009. Vascular endothelial responses to altered shear stress: pathologic implications for atherosclerosis. *Ann. Med.* 41(1): 19–28.

Cines, D.B., E.S. Pollak, C.A. Buck, J. Loscalzo, G.A. Zimmerman, R.P. McEver, J.S. Pober, T.M. Wick, B.A. Konkle, B.S. Schwartz, E.S. Barnathan, K.R. McCrae, B.A. Hug, A.M. Schmidt and D.M. Stern. 1998. Endothelial cells in physiology and in the pathophysiology of vascular disorders. *Blood* 91(10): 3527–3561.

Cotran, R.S., M.A. Gimbrone, Jr., M.P. Bevilacqua, D.L. Mendrick and J.S. Pober. 1986. Induction and detection of a human endothelial activation antigen *in vivo*. *J. Exp. Med.* 164(2): 661–666.

Crosswhite, P. and Z. Sun. 2010. Nitric oxide, oxidative stress and inflammation in pulmonary arterial hypertension. *J. Hypertens.* 28(2): 201–212.

Dinarello, C.A. 2009. Immunological and inflammatory functions of the interleukin-1 family. *Annu. Rev. Immunol.* 27: 519–550.

Dinarello, C.A. 2011. Blocking interleukin-1beta in acute and chronic autoinflammatory diseases. *J. Intern. Med.* 269(1): 16–28.

Du, X. and J.L. Soon. 2011. Mild to moderate aortic stenosis and coronary bypass surgery. *J. Cardiol.* 57(1): 31–35.

Dustin, M.L., R. Rothlein, A.K. Bhan, C.A. Dinarello and T.A. Springer. 1986. Induction by IL 1 and interferon-gamma: tissue distribution, biochemistry, and function of a natural adherence molecule (ICAM-1). *J. Immunol.* 137(1): 245–254.

Glen, K., N.T. Luu, E. Ross, C.D. Buckley, G.E. Rainger, S. Egginton and G.B. Nash. 2011. Modulation of functional responses of endothelial cells linked to angiogenesis and inflammation by shear stress: differential effects of the mechanotransducer CD31. *J. Cell Physiol.*

Griffin, W.S. and R.E. Mrak. 2002. Interleukin-1 in the genesis and progression of and risk for development of neuronal degeneration in Alzheimer's disease. *J. Leukoc. Biol.* 72(2): 233–238.

Hallenbeck, J.M. 2002. The many faces of tumor necrosis factor in stroke. *Nat. Med.* 8(12): 1363–1368.

Hershey, G.K., M.F. Friedrich, L.A. Esswein, M.L. Thomas and T.A. Chatila. 1997. The association of atopy with a gain-of-function mutation in the alpha subunit of the interleukin-4 receptor. *N Engl. J. Med.* 337(24): 1720–1725.

Hill, G.E. and C.W. Whitten. 1997. The role of the vascular endothelium in inflammatory syndromes, atherogenesis, and the propagation of disease. *J. Cardiothorac. Vasc. Anesth.* 11(3): 316–321.

Honda, H.M., T. Hsiai, C.M. Wortham, M. Chen, H. Lin, M. Navab and L.L. Demer. 2001. A complex flow pattern of low shear stress and flow reversal promotes monocyte binding to endothelial cells. *Atherosclerosis* 158(2): 385–390.

Huang, R.B. and O. Eniola-Adefeso. 2012. Shear Stress Modulation of IL-1beta-Induced E-Selectin Expression in Human Endothelial Cells. *PLoS One* 7(2): e31874.

Ikeda, U. 2003. Inflammation and coronary artery disease. *Curr. Vasc. Pharmacol.* 1(1): 65–70.

Javaid, K., A. Rahman, K.N. Anwar, R.S. Frey, R.D. Minshall and A.B. Malik. 2003. Tumor necrosis factor-alpha induces early-onset endothelial adhesivity by protein kinase Czeta-dependent activation of intercellular adhesion molecule-1. *Circ. Res.* 92(10): 1089–1097.

Ji, J.Y., H. Jing and S.L. Diamond. 2008. Hemodynamic regulation of inflammation at the endothelial-neutrophil interface. *Ann. Biomed. Eng.* 36(4): 586–595.

Kluger, M.S., D.R. Johnson and J.S. Pober. 1997. Mechanism of sustained E-selectin expression in cultured human dermal microvascular endothelial cells. *J. Immunol.* 158(2): 887–896.

Kraiss, L.W., N.M. Alto, D.A. Dixon, T.M. McIntyre, A.S. Weyrich and G.A. Zimmerman. 2003. Fluid flow regulates E-selectin protein levels in human endothelial cells by inhibiting translation. *J. Vasc. Surg.* 37(1): 161–168.

Kristiansen, O.P. and T. Mandrup-Poulsen. 2005. Interleukin-6 and diabetes: the good, the bad, or the indifferent? *Diabetes* 54 Suppl 2: S114–124.

Luu, N.T., M. Rahman, P.C. Stone, G.E. Rainger and G.B. Nash. 2010. Responses of endothelial cells from different vessels to inflammatory cytokines and shear stress: evidence for the pliability of endothelial phenotype. *J. Vasc. Res.* 47(5): 451–461.

Maedler, K., G. Dharmadhikari, D.M. Schumann and J. Storling. 2011. Interleukin-targeted therapy for metabolic syndrome and type 2 diabetes. *Handb. Exp. Pharmacol.* (203): 257–278.

Masullo, P., P. Venditti, C. Agnisola and S. Di Meo. 2000. Role of nitric oxide in the reperfusion induced injury in hyperthyroid rat hearts. *Free Radic. Res.* 32(5): 411–421.

Matharu, N.M., H.M. McGettrick, M. Salmon, S. Kissane, R.K. Vohra, G.E. Rainger and G.B. Nash. 2008. Inflammatory responses of endothelial cells experiencing reduction in flow after conditioning by shear stress. *J. Cell Physiol.* 216(3): 732–741.

Methe, H., M. Balcells, C. Alegret Mdel, M. Santacana, B. Molins, A. Hamik, M.K. Jain and E.R. Edelman. 2007. Vascular bed origin dictates flow pattern regulation of endothelial adhesion molecule expression. *Am. J. Physiol. Heart Circ. Physiol.* 292(5): H2167–2175.

Munro, J.M., J.S. Pober and R.S. Cotran. 1991. Recruitment of neutrophils in the local endotoxin response: association with *de novo* endothelial expression of endothelial leukocyte adhesion molecule-1. *Lab. Invest.* 64(2): 295–299.

New, S.E. and E. Aikawa. 2011. Cardiovascular calcification: an inflammatory disease. *Circ. J.* 75(6): 1305–1313.

New, S.E. and E. Aikawa. 2011. Molecular imaging insights into early inflammatory stages of arterial and aortic valve calcification. *Circ. Res.* 108(11): 1381–1391.

Nishimoto, N. 2006. Interleukin-6 in rheumatoid arthritis. *Curr. Opin. Rheumatol.* 18(3): 277–281.

Petersen, A.M. and B.K. Pedersen. 2005. The anti-inflammatory effect of exercise. *J. Appl. Physiol.* 98(4): 1154–1162.

Pober, J.S., M.P. Bevilacqua, D.L. Mendrick, L.A. Lapierre, W. Fiers and M.A. Gimbrone, Jr. 1986. Two distinct monokines, interleukin 1 and tumor necrosis factor, each independently induce biosynthesis and transient expression of the same antigen on the surface of cultured human vascular endothelial cells. *J. Immunol.* 136(5): 1680–1687.

Raab, M., H. Daxecker, S. Markovic, A. Karimi, A. Griesmacher and M.M. Mueller. 2002. Variation of adhesion molecule expression on human umbilical vein endothelial cells upon multiple cytokine application. *Clin. Chim. Acta* 321(1–2): 11–16.

Raetz, C.R. and C. Whitfield. 2002. Lipopolysaccharide endotoxins. *Annu. Rev. Biochem.* 71: 635–700.

Rahman, A., K.N. Anwar and A.B. Malik. 2000. Protein kinase C-zeta mediates TNF-alpha-induced ICAM-1 gene transcription in endothelial cells. *Am. J. Physiol. Cell Physiol.* 279(4): C906–914.

Reinhart-King, C.A. 2008. Endothelial cell adhesion and migration. *Methods Enzymol.* 443: 45–64.

Reinhart-King, C.A., M. Dembo and D.A. Hammer. 2008. Cell-cell mechanical communication through compliant substrates. *Biophys. J.* 95(12): 6044–6051.

Reinhart-King, C.A., K. Fujiwara and B.C. Berk. 2008. Physiologic stress-mediated signaling in the endothelium. *Methods Enzymol.* 443: 25–44.

Secomb, T.W., R. Hsu and A.R. Pries. 2001. Effect of the endothelial surface layer on transmission of fluid shear stress to endothelial cells. *Biorheology* 38(2–3): 143–150.

Sheikh, S., M. Rahman, Z. Gale, N.T. Luu, P.C. Stone, N.M. Matharu, G.E. Rainger and G.B. Nash. 2005. Differing mechanisms of leukocyte recruitment and sensitivity to conditioning by shear stress for endothelial cells treated with tumour necrosis factor-alpha or interleukin-1beta. *Br. J. Pharmacol.* 145(8): 1052–1061.

Sheikh, S., G.E. Rainger, Z. Gale, M. Rahman and G.B. Nash. 2003. Exposure to fluid shear stress modulates the ability of endothelial cells to recruit neutrophils in response to tumor necrosis factor-alpha: a basis for local variations in vascular sensitivity to inflammation. *Blood* 102(8): 2828–2834.

Smith, P.C., A. Hobisch, D.L. Lin, Z. Culig and E.T. Keller. 2001. Interleukin-6 and prostate cancer progression. *Cytokine Growth Factor Rev.* 12(1): 33–40.

Swerlick, R.A., K.H. Lee, T.M. Wick and T.J. Lawley. 1992. Human dermal microvascular endothelial but not human umbilical vein endothelial cells express CD36 *in vivo* and *in vitro*. *J. Immunol.* 148(1): 78–83.

Tarbell, J.M. and M.Y. Pahakis. 2006. Mechanotransduction and the glycocalyx. *J. Intern. Med.* 259(4): 339–350.

Tsou, J.K., R.M. Gower, H.J. Ting, U.Y. Schaff, M.F. Insana, A.G. Passerini and S.I. Simon. 2008. Spatial regulation of inflammation by human aortic endothelial cells in a linear gradient of shear stress. *Microcirculation* 15(4): 311–323.

Urschel, K., C.D. Garlichs, W.G. Daniel and I. Cicha. 2011. VEGFR2 signalling contributes to increased endothelial susceptibility to TNF-alpha under chronic non-uniform shear stress. *Atherosclerosis* 219(2): 499–509.

Urschel, K., A. Worner, W.G. Daniel, C.D. Garlichs and I. Cicha. 2010. Role of shear stress patterns in the TNF-alpha-induced atherogenic protein expression and monocytic cell adhesion to endothelium. *Clin. Hemorheol. Microcirc.* 46(2–3): 203–210.

van Horssen, R., T.L. Ten Hagen and A.M. Eggermont. 2006. TNF-alpha in cancer treatment: molecular insights, antitumor effects, and clinical utility. *Oncologist* 11(4): 397–408.

The Role of Cholesterol and Lipoproteins in Control of Endothelial Biomechanics

Irena Levitan,[1,]* Tzu-Pin Shentu,[2] Mingzhai Sun[3] and Gabor Forgacs[4]

Introduction

Cholesterol is one of the major lipid components of the plasma membrane in all mammalian cells where cholesterol:phospholipids molar ratio may be as high as 1:1 (Yeagle 1985). It is well known that maintaining normal levels of cholesterol is essential for cell function and growth but that the excess of cholesterol above the physiological level is cytotoxic (Kellner-Weibel et al. 1999, Simons and Ikonen 2000, Yeagle 1985, Yeagle 1991). Numerous studies have shown that one of the major properties of cholesterol is its ability to alter the physical properties of the lipid membrane bilayer including ordering of the phospholipids (e.g., Demel et al. 1972, Demel and De Kruyff 1976, Stockton and Smith 1976), membrane fluidity

[1]Section of Pulmonary, Critical Care and Sleep Medicine, Department of Medicine, University of Illinois at Chicago, Chicago, IL 60612.
[2]College of Medicine, Department of Cardiology, UCSD, Dan Diego, CA 92093.
[3]Davis Heart and Lung Research Institute, The Ohio State University, Columbus, OH 43210.
[4]Department of Physics and Astronomy, University of Missouri, Columbia, MO 65211.
*Corresponding author: levitan@uic.edu

(e.g., Brulet and McConnell 1976, Cooper 1978, Xu and London 2000), and membrane elastic modulus (Evans and Needham 1987, Needham and Nunn 1990). However, it is much less clear how changes in the level of membrane cholesterol affect the physical properties of the cellular envelope, a bi-component system composed of the membrane lipid bilayer and the underlying cortical cytoskeleton. Our studies focus on the impact of cholesterol on the stiffness, contractility and membrane-cytoskeleton adhesion of vascular endothelial cells. Our first unexpected and surprising observation was that it is cholesterol depletion rather than cholesterol enrichment that increases endothelial stiffness, enhances contractility and strengthens membrane-cytoskeleton adhesion. As described in more detail below, this was unexpected because an increase in membrane cholesterol typically results in increased stiffness of the lipid bilayer, contrary to our observations in endothelial cells. It was a further surprise that exposing endothelial cells to oxidized modifications of low density lipoproteins (oxLDL) had the same effect on endothelial biomechanics as cholesterol depletion. In this chapter, we will discuss these findings in the context of endothelial function in health and disease.

Cholesterol Regulates Endothelial Stiffness: Membrane Bilayer vs. Cellular Envelope

Cholesterol increases elastic modulus of lipid membrane bilayer

The cholesterol molecule has a planar configuration with four fused steroid rings and a hydrophobic tail forming a highly hydrophobic structure which orients within the phospholipid bilayer with the steroid ring perpendicular to the membrane surface and parallel to the hydrocarbon chains of the phospholipids, thus restricting the motion of the phospholipid hydrocarbon chains within the bilayer (Ohvo-Rekilä et al. 2002, Yeagle 1985). Hindering the motion of the phospholipids results in an increase in lipid packing and decrease in membrane fluidity. These effects may also result in an increase in bilayer thickness because inclusion of a cholesterol molecule within the hydrophobic core of the membrane is expected to promote an increase in *trans* configurations of the hydrocarbon chains, which would make them effectively longer. Furthermore, earlier studies have shown that elevation of membrane cholesterol increases the stiffness of membrane lipid bilayers in artificial membrane vesicles, as evaluated by measuring elastic area compressibility modulus, the slope of the curve of membrane tension plotted as a function of membrane area expansion (Needham and Nunn 1990). In the cellular environment, however, the rigidity of lipid bilayer is not expected to contribute significantly to the overall stiffness of the membrane-cytoskeleton complex whose rigidity is dominated by the cortical cytoskeleton (Pourati et al. 1998, Rotsch and Radmacher 2000, Zhang et al. 2002). Our first goal, therefore, was to test how the stiffness of vascular endothelial cells is affected by changes in the membrane cholesterol.

Loss of cholesterol results in endothelial stiffening but cholesterol enrichment has no effect

Our initial expectation was that depleting membrane cholesterol would decrease endothelial stiffness because growing evidence showed that cholesterol-rich membrane domains are the focal points for coupling between the plasma membrane and the cortical cytoskeleton (Brown and London 2000, Edidin 2003, Simons and Ikonen 1997). Thus, we expected that the loss of membrane cholesterol that results in the disruption of these domains should also result in a dissociation of the membrane from the cytoskeleton, leading to a decrease in membrane stiffness and increase in its deformability. To test this hypothesis, we modulated the level of endothelial membrane cholesterol by exposing the cells to methyl-β-cyclodectrin (MβCD), a cyclic oligosaccharide that has high affinity to cholesterol and that can be used both as cholesterol acceptor to deplete cells of cholesterol and as cholesterol donor to increase cellular cholesterol level (Levitan et al. 2000, Zidovetzki and Levitan 2007). Our studies, however, showed that in contrast to expectation, cholesterol depletion increases rather than decreases membrane stiffness in aortic endothelial cells (Fig. 1), (Byfield et al. 2004a). In this study, endothelial stiffness was estimated by measuring progressive

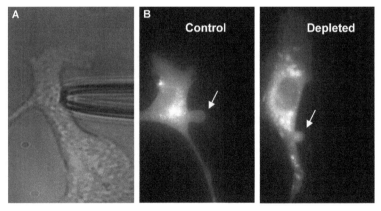

Figure 1. Endothelial stiffening induced by cholesterol depletion. A: an image of a cell with a micropipette, B: typical images of membrane deformation for control and cholesterol-depleted cells (from Byfield et al. 2004b).

membrane deformation in response to negative pressure applied to the cell surface by a glass micropipette, a technique that is called microaspiration, the same approach that was used earlier to determine the effect of cholesterol on membrane stiffness of liposomes (Needham and Nunn 1990). The basic principle of this experimental approach is that the stiffer are the membranes, the slower the rate of membrane deformation and the shorter the steady-state aspiration length. It is also important to note that endothelial stiffness was affected only by cholesterol depletion whereas cholesterol enrichment had no effect at all.

Endothelial stiffening in response to cholesterol depletion was not accompanied by any detectable increase in F-actin suggesting that an increase in actin polymerization is unlikely to underlie the mechanism of this effect. At the same time, disrupting F-actin by latrunculin A, a drug well known to depolymerize actin filaments, resulted in abrogating the stiffening effect of cholesterol depletion (Byfield et al. 2004a). These observations demonstrated that integrity of F-actin is essential for cholesterol depletion-induced endothelial stiffening. While the two statements may seem contradictory, they are easily reconcilable by suggesting that it is not the cellular concentration of F-actin that is most important for this effect but rather its biomechanical properties, such as its connectivity or extent of cross-linking. An intriguing possibility that might underlie an increase in endothelial stiffness is that cholesterol depletion may increase actin stability by sequestering a regulatory phospholipid Phosphatidyl Inositol bi-Phosphate (PIP_2), as was suggested by Kwik et al. (Kwik et al. 2003). Indeed, PIP_2 is considered to be a linker that couples the cytoskeleton to the plasma membrane and plays a major role in the organization of the cytoskeleton (Janmey 1998, Yin and Janmey 2003). It was somewhat unexpected, however, that sequestration of PIP_2 was shown to result in the constrained motion of a class of membrane proteins. This finding in turn led to the hypothesis that this constrained motion might slow down actin turnover resulting in actin stabilization (Kwik et al. 2003). Consistent with this notion, we also found that sequestering PIP_2 simulates the effect of cholesterol depletion on endothelial stiffness (unpublished observations). Clearly, other signaling pathways may also participate/be responsible for these effects and future studies are needed to elucidate these mechanisms. It is also important to note that the stiffening effect in response to cholesterol depletion is not unique to endothelial cells. Our recent study demonstrated that a similar effect is observed in CHO cells, a cell line that has no endothelial phenotype (Kowalsky et al. 2008).

Differential Effect of Sterols on Endothelial Stiffness: Impact of Lipid Packing

Cholesterol stereoisomers have the same effect on endothelial stiffness

A powerful tool to study the mechanism by which cholesterol regulates various cellular properties is the substitution of endogenous cholesterol with sterol analogues that are known to have different effects on the physical properties of lipid bilayers. In our earlier studies, we used this approach to compare the effects of cholesterol (3β-hydroxy-5-cholestene) and its stereoisomer epicholesterol (3α-hydroxy-5-cholestene) that differs from cholesterol in the rotational angle of the hydroxyl group at position 3 (Romanenko et al. 2002). Indeed, stereoisomers (isomers that differ only in the spatial orientation of their component atoms) are widely used to distinguish between specific and non-specific effects of different biological molecules. This is because, stereoisomers, typically, have similar physical properties but are strikingly different in their specific interactions with membrane proteins. Furthermore, earlier studies have shown that the effects of epicholesterol on membrane fluidity and on formation of the lipid domains have been shown to be very similar to those of cholesterol (Gimpl et al. 1997, Xu and London 2000). To substitute endogenous cholesterol with the isomer, cells are first depleted of cholesterol using MβCD and then exposed to MβCD saturated with epicholesterol, which results in epicholesterol to be inserted into the membranes (Romanenko et al. 2002). Using this approach, we demonstrated that replenishing membrane cholesterol in cholesterol-depleted endothelial cells with epicholesterol completely restores cellular stiffness to the control level (Byfield et al. 2006a). These observations suggest that the mechanism by which cholesterol regulates endothelial stiffness is related to its role in defining the physical properties of the lipid bilayer rather than to specific cholesterol-protein interactions.

Sterols that disrupt lipid packing have the same effect on endothelial stiffness as cholesterol depletion

To further test whether endothelial stiffness is regulated by changes in lipid packing of the membrane bilayer, we exposed the cells to two sterols, 7-ketocholesterol or androstenol, both of which have been shown earlier to disrupt lipid order of the membranes (Massey and Pownall 2006, Wang et al. 2004, Xu and London 2000). These sterols were also shown to decrease the rate of lipid raft formation resulting in a decrease in the number and size of membrane domains, effects similar to cholesterol depletion. Our prediction, therefore, was that if cholesterol depletion results in endothelial stiffening because it alters the physical properties of the bilayer/disrupts lipid packing, then both 7-ketocholesterol and androstenol would have the same effect. We found that this is indeed the case. Specifically, we have shown that exposing cells to 7-ketocholesterol or androstenol results in a significant increase in endothelial elastic modulus indicating an increase in cell stiffness (Shentu et al. 2010). We also found that an increase in cell stiffness is inversely correlated with a decrease in lipid packing of the membrane ordered domains, as assessed by Laurdan two photon imaging. The general principle of the technique is that Laurdan is sensitive to the polarity of the local environment and undergoes a red shift as the phase boundary changes from gel to fluid (Gaus et al. 2003, Gaus et al. 2006). Several earlier studies have used this approach to visualize ordered and disordered domains both in liposomes and in cells (Dietrich et al. 2001, Gaus et al. 2003, Gaus et al. 2006). As described earlier, changes in membrane order are estimated by calculating general polarization (GP) ratio, a normalized ratio of fluorescence intensity at 410–490 nm range (gel phase) and 503–553 nm range (fluid phase). Our studies showed a strong correlation between an increase in elastic modulus and a decrease in lipid packing in endothelial cells exposed to 7-ketocholesterol or to androstenol (Fig. 2) (Shentu et al. 2010). An increase in endothelial stiffness in response to 7-ketocholesterol is demonstrated as a shift in the distribution of the elastic modulus of the cells to higher values (compare the histograms of the elastic modulus for control and 7-ketocholesterol-treated cells in Fig. 2A). Average values of the elastic modulus under the two experimental conditions are shown in Fig. 2B.

A decrease in lipid packing is demonstrated by analyzing the GP values for control cells and cells exposed to 7-ketocholesterol: all images are shown in pseudo color with yellow and red corresponding to higher GP values, presumably ordered domains, and green and blue corresponding to lower GP values, presumably disordered domains (Fig. 2C). A decrease in lipid order is manifested by a shift in the GP values from more ordered (red and yellow) to more fluid (green and blue) domains. Analysis of the general polarization (GP) ratios provides the statistically basis on the impact of oxLDL on membrane structure. The histograms of the GP values represent ordered (peak with higher GP values) and disordered (lower GP values) membrane domains, as described in the earlier studies (Gaus et al. 2003, Gaus et al. 2006). A shift from more ordered to more fluid domain structure in response to 7-ketocholesterol is also apparent from the GP histograms that show the abundance and the lipid order properties of both fluid (green curve) and ordered domains (red curve) (Fig. 2D). A shift of the red curve to the right indicates fluidization of the domains whereas a decrease in the area under the curve indicates a decrease in the cell area covered by these domains. Similar effects on both cell stiffness and lipid packing were

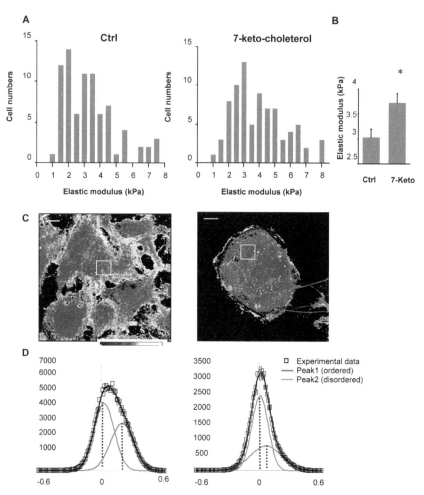

Figure 2. Impact of 7-ketocholesterol on endothelial stiffness and lipid packing. A: Histograms of elastic modulus measured in control (Ctrl) and 7-keto-cholesterol-treated cells. B: Average elastic modulus for control (Ctrl) and 7-keto-cholesterol–treated cells (means± SEM, n=80 cells for each experimental condition). C: Typical GP images of control cells (Ctrl), 7-keto-cholesterol-treated cells. Scale bar is 11.2 μm. D: GP histograms for the corresponding image fitted by a double-Gaussian distribution with the curve shifted to the right representing ordered domains (red) and the curve shifted to the left representing fluid domains (green). The sum of the Gaussians is shown in black (from Shentu et al. 2010).

Color image of this figure appears in the color plate section at the end of the book.

observed when cells were exposed to 7-ketocholesterol, androstenol or when they were cholesterol depleted. Furthermore, enriching the cells with cholesterol, a treatment that is well known to facilitate domain formation rescued the effects of the sterols on endothelial stiffness. Thus, we suggest that a decrease in lipid packing of endothelial membrane domains is a key step in regulation of endothelial stiffness.

Paradoxical Effects of Cholesterol on Membrane-cytoskeleton Adhesion

Next, we addressed the question of how changes in membrane cholesterol affect the adhesion between the plasma membrane and the sub-membrane cytoskeleton, which is critical for keeping the integrity of the membrane-cytoskeleton complex. To investigate how cholesterol affects membrane-cytoskeleton adhesion, we applied Atomic Force Microscopy (AFM) to directly measure the adhesion strength by extraction of membrane tethers (nano-tubular structures) from the surfaces of cells with different cholesterol content (Sun et al. 2005, Sun et al. 2007). Briefly, an AFM cantilever (a force probe) is brought to the cell surface to form a contact, and then pulled away at a constant speed (v) from the cell. Membrane tethers form between the cell and the cantilever, and the tether force (F) is recorded by the deformation of the cantilever. The dependence of the tether force and the pulling speed is described by $F = F_0 + 2\pi\eta_{eff}v$ (Hochmuth et al. 1996). The threshold force F_0 is directly related to the membrane-cytoskeleton interaction; the effective membrane surface viscosity η_{eff} contains contributions associated with the intrinsic material properties of the lipid bilayer, the interlayer slip and the association of the membrane with the cytoskeleton (Hochmuth et al. 1996).

Cholesterol depletion significantly increases membrane-cytoskeleton adhesion. Cholesterol depletion is known to disperse lipid rafts, which contain protein clusters that may connect the lipid bilayer to the underlying cytoskeleton. Thus one would expect cholesterol depletion to decrease the membrane-cytoskeleton interaction and to make it easier to detach the bilayer from the cytoskeleton. However, to our surprise, cholesterol depletion using MβCD significantly increases F_0, whereas, as expected, after latrunculin A treatment, which depolymerizes the actin cytoskeleton network, F_0 significantly decreases. Moreover, there is no difference between values of F_0 of latrunculin A treated control-cells and latrunculin A treated cholesterol-depleted cells. To directly quantify the membrane-cytoskeleton adhesion, we define F_{ad} as the difference of F_0 s before and after latrunculin treatment, i.e., $F_{ad} = F_0 - F_0^{lata}$. As shown in Fig. 3, after cholesterol depletion F_{ad} becomes almost twice as big as that of the control cells. The conclusion that cholesterol depletion increases membrane-cytoskeleton adhesion is further supported by the variance analysis of the tether force. After cholesterol depletion, the variance of tether force becomes much bigger than that of control

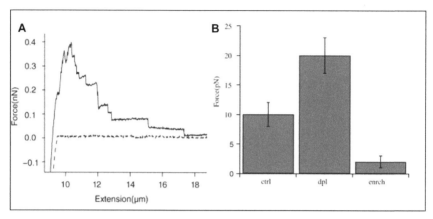

Figure 3. Impact of cholesterol on membrane-cytoskeleton adhesion force: **A:** A typical force vs. extension curve. The dotted line corresponds to the approach curve and the solid line is the retraction curve. On the retraction curve, there are step-like structures, which correspond to the detachment of tethers from the cantilever. The shown retraction curve corresponds to an experiment performed on the control cells at 3 µ/s. **B:** F_{ad}. Values of F_{ad} for cholesterol-depleted (dpl) and cholesterol-enriched (enrch) cells are both significantly different from the control (ctrl) (from Sun et al. 2007).

cells, which suggests that cholesterol depletion makes tethers much more heterogeneous, presumably due to the enhanced membrane-cytoskeleton adhesion (Sun et al. 2007). Thus, the effects of cholesterol depletion on global endothelial stiffness and on membrane-cytoskeleton adhesion are similar because in both cases we see a decrease in membrane deformability and increase in membrane-cytoskeleton adhesion suggesting stabilization and strengthening of cortical cytoskeleton.

Cholesterol enrichment, on the other hand, significantly decreases the threshold tether force F_0 and the effective surface viscosity, η_{eff}. Furthermore, latrunculin treatment after cholesterol enrichment only slightly decreases F_0 and η_{eff} (Sun et al. 2007). As shown in Fig. 3, F_{ad} is significantly smaller, suggesting that cholesterol enrichment reduces the membrane-cytoskeleton adhesion and makes the bilayer easier to detach from the cytoskeleton. It is worth to note that after latrunculin treatment, we did not observe any difference in either F_0 or η_{eff} among control, cholesterol-depleted and cholesterol-enriched cells supporting further the notion that cholesterol enrichment facilitates the detachment of the cytoskeleton from the plasma membrane. Interestingly, the effect of cholesterol enrichment on endothelial membrane viscosity was very different from that observed previously in artificial lipid bilayers, where an increase in the level of bilayer cholesterol is generally associated with a decrease in membrane fluidity and increase in membrane viscosity. This discrepancy is likely to be due to the differences in lipid composition between live cells and artificial bilayers. It is also possible that association with the cortical cytoskeleton affects not only membrane deformability but also membrane viscosity. It is also noteworthy that in contrast to cholesterol depletion, effects of cholesterol enrichment on the global cell stiffness and on membrane-cytoskeleton adhesion are not the same because while cholesterol enrichment weakens membrane-cytoskeleton adhesion, it has no effect on global cell stiffness (Byfield et al. 2004a, Shentu et al. 2010, Sun et al. 2007). These observations suggest that an important component of cell stiffness may be the stiffness of the "inner" or "deep" cytoskeleton of the cell.

Not Only the Membrane: Impact of Cholesterol on the "deep" Cytoskeleton

To estimate the impact of membrane cholesterol on stiffness of the intracellular cytoskeleton, we used particle tracking analysis of correlated motion of fluorescent beads that were phagocytosed by the cells (Byfield et al. 2006a). Earlier studies have shown that motion of internalized particles can be used to describe the mechanics of a material with greater particle motions, typically quantified as mean squared displacements, indicating softer materials (Mason and Weitz 1995). In this study, we have used a method of particle tracking developed by Crocker et al. (2000) and Lau et al. (2003) to analyze correlated motion of the particles within a cell (Crocker and Grier 1996). The general principle of this method is that the motion of individual particles deforms the surrounding material and affects the motion of other particles in its vicinity. Thus, by correlating the motion of two particles we can measure these deformations and infer mechanical information. Larger/smaller deformations in the surrounding particles indicate a softer/harder material, quantified here as a smaller mean square displacement (msd) (Crocker et al. 2000). This technique is sensitive to both underlying mechanical properties of the cell and motor activity (i.e., myosin stroking). Previously, this technique has been shown to measure mechanics consistent with the rheological properties of smooth muscle cells (Lau et al. 2003) measured with magnetic twisting cytommetry (Fabry et al. 2001). Here we show that cholesterol depletion resulted in an approximately 50% reduction in correlated particle motion ($p < 0.05$, when compared at $\tau = 1$ sec) (Fig. 4). This is consistent with either an increase in "deep" cytoskeletal rheology and/or a decrease in a motor activity and suggests that the intracellular milieu is affected by cholesterol depletion. We suggest that changes in the cytoskeleton rheology are induced by disruption of lipid rafts and dissociation or altering membrane-cytoskeleton interactions.

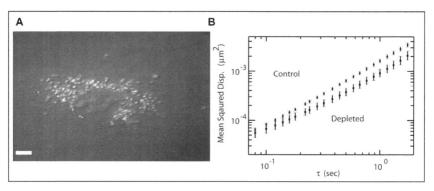

Figure 4. Impact of cholesterol on stiffness of "deep cytoskeleton" in endothelial cells. (A) Typical images of a control cell with engulfed 0.5 μm beads located in the intracellular milieu. Bar is 4 μm. (B) Log-log plot of mean squared displacement of correlated particle motion over time for control (n=14) and cholesterol depleted (n=12) cells. Error bars are generated using log transform statistics and represent the standard error of the mean. $p < 0.05$ at $\tau = 1$ sec. (from Byfield et al. 2006a).

Impact of oxLDL on Endothelial Biomechanics and Lipid Packing

oxLDL and endothelial stiffness: The major physiological cholesterol carrier in the plasma is Low Density Lipoprotein (LDL). It is well known that increased level of LDL is associated with the development of cardiovascular disease and is a major factor in the development of atherosclerosis. Multiple lines of evidence, however, support the hypothesis that it is not LDL by itself but its oxidative modifications (oxLDL) that play the central role in atherogenesis, as discussed in detail in our recent review (Levitan et al. 2010). Indeed, oxLDL is well known to accumulate in atherosclerotic lesions (Yla-Herttuala et al. 1989) and the level of oxLDL increases with hypercholesterolemia both in animal models of atherosclerosis (Hodis et al. 1994, Holvoet et al. 1998) and in humans (Cazzolato et al. 1991, van Tits et al. 2005). In addition, multiple studies have shown that exposure to oxLDL results in endothelial dysfunction including impairment of NO release (Blair et al. 1999), disruption of the endothelial barrier (Gardner et al. 1999) and decrease in endothelial cells (EC) migration (Murugesan et al. 1993). The general mechanism by which oxLDL is believed to exert its pro-atherogenic functions is increased uptake through an array of scavenger receptors resulting in massive cholesterol loading. Our studies demonstrate, however, that the impact of oxLDL on endothelial stiffness strongly resembles the effect of cholesterol depletion and not cholesterol enrichment (Byfield et al. 2006b, Shentu et al. 2010). Specifically, we have shown that exposure to oxLDL stiffens endothelial cells as estimated by two independent experimental approaches, microaspiration (Byfield et al. 2006b) and atomic force microscopy (AFM) (Shentu et al. 2010). Another indication that oxLDL-induced endothelial stiffness is related to cholesterol depletion is that oxLDL-induced effect could be fully recovered by a sequential exposure of the cells to oxLDL and then to MβCD-cholesterol, a treatment that enriches cells with cholesterol. Thus, elevating cholesterol level of the plasma membrane counteracts the effect of oxLDL. It is also important to note that cholesterol enrichment by itself has no effect on endothelial stiffness, indicating reversal of oxLDL-induced endothelial stiffening cannot be attributed to simple cancellation of two opposite effects. Finally, we have also established that while the effects of oxLDL and cholesterol depletion on endothelial biomechanics are very similar, exposure to oxLDL does not cause cholesterol depletion from endothelial cells. These observations led us to hypothesize that oxLDL exposure and cholesterol depletion should initiate a common mechanism that leads to endothelial stiffening.

Both cholesterol depletion and oxLDL increase endothelial force generation: Since earlier studies indicated that increased cellular stiffness is correlated with the magnitude of forces that cells exert on substrates (Wang et al. 2002), we tested whether cholesterol depletion and/or exposure to oxLDL may also increase endothelial force generation. The correlation between endothelial stiffness and force generation on the cell-substrate interface was tested by measuring gel contraction, as described earlier (Sieminski et al. 2004). As expected, seeding the cells into gels resulted in gel contraction under all experimental conditions (Byfield et al. 2006b). Importantly, however, cells that were pretreated with oxLDL or with MβCD showed significantly greater gel contraction, as indicated by a decrease in gel area. More recently, we have also shown that oxLDL-induced increase in endothelial force generation is fully reversed by supplying the cells with a surplus of cholesterol (Shentu et al. 2010).

Cholesterol depletion facilitates traction force generation by individual cells: Impact of membrane cholesterol on endothelial force generation was also analyzed using Traction Force Microscopy (Norman et al. 2010), a common method to measure the forces exerted by cells on compliant substrates (e.g., Dembo and Wang 1999, Reinhart-King et al. 2002). In this study, we showed that cholesterol-depleted cells exert significantly larger force on their substrates than control cells, with an almost two-fold increase compared to control and cholesterol-enriched cells. No statistical differences occur between control and cholesterol-enriched cells. These observations are consistent with the studies described above showing no significant effect of cholesterol enrichment on the membrane stiffness (Byfield et al. 2004a). These results demonstrate that cholesterol plays an important role not only in controlling endothelial stiffness but also in regulating endothelial force generation on the cell-substrate interface. The major question, however, that remained unanswered at that time was: How is it possible that oxLDL and cholesterol depletion have similar effects on endothelial biomechanics?

oxLDL and lipid order: Based on the studies proposed above, we hypothesized that oxLDL and cholesterol depletion may have the same effect on endothelial stiffness because they both disrupt lipid packing. Furthermore, we also propose that oxLDL disrupts lipid packing by inserting oxysterols into the cellular membranes. To test this hypothesis, we compared the impacts of oxLDL and cholesterol depletion on endothelial lipid packing using Laurdan two photon imaging, as described above. As expected, exposing cells to MβCD results in a shift to less ordered membrane structure, as indicated by a relative decrease in highly-ordered (yellow) domains and an increase in less ordered (blue-green) domains (Fig. 5). Most importantly, exposure to oxLDL also resulted in a shift to less ordered membrane structure, as indicated by a loss of highly-ordered (yellow) domains (Fig. 5). Replenishing membrane cholesterol of oxLDL-treated cells with MβCD-cholesterol partially reverses the effect. These changes are also apparent from the GP histograms shown below the images. Specifically, exposure to MβCD resulted in a decrease in peak values/lipid order of both ordered (red curve and green curve both shift to left) indicating that cholesterol depletion decreases lipid packing of the membrane. Exposure to oxLDL, however, only affects the ordered domains, as indicated by a left shift of red curve. Thus, both MβCD and oxLDL induce a shift of ordered domains to a less ordered state. Conversely, replenishing membrane cholesterol by exposing oxLDL-treated cells to MβCD-cholesterol partially reverses the effect of oxLDL by increasing membrane order in a portion of membrane domains. Taken together, our observations show that oxLDL decreases lipid order preferentially of the ordered domains and that this effect can be partially reversed by increasing membrane cholesterol.

OxLDL contains an array of oxidative lipids, which are bioactive under pathological conditions. To determine which of the components are responsible for an increase in endothelial stiffness, we separated the lipid extract of oxLDL into five separate fractions: cholesterol/cholesterol ester farction (Chol/CE) that also included different oxysterols, as well as lysophosphatidylcholine (LPC), sphingomyelin (SM), phosphatidylcholine (PC), phosphatidylethanolamine (PE), and fractions and tested the impact of each fraction on endothelial stiffness. The most prominent effect on endothelial stiffness was observed when cells were exposed to oxidized cholesterol/cholesteryl ester fraction (Shentu et al. 2012). Furthermore,

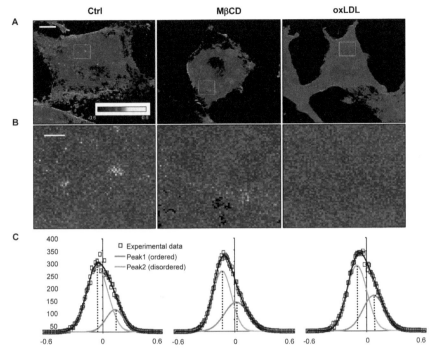

Figure 5. Impact of oxLDL on lipid packing of membrane domains in BAECs. A: Typical GP images of control cells (Ctrl), MβCD-treated cells (MβCD), or oxLDL-treated cells (oxLDL). Scale bar is 5.6μm. **B**: The zoom-in representative regions of the GP images shown above (the zoomed regions are 5.6μm×5.6μm) Scale bar is 1μm. **C**: GP histograms for the three experimental cell populations (dots) fitted by a double-Gaussian distribution with the curve shifted to the right representing ordered domains (red) and the curve shifted to the left representing fluid domains (green). The sum of the Gaussians is shown in black. GP distribution is obtained from the region –0.6 to +0.6 as shown in the x-axis and the number of counts is normalized (sum=10000) as shown in y-axis. (n=23-25 images per experimental condition, 4 independent experiments) (from Shentu et al. 2010).

Color image of this figure appears in the color plate section at the end of the book.

in terms of individual oxysterols, we identified 7-ketocholesterol and 7α-hydroxycholesterol, as having significant effects on endothelial stiffness, whereas several other oxysterols found in oxLDL, such as 5α,6α epoxide and 5β,6β epoxide isomers, had no significant effect on endothelial biomechanics. In addition, we also found that 27-hydroxy-cholesterol that is found in atherosclerotic lesions (Bjrkhem et al. 1994, Crisby et al. 1997) also induces significant endothelial stiffening (Shentu et al. 2012). Furthermore, both 7-ketocholesterol-induced and 27-hydroxycholesterol-induced endothelial stiffening were fully reversed by enriching the cells with cholesterol (Shentu et al. 2012). In summary, our observations suggest that while oxLDL induces endothelial stiffening through the incorporation of specific oxysterols into endothelial membranes, maintaining high level of membrane cholesterol in endothelial cells is actually protective against the effects of oxLDL.

Impact of Endothelial Stiffening on the Sensitivity to Shear Stress

It seems almost obvious that changes in biomechanical properties should be expected to change the mechanosensitivity of the cells. This is particularly important for the endothelium, which is constantly exposed to hemodynamic forces generated by blood flow, which in turn are known to play prominent roles in the acute control of vascular tone, in the regulation of arterial structure and localization of atherosclerotic lesions (Davies et al. 2005, Gimbrone et al. 1997, Ross 1999). It is less clear, however, how exactly an increase in cell stiffness and membrane tension should affect the sensitivity of endothelial cells to shear stress. Our initial expectation was that an increase in cells stiffness should impair their ability to respond to shear stress forces. However, what we have discovered was that oxLDL-induced increase in endothelial stiffness is accompanied with an increased rather than decreased ability of endothelial cells to realign in the direction of the flow, a hallmark of endothelial shear stress response, and that the same effect is observed in response to cholesterol depletion (Kowalsky et al. 2008).

Furthermore, an increased sensitivity to shear stress was observed not only on the cellular level but also at the level of individual actin fibers (Kowalsky et al. 2008). Specifically, the angles of individual stress fibers were measured in cells selected randomly from oxLDL-treated and control cell populations (30–40 individual fibers, which could be clearly identified and generally represented the brightest ones, were analyzed in each cell; curved fibers were excluded from the analysis) after the cells were exposed to flow for 6 hours. The angles were taken relative to the direction of the flow. This analysis shows that, consistent with the changes in cell orientations, exposure to oxLDL also significantly decreased the spread of the orientation angles and facilitated the realignment of the fibers in the direction of flow (Fig. 6). A similar effect was also observed in response to cholesterol depletion. We propose that oxLDL-induced increase in endothelial stiffness may sensitize the cells to a mechanical stimulus generated by shear stress. Further studies are needed to elucidate the mechanism of this effect and the implications for the development of atherosclerosis.

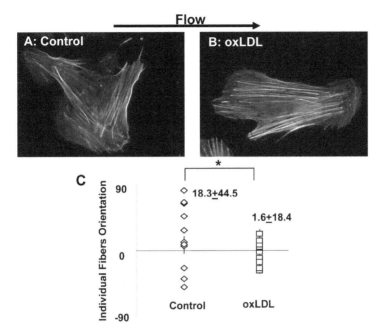

Figure 6. Impact of oxLDL on flow-induced reorientation of actin fibers. (A-B): Typical images of F-actin structure control and oxLDL-treated cells exposed to 6 h of flow. All images were taken with the same contrast levels (while individual cells vary in size under all experimental conditions, there is no significant differences in cell spreading between control and oxLDL-treated cells (the average cell areas normalized to control were 1 ± 0.07 vs. 1.19 ± 0.12 for control and oxLDL-treated cells under static conditions and 1 ± 0.13 vs. 1.20 ± 0.20 for control and oxLDL-treated cells under flow conditions)). **C:** Angles of individual actin fibers in control and oxLDL-treated cells exposed to 6 h of flow. The symbols and the error bars represent mean+SEM for individual cells (30–40 individual fibers were measured in each cell, n=11 and 17 cells for control and oxLDL-treated conditions respectively) (from Kowalsky et al. 2008).

Acknowledgements

We thank Gregory Kowalsky for his help in the preparation of the manuscript and formatting the figures. The work was supported by National Institutes of Health grants HL073965 and HL083298 (to I. Levitan) and by AHA predoctoral fellowship and Max Goldenberg Foundation 10PRE2570025 (to T. P. Shentu).

References

Bjorkhem, I., O. Andersson, U. Diczfalusy, B. Sevastik, R.J. Xiu, C. Duan and E. Lund. 1994. Atherosclerosis and sterol 27-hydroxylase: evidence for a role of this enzyme in elimination of cholesterol from human macrophages. *Proceedings of the National Academy of Sciences* 91(18): 8592–8596.

Blair, A., P.W. Shaul, I.S. Yuhanna, P.A. Conrad and E.J. Smart. 1999. Oxidized low-density lipoprotein displaces endothelial Nitric-oxide synthase from plasmalemmal caveolae and impairs eNOS activation. *The Journal of Biological Chemistry* 274: 32512–32519.

Brown, A.D. and E. London. 2000. Structure and function of sphingolipid- and cholesterol-rich membrane rafts. *The Journal of Biological Chemistry* 275: 17221–17224.

Brulet, P. and H.M. McConnell. 1976. Lateral hapten mobility and immunochemistry of model membranes. *Proc. Natl. Acad. Sci. USA* 73: 2977–2981.

Byfield, F., H. Aranda-Aspinoza, V.G. Romanenko, G.H. Rothblat and I. Levitan. 2004a. Cholesterol depletion increases membrane stiffness of aortic endothelial cells. *Biophys. J.* 87: 3336–3343.

Byfield, F.J., H. Aranda-Espinoza, V.G. Romanenko, G.H. Rothblat and I. Levitan. 2004b. Cholesterol depletion increases membrane stiffness of aortic endothelial cells. *Biophys. J.* 87(5): 3336–3343.

Byfield, F.J., B.D. Hoffman, V.G. Romanenko, Y. Fang, J.C. Crocker and I. Levitan. 2006a. Evidence for the role of cell stiffness in modulation of volume-regulated anion channels. *Acta Physiologica* 187(1–2): 285–294.

Byfield, F.J., S. Tikku, G.H. Rothblat, K.J. Gooch and I. Levitan. 2006b. OxLDL increases endothelial stiffness, force generation and network formation. *J. Lipid Res.* 47: 715–723.

Cazzolato, G., P. Avogaro and G. Bittolo-Bon. 1991. Characterization of a more electronegatively charged ldl subfraction by ion exchange hplc. *Free Radic. Biol. Med.* 11: 247–253.

Cooper, R.A. 1978. Influence of increased membrane cholesterol on membrane fluidity and cell function in human red blood cells. *J. Supramol. Struct.* 8: 413–430.

Crisby, M., J. Nilsson, V. Kostulas, I. Bjökhem and U. Diczfalusy. 1997. Localization of sterol 27-hydroxylase immuno-reactivity in human atherosclerotic plaques. *Biochimica et Biophysica Acta (BBA) - Lipids and Lipid Metabolism* 1344: 278–285.

Crocker, J.C. and D.G. Grier. 1996. Methods of digital video microscopy for colloidal studies. *Journal of Colloid and Interface Science* 179(1): 298–310.

Crocker, J.C., M.T. Valentine, E.R. Weeks, T. Gisler, P.D. Kaplan, A.G. Yodh and D.A. Weitz. 2000. Two-point microrheology of inhomogeneous soft materials. *Physical Review Letters* 85(4): 888–891.

Davies, P.F., J.A. Spaan and R. Krams. 2005. Shear Stress Biology of the Endothelium. *Annals of Biomedical Engineering* V33(12): 1714.

Dembo, M. and Y.L. Wang. 1999. Stresses at the cell-to-substrate interface during locomotion of fibroblasts. *Biophys. J.* 76: 2307–2316.

Demel, R.A., K.R. Bruckdorfer and L.L.M. van Deenen. 1972. The effect of sterol structure on the permeability of liposomes to glucose, glycerol and Rb+. *Biochem. et Biophys. Acta* 255: 321–330.

Demel, R.A. and B. De Kruyff. 1976. The function of sterols in membranes. *Biochim. Biophys. Acta* 457: 109–132.

Dietrich, C., L.A. Bagatolli, Z.N. Volovyk, N.L. Thompson, M. Levi, K. Jacobson and E. Gratton. 2001. Lipid rafts reconstituted in model membranes. *Biophys. J.* 80: 1417–1428.

Edidin, M. 2003. The state of lipid rafts: From model membranes to cells. *Annual Review Biophysical Biomolecular Structure* 32: 257083.

Evans, E. and D. Needham. 1987. Physical properties of surfactant bilayer membranes: thermal transition, elasticity, rigidity, cohesion and colloidal interactions. *Journal of Physical Chemistry* 91: 4219–4228.

Fabry, B., G.N. Maksym, J.P. Butler, M. Glogauer, D. Navajas and J.J. Fredberg. 2001. Scaling the microrheology of living cells. *Physical Review Letters* 87(14): 148102–(148101–148104).

Gardner, G., C.L. Banka, K.A. Roberts, A.E. Mullick and J.C. Rutledge. 1999. Modified LDL–mediated increases in endothelial layer permeability are attenuated with 17ß-estradiol. *Arterioscler. Thromb. Vasc. Biol.* 19(4): 854–861.

Gaus, K., E. Gratton, E.P. Kable, A.S. Jones, I. Gelissen, L. Kritharides and W. Jessup. 2003. Visualizing lipid structure and raft domains in living cells with two-photon microscopy. *Proc. Natl. Acad. Sci. USA* 100(26): 15554–15559.

Gaus, K., T. Zech and T. Harder. 2006. Visualizing membrane microdomains by Laurdan 2-photon microscopy. *Mol. Membr. Biol.* 23(1): 41–48.

Gimbrone, M.A.J., N. Resnick, T. Nagel, L.M. Khachigian, T. Collins and J.N. Topper. 1997. Hemodynamics, endothelial gene expression, and atherogenesis. *Ann. N Y Acad. Sci.* 811: 1–10.

Gimpl, G., K. Burger and F. Fahrenholz. 1997. Cholesterol as modulator of receptor function. *Biochemistry* 36: 10959–10974.

Hochmuth, F.M., J.Y. Shao, J. Dai and M.P. Sheetz. 1996. Deformation and flow of membrane into tethers extracted from neuronal growth cones. *Biophys. J.* 70: 358–369.

Hodis, H.N., D.M. Kramsch, P. Avogaro, G. Bittolo-Bon, G. Cazzolato, J. Hwang, H. Peterson and A. Sevanian. 1994. Biochemical and cytotoxic characteristics of an *in vivo* circulating oxidized low density lipoprotein (LDL-). *J. Lipid Res.* 35: 669–677.

Holvoet, P., G. Theilmeier, B. Shivalkar, W. Flameng and D. Collen. 1998. LDL hypercholesterolemia is associated with accumulation of oxidized LDL, atherosclerotic plaque growth, and compensatory vessel enlargement in coronary arteries of miniature pigs. *Arterioscler. Thromb. Vasc. Biol.* 18: 415–422.

Janmey, P.A. 1998. The Cytoskeleton and Cell Signaling: Component Localization and Mechanical Coupling. *Phys. Rev.* 78(3): 763–781.

Kellner-Weibel, G., Y.J. Geng and G.H. Rothblat. 1999. Cytotoxic cholesterol is generated by the hydrolysis of cytoplasmic cholesteryl ester and transported to the plasma membrane. *Atherosclerosis* 146: 309–319.

Kowalsky, G.B., F.J. Byfield and I. Levitan. 2008. oxLDL facilitates flow-induced realignment of aortic endothelial cells. *Am. J. Physiol. Cell Physiol.* 295(2): C332–340.

Kwik, J., S. Boyle, D. Fooksman, L. Margolis, M.P. Sheetz and M. Edidin. 2003. Membrane cholesterol, lateral mobility, and the phosphatidylinositol 4,5-bisphosphate-dependent organization of cell actin. *Proc. Natl. Acad. Sci. USA* 100(24): 13964–13969.

Lau, A.W.C., B.D. Hoffman, A. Davies, J.C. Crocker and T.C. Lubensky. 2003. Microrheology, stress fluctuations, and active behavior of living cells. *Physical Review Letters* 91(19): 198101–(198101–198104).

Levitan, I., A.E. Christian, T.N. Tulenko and G.H. Rothblat. 2000. Membrane cholesterol content modulates activation of volume-regulated anion current (VRAC) in bovine endothelial cells. *Journal of General Physiology* 115: 405–416.

Levitan, I., S. Volkov and P.V. Subbaiah. 2010. Oxidized LDL: Diversity, patterns of recognition and pathophysiology. *Antioxidants & Redox Signaling* 13: 39–75.

Mason, T.G. and D.A. Weitz. 1995. Optical Measurements of Frequency-Dependent Linear Viscoelastic Moduli of Complex Fluids. *Physical Review Letters* 74(7): 1250–1253.

Massey, J.B. and H.J. Pownall. 2006. Structures of biologically active oxysterols determine their differential effects on phospholipid membranes. *Biochemistry* 45(35): 10747–10758.

Murugesan, G., G.M. Chisolm and P.L. Fox. 1993. Oxidized low density lipoprotein inhibits the migration of aortic endothelial cells *in vitro*. *J. Cell Biol.* 120: 1011–1019.

Needham, D. and R.S. Nunn. 1990. Elastic deformation and failure of lipid bilayer membranes containing cholesterol. *Biophys. J.* 58: 997–1009.

Norman, L.L., R.J. Oetama, M. Dembo, F. Byfield, D.A. Hammer, I. Levitan and H. Aranda-Espinoza. 2010. Modification of Cellular Cholesterol Content Affects Traction Force, Adhesion and Cell Spreading. *Cell Mol. Bioeng.* 3: 151–162.

Ohvo-Rekilä, H., B. Ramstedt, P. Leppimäki and J.P. Slotte. 2002. Cholesterol interactions with phospholipids in membranes. *Prog. Lipid Res.* 41: 66–97.

Pourati, J., A. Maniotis, D. Spiegel, J.L. Schaffer, J.P. Butler, J.J. Fredberg, D.E. Ingber, D. Stamenovic and N. Wang. 1998. Is cytoskeletal tension a major determinant of cell deformability in adherent endothelial cells? *Am. J. Physiol. Cell Physiol.* 274(5): C1283–1289.

Reinhart-King, C., M. Dembo and D. Hammer. 2002. Endothelial cell traction forces on RGD-derivatized polyacrylamide substrata. *Langmuir* 19: 1573–1579.

Romanenko, V.G., G.H. Rothblat and I. Levitan. 2002. Modulation of endothelial inward rectifier K+ current by optical isomers of cholesterol. *Biophys. J.* 83: 3211–3222.

Ross, R. 1999. Atherosclerosis—an inflammatory disease. *N Engl. J. Med.* 340: 115–126.

Rotsch, C. and M. Radmacher. 2000. Drug-Induced changes of cytoskeletal structure and mechanics in fibroblasts: an atomic force microscopy study. *Biophys. J.* 78(1): 520–535.

Shentu, T.P., D.K. Singh, M.-J. Oh, S. Sun, L. Sadaat, A. Makino, T. Mazzone, P.V. Subbaiah, M. Cho and I. Levitan. 2012. The role of oxysterols in control of endothelial stiffness. *Journal of Lipid Research*.

Shentu, T.P., I. Titushkin, D.K. Singh, K.J. Gooch, P.V. Subbaiah, M. Cho and I. Levitan. 2010. oxLDL-induced decrease in lipid order of membrane domains is inversely correlated with endothelial stiffness and network formation. *Am. J. Physiol. Cell Physiol.* 299(2): C218–229.

Sieminski, A.L., R.P. Hebbel and K.J. Gooch. 2004. The relative magnitudes of endothelial force generation and matrix stiffness modulate capillary morphogenesis *in vitro*. *Exp. Cell Res.* 297: 574–584.

Simons, K. and E. Ikonen. 1997. Functional rafts in cell membranes. *Nature* 387: 569–572.

Simons, K. and E. Ikonen. 2000. How cells handle cholesterol. *Science* 290: 1721–1726.

Stockton, B.W. and I.C.P. Smith. 1976. A deuterium NMR study of the condensing effect of cholesterol on egg phosphatidylcholine bilayer membranes. *Chem. Phys. Lipids* 17: 251–263.

Sun, M., J.S. Graham, B. Hegedus, F. Marga, Y. Zhang, G. Forgacs and M. Grandbois. 2005. Multiple Membrane Tethers Probed by Atomic Force Microscopy. *Biophys. J.* 89(6): 4320–4329.

Sun, M., N. Northup, F. Marga, F.J. Byfield, I. Levitan and G. Forgacs. 2007. Cellular cholesterol effects on membrane-cytoskeleton adhesion. *J. Cell Sci.* 120: 2223–2231.

van Tits, L.J., T.M. van Himbergen, H.L. Lemmers, J. de Graaf and A.F. Stalenhoef. 2005. Proportion of oxidized ldl relative to plasma apolipoprotein b does not change during statin therapy in patients with heterozygous familial hypercholesterolemia. *Atherosclerosis* [Epub ahead of print].

Wang, J., Megha and E. London. 2004. Relationship between sterol/steroid structure and participation in ordered lipid domains (lipid rafts): implications for lipid raft structure and function. *Biochemistry* 43: 1010–1018.

Wang, N., I.M. Tolic-Norrelykke, J. Chen, S.M. Mijailovich, J.P. Butler, J.J. Fredberg and D. Stamenovic. 2002. Cell prestress. I. Stiffness and prestress are closely associated in adherent contractile cells. *Am. J. Physiol. Cell Physiol.* 282: C606–616.

Xu, X. and E. London. 2000. The effect of sterol structure on membrane lipid domains reveals how cholesterol can induce lipid domain formation. *Biochemistry* 39: 843–849.

Yeagle, P.L. 1985. Cholesterol and the cell membrane. *Biochimica et Biophysica Acta* 822: 267–287.

Yeagle, P.L. 1991. Modulation of membrane function by cholesterol. *Biochimie* 73: 1303.

Yin, H.L. and P.A. Janmey. 2003. Phosphoinositide regulation of the actin cytoskeleton. *Annu. Rev. Physiol.* 65: 761–789.

Yla-Herttuala, S., W. Palinski, M.E. Rosenfeld, S. Parthasarathy, T.E. Carew, S. Butler, J.L. Witztum and D. Steinberg. 1989. Evidence for the presence of oxidatively modified low density lipoprotein in atherosclerotic lesions of rabbit and man. *J. Clin. Invest.* 84: 1086–1095.

Zhang, G., M. Long, Z.Z. Wu and W.Q. Yu. 2002. Mechanical properties of hepatocellular carcinoma cells. *World J. Gastroenterol.* 8: 243–246.

Zidovetzki, R. and I. Levitan. 2007. Use of cyclodextrins to manipulate plasma membrane cholesterol content: Evidence, misconceptions and control strategies. *Biochimica et Biophysica Acta (BBA) - Biomembranes* 1768(6): 1311.

Implications of Fluid Shear Stress in Capillary Sprouting during Adult Microvascular Network Remodeling

Walter L. Murfee

Introduction: The importance of understanding how microvascular networks remodel in the adult

A key requirement for tissue function is an adequate blood supply. Consequently, the microcirculation is a common denominator for multiple pathological conditions including diabetic retinopathy, myocardial ischemia, and cancer. Microvascular remodeling is a complex continuum of molecular and cellular events and is a term used to describe any type of structural adaptation associated with a vascular network. This general phenomenon is commonly studied as one of three specific sub-processes: vasculogenesis, arteriogenesis, and angiogenesis. Vasculogenesis is defined

Department of Biomedical Engineering, Tulane University, Lindy Boggs Center, Suite 500, New Orleans, LA 70118-5698.
Email: wmurfee@tulane.edu

as the *de novo* formation of new vessels. Arteriogenesis refers to as the maturation of new vessels and angiogenesis is defined as the growth of new vessels from existing ones. Work over the past twenty years has focused on understanding the major genetic and molecular factors involved in these sub-processes (Carmeliet 2004, Peirce and Skalak 2003). Advancing both our understanding of how a microvascular network changes in response to its local environment and our ability to therapeutically apply this understanding requires the identification of environmental cues at specific locations across the hierarchy of a network. Consider an intact adult network (Fig. 1) consisting of arterioles, capillaries, and venules. The structural heterogeneity between endothelial cells along different vessels types (Fig. 1) emphasizes the importance of identifying: 1) vessel specific environments, 2) how these environments change during remodeling; and 3) how these changes direct endothelial cell behavior. Now consider an adult microvascular network that has undergone remodeling (Fig. 2).

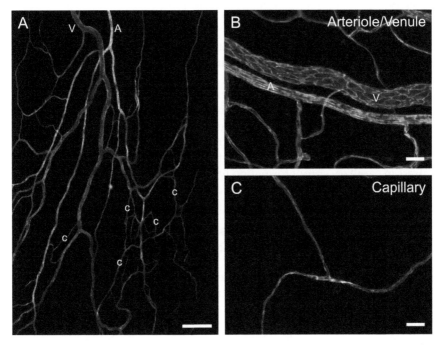

Figure 1. Representative endothelial cell structures along an adult rat mesenteric microvascular network. Antibody labeling against PECAM identifies all endothelial cells across the hierarchy of microvascular networks. A) A typical microvascular network with a feeding arteriole, "A", capillaries, "c", and a draining venule, "V". PECAM labeling of endothelial cell junctions identifies different cellular morphologies per vessel type. B) Arterioles versus venules display a smaller relative diameter and more elongated endothelial cells. C) Capillaries are typically lined by single elongated cells. Scale bars = 200 μm (A), 20 μm (B, C).

Unstimulated   **Stimulated**

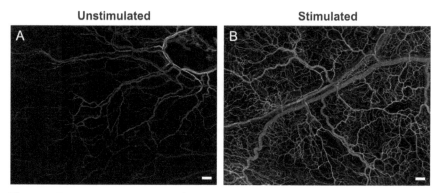

Figure 2. Representative microvascular networks before and after remodeling. Antibody labeling against PECAM identifies all endothelial cells across the hierarchy of microvascular networks. A) Unstimulated adult rat mesenteric microvascular network. B) Adult rat mesenteric microvascular network 10 days post mast cell granulation stimulation. Microvascular network growth is apparent by the increase in vessel density. Faint PECAM labeling in (A) identifies lymphatic vessels, which are distinguishable from blood vessels based their increased diameters and blunt ends. Scale bars = 200 μm.

Dramatic changes in network size and vessel density further emphasize the need to consider how local environmental cues might differ over the time course of remodeling.

This chapter provides an overview of what we know regarding the relationships between fluid shear stress and the endothelial cell dynamics involved inadult microvascular remodeling. First, the implications for shear stresses in vasculogenesis, arteriogenesis, and angiogenesis will be summarized. Then, experimental results that support a role for shear stress in capillary sprout formation will be considered and used to generate unanswered questions to motivate future research.

The Role of Shear Stress in Adult Microvascular Network Remodeling

Hemodynamics, including wall shear stress and circumferential stress, undoubtedly play a role in the structural adaptions associated with microvascular network remodeling (Skalak and Price 1996). The relative importance of both mechanical stimuli is unclear, yet for the purposes of this chapter, a focus is placed on wall shear stress, which is assumed to be sensed by the endothelial cells along microvessels. Wall shear stress is due to the viscous forces associated with the fluid flow through the lumen of a vessel and is the product of viscosity and shear rate at the wall (the radial component of the change in fluid velocity). Most commonly, shear stress, τ, in a microvessel is approximated by Poiseuille flow through a cylindrical tube, $\tau = \mu 8 \, (V_{mean}/d)$. μ represents viscosity, V_{mean} represents the average velocity and d is the inner tube diameter. This equation does not account for non-Newtonian properties of blood or local cell geometries. Nonetheless, application of this relationship has proven useful for approximating relative shear stress values in different vessel types.

The presence of injected dextran molecules along each vessel segment in a growing network suggests even the newest capillaries and capillary sprouts experience fluid flow (Fig. 3). Thus, presumably each endothelial cell along the network experiences a shear stress. Typically endothelial cells are thought to experience wall shear stresses in the range of 5–150 dyne/cm^2, with arterioles experiencing increased magnitudes compared to venules (Pries et al. 1995). The actual shear stress magnitudes along capillary sprouts remain unknown.

The research related to endothelial cells and shear stress generally addresses one of two questions: 1) How does an endothelial cell sense and transduce shear stress? and 2) How does an endothelial cell respond to shear stress? For a comprehensive review of the molecular basis for endothelial cell mechanotransduction see the review by Li et al. (Li et al. 2005). Mechano-sensing elements on the endothelial cell surface include cell-cell junction, G-protein-coupled receptors, integrins, tyrosine kinase receptors, and voltage-gated ion channels. Looking to the future, systematic identification of the interrelationships between these sensing elements will be critical for understanding which elements are most important for triggering a response. The importance for this new area of research is supported by results by Tzima et al.'s discovery that a complex including PECAM, VE-cadherin, and VEGFR-2 is necessary for an endothelial cell response to shear stress (Tzima et al. 2005). Another area of investigation critical for advancing our understanding of endothelial cell mechanotransduction will be the identification of how sensing elements interact with the glycocalyx, which is a glycoprotein structure that extends into the lumen from the surface of endothelial cells (Curry and Adamson 2012, Tarbell and Ebong 2008).

The rest of this section presents the direct evidence for shear stress involvement in each remodeling sub-process involved in microvascular network growth.

Vasculogenesis

Vasculogenesis refers to the formation of blood vessels from undifferentiated precursor cells. This process is most often associated with vascular development during embryonic development, yet recent work has suggested a potential role for the recruitment of vascular precursor cells to neo-vessels in the adult (Carmeliet 2004). The capability of circulating cells

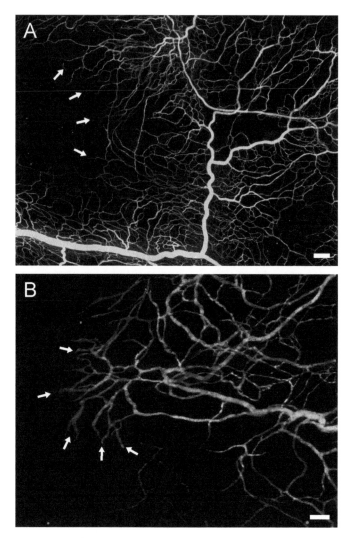

Figure 3. The presence of fluid flow in microvessels across the hierarchy of angiogenic adult rat mesenteric microvascular networks. Antibody labeling against PECAM (red) identifies all endothelial cells across the hierarchy of microvascular networks. Prior to tissue harvesting, vessel lumens were identified via intra-femoral vein injection of a FITC-conjugated fixable 40 kDa dextran (green). A, B) Representative images of growing networks 10 days post mast cell degranulation stimulation. Network growth is apparent by the increase in capillary density. Arrows indicate capillary sprouts. Dextran is present along capillaries in high vessel density regions and capillary sprouts indicating that new capillaries and sprouts have lumens. Scale bars = 200 μm (A), 100 μm (B).

Color image of this figure appears in the color plate section at the end of the book.

to incorporate into remodeling microvessels in adult tissues has generated much excitement and introduced the potential use of adult precursor cells for therapeutic stimulation of vascular growth (Asahara et al. 1997, Lyden et al. 2001, Majka et al. 2003). Depending on the angiogenic stimulus, precursor cell incorporation rates vary from approximately 10% to 90% (Asahara et al. 1997, Lyden et al. 2001). The discrepancies in incorporation rates bring to question whether incorporation is even important. Recent work suggests that bone marrow-derived cells might rather contribute to vascular remodeling through paracrine mechanisms and do not necessarily incorporate into vessels (Kinnaird et al. 2004, O'Neill et al. 2005, Ziegelhoeffer et al. 2004). Regardless of the incorporation controversy, circulating vascular progenitor cells seem to play a functional role in adult microvascular remodeling.

Given their trafficking through the circulation, vascular progenitor cells can be assumed to experience shear stress. *In vitro* experiments have shown that progenitor cell phenotype can be directly influenced by a shear stress stimulus (Maul et al. 2011, Obi et al. 2009, Yamamoto et al. 2003). In the study by Obi et al., cultured endothelial progenitor cells were shown to increase their ephrin B2 expression in a shear stress dependent manner over the range of $0.1–2.5$ dyne/cm^2 suggesting that fluid shear stress is capable of inducing an arterial phenotype (Obi et al. 2009). Interestingly, this magnitude range might be more relevant for a capillary sprout scenario (Stapor et al. 2011). Thus, the interpretation of the results obtained by Obi et al. depends on the application of what is known regarding shear stress magnitudes at specific vessel locations and the correlation between location and cell phenotype. In adult microvascular networks, direct evidence for a putative role of shear stress in the cell dynamics necessary for vasculogenesis still remains to be identified.

Arteriogenesis: A/V determination

Arteriogenesis refers to arterialization, described as the capillary acquisition of perivascular cells, and subsequent vessel enlargement, including collateral development. The source of new perivascular cells during capillary arterialization can be attributed to the migration of upstream smooth muscle cells (SMCs), differentiation of existing perivascular cells or the recruitment of circulating progenitor cells (Au et al. 2008, Hellstrom et al. 1999, Majka et al. 2003, Van Gieson et al. 2003). Trying to understand this lineage leads us to the related question: How do local environmental cues regulate cell fate?

Shear stress has been implicated in both arterialization and collateral development *in vivo*. Wang and Prewitt demonstrated that a chronic flow reduction can lead to a loss of arterioles (Wang and Prewitt 1993). In regards to collateral development, Schaper and coworkers have correlated increases in fluid flow with increases in vessel diameter, wall thickening, macrophage recruitment, and growth factor production following the re-direction of flow through pre-existing collateral vessels due to femoral artery ligation (Heil and Schaper 2004). In spite of these examples, the isolated effects of shear stress on endothelial cells during either arterialization or collateral development remain understudied.

Since the identification of ephrinB2 as an arterial marker during embryonic development (Wang et al. 1998), much attention has been given to arterial/venous (A/V) identity. Reported A/V markers include the eph receptor tyrosine kinases and their ephrin ligands (Adams et al. 1999, Gale et al. 2001, Shin et al. 2001, Wang et al. 1998), Neuropilin-1 (NP-1) and Neuropilin-2 (NP-2) (Herzog et al. 2001, Moyon et al. 2001), Notch receptors (Villa et al. 2001), and gridlock (Zhong et al. 2001). These molecules have been shown to exhibit differential A/V expression and effects on vascular patterning during embryonic development (Lawson and Weinstein 2002). The plasticity of vascular cells prompts the question of whether A/V markers are downstream of local environmental signaling or causal regulators of A/V vessel phenotype. Given the shear stress magnitudes seen on the arterial side of a network versus the venous side (Pries et al. 1995), shear stress represents a logical candidate for a regulator of A/V identity. The regulating potential for shear stress is supported during chicken development as endothelial cells along the arterial side of a network switch their A/V phenotype after a main feeding artery is ligated (le Noble et al. 2004). However, this paradigm is challenged by expression of gridlock and ephrinB2 by arterial vessels before the onset of flow during vascular development (Wang et al. 1998, Zhong et al. 2001). The question remains whether A/V identity is regulated by shear stress or genetic programing.

Angiogenesis

Angiogenesis is the process that describes the formation of capillaries from pre-existing vessels and involves endothelial cell interactions with the local environment and other cell types, including vascular pericytes (Eilken and Adams 2010, Gerhardt and Betsholtz 2003). The endothelial cell dynamics involved in angiogenesis are generally associated with two modes: capillary sprouting and intussusception (Fig. 4). Capillary sprouting, the most commonly described form of angiogenesis, includes endothelial cell proliferation and migration away from an existing vessel. Intussusception, also referred to as capillary splitting, can be conceptualized as the pinching off and separating of an existing vessel into two vessels. During this process, endothelial cells extend intralumenal filopodia that separate the opposite sides of the lumen and lead the actual splitting process. Work by Kelly-Goss

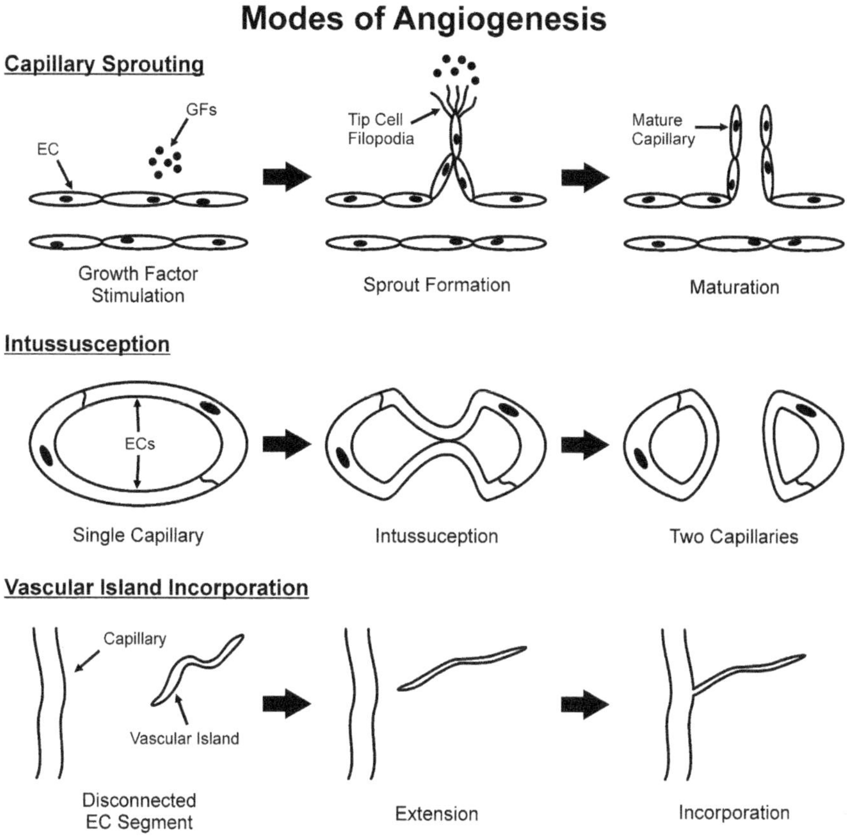

Modes of Angiogenesis

Capillary Sprouting

Growth Factor Stimulation

Sprout Formation

Maturation

Intussusception

Single Capillary

Intussuception

Two Capillaries

Vascular Island Incorporation

Disconnected EC Segment

Extension

Incorporation

Figure 4. Endothelial cell dynamics associated with three models of angiogenesis. A) Capillary sprouting. Endothelial cells (ECs) stimulated by a local growth factor (GF) gradient proliferate and migrate away from an existing host vessel. A tip cell extends filopodia and guides sprout extension into the avascular space. B) Intussusception (capillary splitting). Endothelial cells extend across the lumen of a single capillary leading to its separation into two capillaries. C) Vascular island incorporation. Vascular islands, defined as endothelial cell segments initially disconnected from a network, undergo proliferation, extension, and incorporation to nearby vessels.

et al. suggests a third mode, vascular island incorporation (Kelly-Goss et al. 2012), might also exist (Fig. 4). In this mode, endothelial cell segments initially disconnected from a network undergo proliferation, extension and connection to nearby vessels (Kelly-Goss et al. 2012, Stapor et al. 2012).

The relative contribution of capillary sprouting, intussusception and vascular island incorporation to cumulative angiogenic growth is difficult to assess. Quantification of angiogenesis in a microvascular network post stimulation is usually limited by the analyses of images captured at a specific time point (Fig. 2) and an increase in the number of capillaries could be attributed to each mode of angiogenesis. The majority of the evidence that links shear stress and angiogenesis is based on *in vitro* experiments performed on endothelial cells far removed from the context of the *in vivo* scenario. A few studies that do link shear stress to angiogenesis *in vivo* show that shear stress increases the number of capillaries (Ichioka et al. 1997, Milkiewicz et al. 2001, Nasu et al. 1999, Zhou et al. 1998). While these studies motivate the importance of understanding the role of shear stress, the interpretation of their results is limited by the challenge of decoupling the shear stress stimuli from other potential factors and the difficulty in spatially correlating changes in shear stress to specific endothelial cell dynamics.

Endothelial cell dynamics during capillary sprouting

Capillary sprouting is the most commonly studied form of angiogenesis and, as mentioned above, involves endothelial cell proliferation and migration. Capillary sprouts typically originate from existing capillaries or venules in response to a local growth factor gradient. The presence of blood flow in capillary sprouts (Figs. 3 and 5; Guerreiro-Lucas et al. 2008) prompts the discussion of whether shear stresses locally influence endothelial cell behavior.

As a new sprout is initiated, endothelial cells loosen their adhesion to neighboring cells and degrade their basement membrane. Concurrently, a tip cell phenotype is thought to emerge (Eichmann et al. 2005, Eilken and Adams 2010). Tip cells are specialized endothelial cells that extend numerous filopodia and help guide the directional migration of a sprout. As endothelial cells invade an avascular tissue space, they actively degrade the surrounding extracellular matrix to permit migration and assemble new basement membrane to stabilize the sprout. Stabilization is also influenced by the recruitment of pericytes (Gerhardt and Betsholtz 2003). The tip cell is spatially followed by a phenotypically different endothelial cell type, named stalk cells, which do not extend filopodia. Stalk cells undergo proliferation and can be viewed as the pushing force responsible for sprout elongation (Gerhardt et al. 2003). Tip cells, on the other hand, express high levels of VEGFR-2 enabling them to be more sensitive to local growth factor gradients (Eilken and Adams 2010) and can be viewed as the cells that guide sprout direction. Most of the evidence for the molecular mechanisms involved in tip cell versus stalk cell dynamics has been generated from time lapse observation in developing zebrafish and observations made at discrete time points during post natal retinal angiogenesis in mice. While future studies will be required to generalize these characteristics to all adult scenarios, the concept of tip cells indicates the specialization of endothelial cell phenotypes along a single capillary sprout.

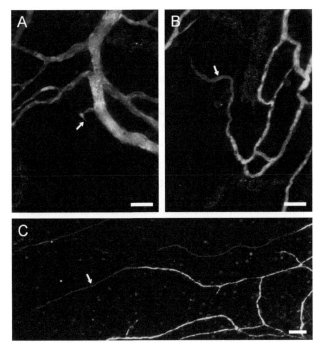

Figure 5. Representative images that support the presence of lumens along capillary sprouts in adult rat mesenteric microvascular networks. Antibody labeling against PECAM (red) identified all endothelial cells across the hierarchy of adult rat mesenteric microvascular networks. Prior to tissue harvesting, vessel lumens were identified via intra-femoral vein injection of a FITC-conjugated fixable 40 kDa dextran (green). Networks were stimulated by mast cell degranulation. Arrows indicate capillary sprouts. A) Example of a short capillary sprout off a venule. B) Example of a capillary sprout off an existing capillary. C) Example of a longer capillary sprout off a capillary. Faint PECAM labeling in (C) identifies blunt ended lymphatic vessels. In some cases the presence of the injected dextran extended the length of the sprout and in some cases stopped before the end of the sprout. Scale bars = 50 μm (A), 50 μm (B), 100 μm (C).

Color image of this figure appears in the color plate section at the end of the book.

Now let's consider the hypothesis–low levels of shear stress cause stalk cell proliferation. This hypothesis is supported by *in vitro* experiments (Davies 1995, Kaunas et al. 2011, Skalak and Price 1996). However, an *in vivo* capillary sprout scenario suggests that tip cells would also experience low shear stress magnitudes and, thus, also undergo proliferation. A conceptual issue is that proliferation is not typically associated with the tip cell phenotype (Gerhardt et al. 2003). In addition, endothelial cell proliferation during angiogenesis can be observed along vessel segments upstream and downstream of capillary sprout locations and these vessels experience relatively higher levels of shear stress (Murfee et al. 2006, Stapor et al. 2011). Anderson et al. discovered that endothelial cells along capillary sprouts downregulate their expression of CD36, a thrombospondin receptor (Anderson et al. 2008). Since thrombospondin is considered to anti-angiogenic, this phenotypic switch would make sense. Anderson et al. further demonstrated *in vitro* that reduced shear stress regulates this phenotype, yet endothelial cells along capillary segments that lack CD36 expression *in vivo* possibly experience high and low shear stress magnitudes. Consideration of stalk cell proliferation or the CD36 phenotypic switch during angiogenesis emphasizes the need to identify the actual shear stress magnitudes along a capillary sprout.

The effects of shear stress on endothelial cell dynamics

In vitro flow chamber experiments, motivated by understanding the causes of endothelial dysfunction at sites of atherosclerotic plaque formation, have demonstrated that shear stress can influence endothelial cell dynamics associated with capillary sprouting (Davies 1995, Skalak and Price 1996). Typical parallel plate flow chamber studies involve passing media over a two-dimensional monolayer of endothelial cells on a rigid substrate. Endothelial cell responses to shear stress include proliferation, migration, cytoskeletal reorganization, cell-matrix adhesion, protein phosphorylization and growth factor production (Davies 1995, Helmke 2005, Skalak and Price 1996). Shear stresses as low as 0.2 dyne/cm^2 have been shown to activate endothelial ion channels (Cooke et al. 1991). This example indicates the need to determine specific thresholds for each potential response. Insight into the role of shear stress on capillary sprouting can be gained by identifying which responses are associated with the capillary sprout process and spatially mapping these responses to local shear stress magnitudes experienced by the endothelial cells.

As mentioned in the previous sections, endothelial cells exhibit structural and phenotypic differences at different locations within a microvascular network. These differences, along with apparent vessel specific flow conditions, motivate the need for more site specific endothelial cell experiments, in which endothelial cells derived from arterial, venous, and capillary locations, are exposed to relevant shear stress magnitudes. A need also exists for experimental methods to bridge the gap between *in vivo* and *in vitro* environments. In the past few years, a modification of the common parallel plate flow chamber has offered such a tool (Bayless and Davis 2003, Kang et al. 2008, Ueda et al. 2004). For this model endothelial cells are seeded onto the surface of a three-dimensional collagen matrix rather than a rigid surface. Replicating the scenario of endothelial sprouting away from the lumen of a vessel, endothelial cells are simultaneously exposed to fluid flow and allowed to migrate into the three-dimensional matrix. Using this system, Kang et al. demonstrated that a wall shear stress equal to 3 dyne/cm^2 promotes the invasion of human microvascular endothelial cells into the collagen matrix. Interestingly, the presence of a chemotactant was necessary for the shear stress mediated sprouting. Sprouting into the three-dimensional matrix was also dependent on shear stress magnitude, as the number of invading cells was greatest at intermediate levels of shear stress, approximately 5 dynes/cm^2 (compared to 0.12 and 12 dynes/cm^2). These results emphasize the necessity for future experiments to examine the integrated effects of mechanical forces and local biochemical factors. They also suggest that specific shear stress magnitudes can be a positive modulator of capillary sprouting.

Work by Song and Munn additionally support the need to define the interrelationships between shear stress and biochemical signals (Song and Munn 2011). Using a microfluidic analog of capillary sprouting, the authors demonstrated that a wall shear stress of 3 dyne/cm^2 attenuated VEGF induced endothelial sprouting. In their model, endothelial cells were initially seeded along the sides of a microchannel fabricated from PDMS using soft lithography techniques. Gaps along the channel wall permitted endothelial cell migration into a three-dimensional extracellular collagen matrix. Similar to the results from Kang et al., shear stress alone did not induce capillary sprouting. However, in contrast to the results from Kang et al., a shear stress equal to 3 dyne/cm^2 was a negative regulator of capillary sprouting. While an explanation for these different results is unclear, the future use of both models serves to advance our understanding of the exact role of shear stress on endothelial cell dynamics involved in the initial sprout formation of host vessels.

The presence of shear stress along a capillary sprout

Application of the results from *in vitro* studies of shear stress effects on endothelial cell behavior necessitates knowing the shear stress magnitudes present along a capillary sprout. Computational fluid dynamic models have provided valuable insight about flow profiles within capillaries. For example, work by Skalak and coworkers predicted the effects of rigid and elastic spheres on local flow in straight tubes (Sugihara-Seki and Skalak 1988, Tözeren and Skalak 1978, Wang and Skalak 1969). Secomb and co-workers investigated how red blood cells deform during flow through a cylindrical vessel and at bifurcations (Barber et al. 2008, Secomb et al. 1986). More recently, Xiong and Zhang simulated the shear stress distribution along capillaries, taking into account the flow of deformable red blood cells (Xiong and Zhang 2010). Extension of these computational approaches to capillary sprout scenarios will help address the issue of whether shear stresses above physiologically relevant thresholds are be present along a capillary sprout.

Work by Stapor et al. recently attempted to provide the first estimates of shear stresses in a sprout (Fig 6; Stapor et al. 2011). Based on average geometrical measurements from real adult rat mesenteric networks, the authors simulated flow through capillary sprouts for three vessel wall cases: 1) non-permeable, 2) uniformly permeable, and 3) permeable at distinct locations. The open or permissive slots for the third case represented endothelial cell clefts. For plasma flow through a blind-ended sprout (6 μm in diameter and 50 μm in length) originating from a 11 μm host vessel,

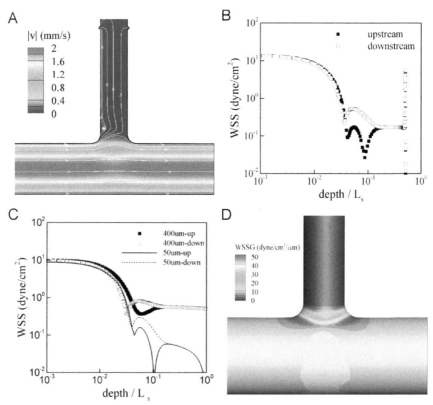

Figure 6. Computational predictions of hemodynamics within a capillary sprout. A) Plasma velocity profile for a 50 μm long, 6 μm diameter capillary sprout permeable at endothelial cell cleft regions. B) Predicted wall shear stress magnitudes along the length (depth) of the sprout (A). The shear stress distribution was normalized by the length of the sprout, "L_s", and shown for the upstream and downstream walls of the sprout. C) Wall shear stress distribution along the walls of 50 μm and 400 μm long capillary sprouts with a 6 μm diameter uniformly permeable walls. Notice the effect of sprout length on the elevated shear stress magnitudes along the length of the sprout. D) Wall shear stress gradient distribution for a 50 μm long, 6 μm diameter capillary sprout with uniform wall permeability. High shear stress gradients appear at the sprout entrance. Images were adapted from Stapor et al. 2011.

Color image of this figure appears in the color plate section at the end of the book.

shear stresses were locally increased at the sprout entrance and dropped below 0.2 dyne/cm² within approximately 10% of the sprout length for the non-permeable and uniformly permeable cases. The presence of endothelial cell clefts caused shear stress magnitudes as high as 5.9 dyne/cm² at the sites of fluid outflow from the sprout into the interstitial space. For longer sprouts (400 μm in length) with uniform permeability, shear stresses were predicted to be greater than 0.2 dyne/cm² along the full length of the sprout. This effect of sprout length can be explained by the proportional increase intransmural flux due to the increase in wall surface area, and indicates the potential influence wall permeability might have on local shear stress experienced by an endothelial cell. Considering a stationary red blood cell in a capillary further suggested the potential for an increase in local shear stress magnitudes above physiologically relevant levels.

The results by Stapor et al. suggest that endothelial cells along a capillary sprout can experience both relatively low and high stress magnitudes. Within the lumen of a uniformly permeable short sprout, an endothelial cell might experience a low shear stress magnitude that promotes proliferation. For later stages of the process when capillary sprouts are longer, endothelial cells could experience a shear stress high enough to limit proliferation and, based on the work by Song and Munn, attenuate further sprout elongation. In another case based on the application of the work by Kang et al., endothelial cell regulation of junctional adhesions could serve to positively regulate sprouting. Local maximum shear stress magnitudes at more open cell-cell junctions would promote endothelial cell sprout formation. These attempts to critically apply experimental evidence regarding how endothelial cells respond to shear stresses based on *in vitro* models highlight the need for interdisciplinary studies to determine the actual shear stress magnitudes experienced by endothelial cells during the capillary sprout process.

Summary: Unanswered questions

The presence of shear stress throughout a network implicates its roles in vasculogenesis, arteriogenesis, and angiogenesis. However, investigations aimed at understanding the roles of shear stress depend on the ability to identify the relevant shear stress changes at specific locations within a network. In the context of capillary sprouting, the shear stresses could be due to blood flow within the sprout lumen, transmural flow through endothelial cell clefts, or even local interstitial flow (Fig. 7). Yet without knowing the shear stress magnitudes, the application of results derived from reductionist *in vitro* flow chamber experiments is difficult. While model systems that more closely replicate *in vivo* scenarios provide new insight regarding the integrative effects of shear stress on endothelial cell dynamics, the questions remain: *What are the wall shear stress magnitudes experienced*

Flow Along a Capillary Sprout

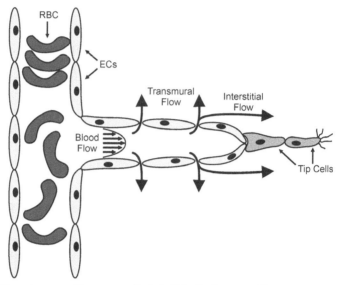

Figure 7. Flow along a capillary sprout. Endothelial cells along a capillary sprout can experience fluid flow induced wall shear stress due to flow through the capillary sprout lumen (blood flow), flow between cell-cell junctions (transmurall flow), or interstitial flow. Identification of shear stress magnitudes experienced by endothelial cells requires a detailed analysis of endothelial cell structures and associated flow profiles.

by endothelial cells during capillary sprouting? and *How do these wall shear stress magnitudes spatially correlate with endothelial cell function?* Combined experimental and computational studies aimed at providing answers to these questions will offer a context in which to apply what is already known regarding endothelial cell responses to shear stress. Finally, this chapter has focused on the role of shear stress in capillary sprouting. Wall shear stress gradients have also been implicated as a positive regulator of endothelial cell migration and proliferation. While less attention has been given to a role for shear stress gradients in capillary sprouting, the potential for high gradients along a sprout (Fig. 6) suggests that cell-gradient relationships warrant consideration.

Acknowledgements

The author would like to acknowledge Peter Stapor and Rick Sweat for their contributions to the figures and numerous discussions. The author would also like to acknowledge Sadegh Azimi, Molly Kelly-Goss and Elizabeth Townsend for their feedback on the manuscript. This work was supported by the Board of Regents of the State of Louisiana LEQSF (2009-12)-RD-A-

19 (PI: W. L. Murfee) and by the Tulane Hypertension and Renal Center of Excellence—NIH grant P20RR017659-10 (PI: L. Gabriel Navar).

References

Adams, R.H., G.A. Wilkinson, C. Weiss, F. Diella, N.W. Gale, U. Deutsch, W. Risau and R. Klein. 1999. Roles of ephrinB ligands and EphB receptors in cardiovascular development: demarcation of arterial/venous domains, vascular morphogenesis, and sprouting angiogenesis. *Genes Dev.* 13: 295–306.

Anderson, C.R., N.E. Hastings, B.R. Blackman and R.J. Price. 2008. Capillary sprout endothelial cells exhibit a CD36 low phenotype: regulation by shear stress and vascular endothelial growth factor-induced mechanism for attenuating anti-proliferative thrombospondin-1 signaling. *Am. J. Pathol.* 173: 1220–1228.

Asahara, T., T. Murohara, A. Sullivan, M. Silver, R. van der Zee, T. Li, B. Witzenbichler, G. Schatteman and J.M. Isner. 1997. Isolation of putative progenitor endothelial cells for angiogenesis. *Science* 275: 964–967.

Au, P., J. Tam, D. Fukumura and R.K. Jain. 2008. Bone marrow-derived mesenchymal stem cells facilitate engineering of long-lasting functional vasculature. *Blood* 111: 4551–4558.

Barber, J.O., J.P. Alberding, J.M. Restrepo and T.W. Secomb. 2008. Simulated two-dimensional red blood cell motion, deformation, and partitioning in microvessel bifurcations. *Ann. Biomed. Eng.* 36: 1690–1698.

Bayless, K.J. and G.E. Davis. 2003. Sphingosine-1-phosphate markedly induces matrix metalloproteinase and integrin-dependent human endothelial cell invasion and lumen formation in three-dimensional collagen and fibrin matrices. *Biochem. Biophys. Res. Commun.* 312: 903–913.

Carmeliet, P. 2004. Manipulating angiogenesis in medicine. *J. Intern. Med.* 255: 538–561.

Cooke, J.P., E. Rossitch Jr., N.A. Andon, J. Loscalzo and V.J. Dzau. 1991. Flow activates an endothelial potassium channel to release an endogenous nitrovasodilator. *J. Clin. Invest.* 88: 1663–1671.

Curry, F.E. and R.H. Adamson. 2012. Endothelial glycocalyx: permeability barrier and mechanosensor. *Ann. Biomed. Eng.* 40: 828–839.

Davies, P.F. 1995. Flow-mediated endothelial mechanotransduction. *Physiol. Rev.* 75: 519–560.

Eichmann, A., F. Le Noble, M. Autiero and P. Carmeliet. 2005. Guidance of vascular and neural network formation. *Curr. Opin. Neurobiol.* 15: 108–115.

Eilken, H.M. and R.H. Adams. 2010. Dynamics of endothelial cell behavior in sprouting angiogenesis. *Curr. Opin. Cell Biol.* 22: 617–625.

Gale, N.W., P. Baluk, L. Pan, M. Kwan, J. Holash, T.M. DeChiara, D.M. McDonald and G.D. Yancopoulos. 2001. Ephrin-B2 selectively marks arterial vessels and neovascularization sites in the adult, with expression in both endothelial and smooth-muscle cells. *Dev. Biol.* 230: 151–160.

Gerhardt, H. and C. Betsholtz. 2003. Endothelial-pericyte interactions in angiogenesis. *Cell Tissue Res.* 314: 15–23.

Gerhardt, H., M. Golding, M. Fruttiger, C. Ruhrberg, A. Lundkvist, A. Abramsson, M. Jeltsch, C. Mitchell, K. Alitalo, D. Shima and C. Betsholtz. 2003. VEGF guides angiogenic sprouting utilizing endothelial tip cell filopodia. *J. Cell Biol.* 161: 1163–1177.

Guerreiro-Lucas, L.A., S.R. Pop, M.J. Machado, Y.L. Ma, S.L. Waters, G. Richardson, K. Saetzler, O.E. Jensen and C.A. Mitchell. 2008. Experimental and theoretical modelling of blind-ended vessels within a developing angiogenic plexus. *Microvasc. Res.* 76: 161–168.

Heil, M. and W. Schaper. 2004. Influence of mechanical, cellular, and molecular factors on collateral artery growth (arteriogenesis). *Circ. Res.* 95: 449–458.

Hellstrom, M., M. Kalen, P. Lindahl, A. Abramsson and C. Betsholtz. 1999. Role of PDGF-B and PDGFR-beta in recruitment of vascular smooth muscle cells and pericytes during embryonic blood vessel formation in the mouse. *Development* 126: 3047–3055.

Helmke, B.P. 2005. Molecular control of cytoskeletal mechanics by hemodynamic forces. *Physiology (Bethesda)* 20: 43–53.

Herzog, Y., C. Kalcheim, N. Kahane, R. Reshef and G. Neufeld. 2001. Differential expression of neuropilin-1 and neuropilin-2 in arteries and veins. *Mech. Dev.* 109: 115–119.

Ichioka, S., M. Shibata, K. Kosaki, Y. Sato, K. Harii and A. Kamiya. 1997. Effects of shear stress on wound-healing angiogenesis in the rabbit ear chamber. *J. Surg. Res.* 72: 29–35.

Kang, H., K.J. Bayless and R. Kaunas. 2008. Fluid shear stress modulates endothelial cell invasion into three-dimensional collagen matrices. *Am. J. Physiol. Heart Circ. Physiol.* 295: H2087–97.

Kaunas, R., H. Kang and K.J. Bayless. 2011. Synergistic Regulation of Angiogenic Sprouting by Biochemical Factors and Wall Shear Stress. *Cell. Mol. Bioeng.* 4: 547–559.

Kelly-Goss, M.R., E.R. Winterer, P.C. Stapor, M. Yang, R.S. Sweat, W.B. Stallcup, G.W. Schmid-Schonbein and W.L. Murfee. 2012. Cell proliferation along vascular islands during microvascular network growth. *BMC Physiol.* 12: 7.

Kinnaird, T., E. Stabile, M.S. Burnett, C.W. Lee, S. Barr, S. Fuchs and S.E. Epstein. 2004. Marrow-derived stromal cells express genes encoding a broad spectrum of arteriogenic cytokines and promote *in vitro* and *in vivo* arteriogenesis through paracrine mechanisms. *Circ. Res.* 94: 678–685.

Lawson, N.D. and B.M. Weinstein. 2002. Arteries and veins: making a difference with zebrafish. *Nat. Rev. Genet.* 3: 674–682.

le Noble, F., D. Moyon, L. Pardanaud, L. Yuan, V. Djonov, R. Matthijsen, C. Breant, V. Fleury and A. Eichmann. 2004. Flow regulates arterial-venous differentiation in the chick embryo yolk sac. *Development* 131: 361–375.

Li, S., N.F. Huang and S. Hsu. 2005. Mechanotransduction in endothelial cell migration. *J. Cell. Biochem.* 96: 1110–1126.

Lyden, D., K. Hattori, S. Dias, C. Costa, P. Blaikie, L. Butros, A. Chadburn, B. Heissig, W. Marks, L. Witte, Y. Wu, D. Hicklin, Z. Zhu, N.R. Hackett, R.G. Crystal, M.A. Moore, K.A. Hajjar, K. Manova, R. Benezra and S. Rafii. 2001. Impaired recruitment of bone-marrow-derived endothelial and hematopoietic precursor cells blocks tumor angiogenesis and growth. *Nat. Med.* 7: 1194–1201.

Majka, S.M., K.A. Jackson, K.A. Kienstra, M.W. Majesky, M.A. Goodell and K.K. Hirschi. 2003. Distinct progenitor populations in skeletal muscle are bone marrow derived and exhibit different cell fates during vascular regeneration. *J. Clin. Invest.* 111: 71–79.

Maul, T.M., D.W. Chew, A. Nieponice and D.A. Vorp. 2011. Mechanical stimuli differentially control stem cell behavior: morphology, proliferation, and differentiation. *Biomech. Model. Mechanobiol.* 10: 939–953.

Milkiewicz, M., M.D. Brown, S. Egginton and O. Hudlicka. 2001. Association between shear stress, angiogenesis, and VEGF in skeletal muscles *in vivo*. *Microcirculation* 8: 229–241.

Moyon, D., L. Pardanaud, L. Yuan, C. Breant and A. Eichmann. 2001. Plasticity of endothelial cells during arterial-venous differentiation in the avian embryo. *Development* 128: 3359–3370.

Murfee, W.L., M.R. Rehorn, S.M. Peirce and T.C. Skalak. 2006. Perivascular cells along venules upregulate NG2 expression during microvascular remodeling. *Microcirculation* 13: 261–273.

Nasu, R., H. Kimura, K. Akagi, T. Murata and Y. Tanaka. 1999. Blood flow influences vascular growth during tumour angiogenesis. *Br. J. Cancer* 79: 780–786.

Obi, S., K. Yamamoto, N. Shimizu, S. Kumagaya, T. Masumura, T. Sokabe, T. Asahara and J. Ando. 2009. Fluid shear stress induces arterial differentiation of endothelial progenitor cells. *J. Appl. Physiol.* 106: 203–211.

O'Neill, T.J., B.R. Wamhoff, G.K. Owens and T.C. Skalak. 2005. Mobilization of bone marrow-derived cells enhances the angiogenic response to hypoxia without transdifferentiation into endothelial cells. *Circ. Res.* 97: 1027–1035.

Peirce, S.M. and T.C. Skalak. 2003. Microvascular remodeling: a complex continuum spanning angiogenesis to arteriogenesis. *Microcirculation* 10: 99–111.

Pries, A.R., T.W. Secomb and P. Gaehtgens. 1995. Design principles of vascular beds. *Circ. Res.* 77: 1017–1023.

Secomb, T.W., R. Skalak, N. Ozkaya and J.F. Gross. 1986. Flow of axisymmetric red blood cells in narrow capillaries. *J. Fluid Mech.* 163: 405–423.

Shin, D., G. Garcia-Cardena, S. Hayashi, S. Gerety, T. Asahara, G. Stavrakis, J. Isner, J. Folkman, M.A. Gimbrone Jr. and D.J. Anderson. 2001. Expression of ephrinB2 identifies a stable genetic difference between arterial and venous vascular smooth muscle as well as endothelial cells, and marks subsets of microvessels at sites of adult neovascularization. *Dev. Biol.* 230: 139–150.

Skalak, T.C. and R.J. Price. 1996. The role of mechanical stresses in microvascular remodeling. *Microcirculation* 3: 143–165.

Song, J.W. and L.L. Munn. 2011. Fluid forces control endothelial sprouting. PNAS 108(37): 15342–7.

Stapor, P.C., M.S. Azimi, T. Ahsan and W.L. Murfee. 2013. An angiogenesis model for investigating multi-cellular interactions across intact microvascular networks. *Am. J. Physiol. Heart Circ. Physiol.* 304(2): H235–45.

Stapor, P.C., W. Wang, W.L. Murfee and D.B. Khismatullin. 2011. The Distribution of Fluid Shear Stresses in Capillary Sprouts. *Cardiovasc. Eng. Technol.* 2: 124–136.

Sugihara-Seki, M. and R. Skalak. 1988. Numerical study of asymmetric flows of red blood cells in capillaries. *Microvasc. Res.* 36: 64–74.

Tarbell, J.M. and E.E. Ebong. 2008. The endothelial glycocalyx: a mechano-sensor and -transducer. *Sci. Signal.* 1: pt8.

Tözeren, H. and R. Skalak. 1978. The steady flow of closely fitting incompressible elastic spheres in a tube. *J. Fluid Mech.* 87: 1–16.

Tzima, E., M. Irani-Tehrani, W.B. Kiosses, E. Dejana, D.A. Schultz, B. Engelhardt, G. Cao, H. DeLisser and M.A. Schwartz. 2005. A mechanosensory complex that mediates the endothelial cell response to fluid shear stress. *Nature* 437: 426–431.

Ueda, A., M. Koga, M. Ikeda, S. Kudo and K. Tanishita. 2004. Effect of shear stress on microvessel network formation of endothelial cells with *in vitro* three-dimensional model. *Am. J. Physiol. Heart Circ. Physiol.* 287: H994–1002.

Van Gieson, E.J., W.L. Murfee, T.C. Skalak and R.J. Price. 2003. Enhanced smooth muscle cell coverage of microvessels exposed to increased hemodynamic stresses *in vivo*. *Circ. Res.* 92: 929–936.

Villa, N., L. Walker, C.E. Lindsell, J. Gasson, M.L. Iruela-Arispe and G. Weinmaster. 2001. Vascular expression of Notch pathway receptors and ligands is restricted to arterial vessels. *Mech. Dev.* 108: 161–164.

Wang, H. and R. Skalak. 1969. Viscous flow in a cylindrical tube containing a line of spherical particles. *J. Fluid Mech.* 38: 75–96.

Wang, D.H. and R.L. Prewitt. 1993. Alterations of mature arterioles associated with chronically reduced blood flow. *Am. J. Physiol.* 264: H40–4.

Wang, H.U., Z.F. Chen and D.J. Anderson. 1998. Molecular distinction and angiogenic interaction between embryonic arteries and veins revealed by ephrin-B2 and its receptor Eph-B4. *Cell* 93: 741–753.

Xiong, W. and J. Zhang. 2010. Shear stress variation induced by red blood cell motion in microvessel. *Ann. Biomed. Eng.* 38: 2649–2659.

Yamamoto, K., T. Takahashi, T. Asahara, N. Ohura, T. Sokabe, A. Kamiya and J. Ando. 2003. Proliferation, differentiation, and tube formation by endothelial progenitor cells in response to shear stress. *J. Appl. Physiol.* 95: 2081–2088.

Zhong, T.P., S. Childs, J.P. Leu and M.C. Fishman. 2001. Gridlock signalling pathway fashions the first embryonic artery. *Nature* 414: 216–220.

Zhou, A., S. Egginton, O. Hudlicka and M.D. Brown. 1998. Internal division of capillaries in rat skeletal muscle in response to chronic vasodilator treatment with alpha1-antagonist prazosin. *Cell Tissue Res.* 293: 293–303.

Ziegelhoeffer, T., B. Fernandez, S. Kostin, M. Heil, R. Voswinckel, A. Helisch and W. Schaper. 2004. Bone marrow-derived cells do not incorporate into the adult growing vasculature. *Circ. Res.* 94: 230–238.

Endothelial Cell Adhesion Molecules and Drug Delivery Applications

Daniel Serrano and *Silvia Muro**

Introduction

The vascular endothelium is an extensive and actively functional surface lining the luminal side of all blood vessels throughout the body. It serves as a regulatory interface between the blood and tissues, and is an important contributor to controlling vascular physiology and overall maintenance of the body homeostasis (Aird 2008). Vascular endothelial cells control the transport of fluids, molecules, and blood cells between the circulation and tissues, contribute to generating vasoactive substances that modulate the vascular tone, regulate the hemostatic balance by producing either pro- or anti-thrombotic mediators, and participate in innate and adaptive immunity during infection and tissue injury (Aird 2008). Therefore, the endothelium represents an important target for clinical intervention as well as a gate for accessing other tissues in the body, and numerous endothelial surface markers are being explored from the perspective of diagnostic and therapeutic targeting (Muzykantov 2011).

University of Maryland College Park, 5115 Plant Sciences Building, College Park, MD 20742.
*Corresponding author: muro@umd.edu

A key function of the endothelial lining, which will be reviewed on this chapter, is that of contributing to extravasation of leukocytes from the bloodstream into tissues at areas of inflammation (Ley et al. 2007). This process takes place in capillaries and, most typically, postcapillary venules where the blood shear stress is relatively low and only a thin layer of endothelial cells separates the circulation from the tissue parenchyma. During this process, inflammatory cells are recruited to the vasculature in the affected areas via a series of interactions, including leukocyte tethering and rolling, followed by firm adhesion and lateral crawling on the endothelial surface, culminating with transendothelial migration into the tissue (Ley et al. 2007). These events are governed by low and high affinity interactions between complementary adhesion molecules on the endothelial and the leukocyte surfaces (Figs. 1–2). First, endothelial generation of vasoactive

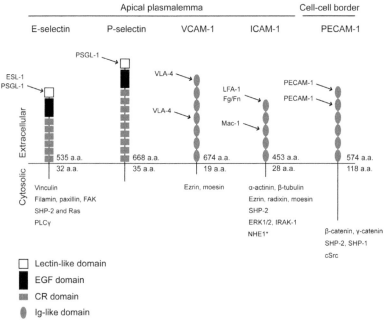

Figure 1. Structural features of leukocyte-interacting endothelial cell adhesion molecules. The main endothelial cell adhesion molecules (CAMs) involved in leukocyte extravasation are displayed on the endothelial plasmalemma. Based on approximate lengths, the extracellular and cytoplasmic domains of these molecules are drawn to scale with respect to each other; the number of amino acids (a.a.) for their extracellular and cytosolic domains are shown above and below, respectively, the cell membrane (horizontal black line). E- and P-selectins contain three types of domains on their extracellular region (rectangles), while immunoglobulin-(Ig)-like CAMs contain one type of domain (ovals). Approximate binding sites for major extracellular ligands are noted with arrows. Representative molecular interactions with endothelial signaling or cytoskeletal partners are listed at the cytosolic region of each adhesion molecule. *NHE1 is a transmembrane (not cytosolic) molecule.

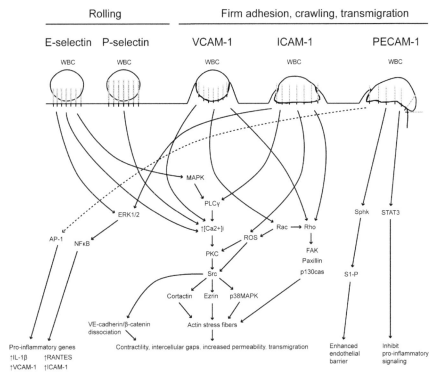

Figure 2. Signaling pathways associated with leukocyte-interacting endothelial cell adhesion molecules. Leukocytes are shown binding to each of the main endothelial cell adhesion molecules (CAMs) involved in white blood cell (WBC) extravasation. Key endothelial players in this event are: (1) E- and P-selectin, involved in the initial stage of rolling interactions via weak bonds of high association-dissociation constants and (2) Ig-like CAMs (ICAM-1, VCAM-1 and PECAM-1), involved in the stages of firm adhesion, crawling, and transendothelial migration via strong bonds and signaling events linked to structural changes in endothelial cells and leukocytes. These pathways often intersect with expression of endothelial inflammatory mediators and rearrangements of the actin cytoskeleton. This reorganization is characterized by assembly of actin stress fibers and special plasmalemma domains, which lead to formation of leukocyte docking structures on the endothelial surface (depicted during the steps of firm adhesion and crawling), mainly mediated by VCAM-1 and ICAM-1. In addition, these events lead to transient disassembly of endothelial junctions, helping in leukocyte extravasation across the endothelial barrier. PECAM-1, found associated with cell-cell borders, might allow for fine-tuning of the endothelial barrier opening, as some of its signaling events are linked to maintenance of cell-cell junctions. Although leukocytes are depicted binding individually to each adhesion molecule for simplicity, physiologically, the binding interactions and resulting signaling shown here are known to overlap.

compounds results in smooth muscle relaxation at the inflamed area, leading to vasodilatation and enhanced blood flow to the site of injury. This augments the arrival of leukocytes to the area, which can initially interact with endothelial cells through low affinity receptors known as selectins

(E- and P-selectin). These endothelial adhesion molecules recognize P-selectin glycoprotein ligand 1 (PSGL1) and other carbohydrate-based ligands on the leukocyte surface. The combined effects of blood flow and leukocyte binding to selectins result in their rolling on the endothelial surface. Then, chemotactic agents produced by endothelial cells facilitate firm adhesion by activating integrins on the leukocyte surface and these integrins bind to high affinity co-receptors on the endothelial counterpart. The most commonly studied interactions with respect to firm adhesion are those of leukocyte integrins $\alpha_4\beta_1$ (very late antigen 4, VLA-4) and $\alpha_L\beta_2$ (lymphocyte function-associated antigen 1, LFA1) with the endothelial co-receptors vascular cell adhesion molecule (VCAM)-1 and intercellular adhesion molecule (ICAM)-1, respectively (Ley et al. 2007). These interactions support not only leukocyte arrest on the endothelial surface but also the "crosstalk" between these two cells types, which regulates subsequent migratory events. This contributes to mediate leukocyte lateral crawling on the endothelium and "probing" for sites suitable for transmigration, which involves extension of leukocyte protrusions against the endothelial plasmalemma and simultaneous formation of leukocyte docking structures and transmigratory cups on the endothelium (Barreiro et al. 2002, Carman 2009). Finally, adherent leukocytes migrate across the endothelial lining. This can occur via the paracellular pathway, where leukocytes migrate through the space between adjacent endothelial cells that have transiently disassembled their cell junctions, or via the transcellular pathway, where the leukocyte moves across the body of an individual endothelial cell in a process that involves fusion of endothelial vesicles into a so-called transcellular pore (Yang et al. 2005, Millán et al. 2006, Carman 2009). Endothelial adhesion molecules that mediate firm leukocyte adhesion and also contribute to these late steps of extravasation include VCAM-1 and, mainly, ICAM-1, but also additional endothelial partners such as PECAM-1 and CD99 (Ley et al. 2007).

Regulation of the expression patterns and functions of all these endothelial markers represents a complex and dynamic process, with these molecules serving not only as adhesive surfaces permitting leukocyte anchoring to the endothelium, but also as signaling platforms contributing to multiple physiological and pathological processes. Indeed, endothelial cell adhesion molecules associate closely with a variety of diseases, including inflammatory and autoimmune conditions, infections and septic shock, ischemia-reperfusion injury, atherosclerosis and thrombosis, metabolic and genetic disorders, asthma and acute lung injury, cancer and tumor metastasis, and many others (Aird 2008). As a consequence, endothelial molecules involved in leukocyte adhesion have been proposed as targets for clinical intervention in these pathologies, which includes the use of agents that block their function, as well as the use of these molecules as targets to deliver imaging or therapeutic compounds to pathologically altered endothelium (Muzykantov 2011). Although less explored, targeting to these molecules could potentially enhance transport of therapeutics across the vascular lining, as some of these markers contribute to leukocyte extravasation and, thus, could be manipulated to modulate the para- and/ or trans-cellular transport routes that separate the bloodstream and the tissue compartment (Ghaffarian et al. 2012, Muro 2012).

In this context, blocking as well as targeting of imaging agents or drugs to endothelial cell adhesion molecules can be achieved by using affinity moieties addressed to these markers, such as antibodies and their fragments, proteins and peptides, and aptamers (Muzykantov 2011). These motifs can be attached to the desired cargo by direct synthesis or a variety of chemical conjugation modalities (e.g., through covalent linkage or interactive biological molecules), or coupling can be indirect by utilizing drug delivery systems that display said targeting motifs. In particular, design of drug delivery systems has become a very active area of research holding great promise, as these systems permit modulation of a variety of parameters pertaining to efficacy and safety of drug delivery, including not only targeting but also improving solubility in body fluids, circulation half-life, clearance, and degradation of treatment agents (Langer 1998, Duncan 2003, Panyam and Labhasetwar 2003). These systems can be fabricated using biological, synthetic, or semi-synthetic materials (Moghimi et al. 2001, El-Sayed et al. 2003, Stayton et al. 2005, Torchilin 2006). Liposomes are phospholipid-shelled capsules that can be loaded with hydrophilic compounds in their inner lumen or hydrophobic drugs within the lipid bilayer (Torchilin 2006). Polymerosomes are the polymer analog of liposomes, formed from block co-polymers whose polymer chains consist of a hydrophobic and a hydrophilic moiety (Discher and Eisenberg 2002). Polymers can also be formulated as linear or branched structures. For instance, dendrimers are tree-like hyperbranched carriers with a very high surface-to-volume ratio (El-Sayed et al. 2003). Nanoparticles of different sizes and shapes can also be fabricated to display solid, hollow, or multi-porous structures (Moghimi et al. 2001, Panyam and Labhasetwar 2003).

These carriers can be further functionalized to prolong circulation of drugs and/or enhance biodistribution in the body at the tissue, cellular, and sub-cellular level. For instance, to avoid rapid recognition and removal by the reticulo-endothelial system of imaging agents or therapeutics injected in the circulation, drug delivery systems can be built to display stealth elements that minimize interactions with plasma opsonins, complement, phagocytic cells, and lymphocytes related to specific immunity (Moghimi et al. 2001, Panyam and Labhasetwar 2003). Also, when carriers are targeted to cell surface molecules involved in vesicular transport or coupled to cell-penetrating peptides, or if they are designed to modify the permeability of cellular layers, they can provide delivery to intracellular compartments or, furthermore, be transported across cellular barriers, both of which represent important requirements for most clinical goals (Duncan 2003, El-Sayed et al. 2003, Torchilin 2006, Muro 2012). As an example, many systems that target endothelial cell adhesion molecules have been observed to gain access into cells by inducing uptake within endocytic vesicles (Fig. 3). Endocytosis refers to a group of processes by which cells engulf extracellular material with their plasma membrane, followed by pinching off of the resulting vesicles into the cytosol (Muro et al. 2004). This can be regulated by various pathways, including clathrin- and caveolae-mediated mechanisms, macropinocytosis, phagocytosis, and other less conventional pathways. Most commonly, endocytosis results in transport of the internalized materials to endosomes and lysosomes, but carriers can be designed to escape these compartments for delivery to the cytosol and other cellular placements (Torchilin 2006).

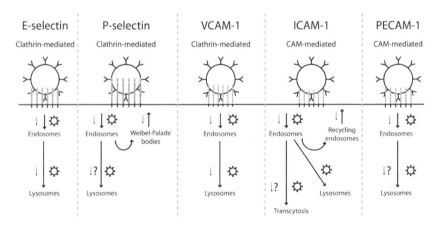

Figure 3. Endocytic pathways associated with leukocyte-interacting endothelial cell adhesion molecules. E-selectin, P-selectin and VCAM-1 have been associated with clathrin-mediated uptake and, as expected, binding of affinity molecules and/or drug carriers to these molecules leads to a similar endocytic pathway. ICAM-1 and PECAM-1, however, have not been associated with endocytosis (although PECAM-1 associates with a recycling process), yet binding of affinity molecules and/or drug carriers to these molecules induces an unconventional endocytic pathway known as cell adhesion molecule- (CAM)-mediated endocytosis. After endocytosis, endosome-to-lysosome trafficking is common among all these molecules. In addition, a fraction of P-selectin recycles to Weibel-Palade bodies, a fraction of ICAM-1 recycles to the cell surface, and a fraction of ICAM-1-targeted carriers traffic across the cell by transcytosis. In some cases (indicated by question marks) the fate of receptors versus carriers has not been characterized.

In addition, effective accumulation of imaging and therapeutic agents within particular tissues in the body may require penetration across the endothelial lining in the microcirculation, which may be accomplished by exploiting endocytic transport via transcytosis, as observed in some examples of targeting endothelial cell adhesion molecules (Ghaffarian et al. 2012, Muro 2012).

Some of the most relevant findings related to the biology of endothelial cell adhesion molecules involved in leukocyte adhesion/extravasation and their potential application in the context of drug delivery are reviewed in this chapter.

E- and P-selectin

Biology

As described above, the initial rolling interactions of leukocytes over the endothelial surface are mediated by proteins in the lectin family known as selectins. Selectins are expressed by both leukocytes and endothelial cells, but more specifically, endothelial cells express endothelial (E)-selectin (CD62E) and platelet (P)-selectin (CD62P). E-selectin was first identified as a 65–115 kDa (depending on its glycosylation state) antigen to antibodies that blocked neutrophil binding to cytokine-activated endothelial cells (Bevilacqua et al. 1987). P-selectin (140 kDa) was first identified as a protein contained in storage granules of platelets and later, the endothelium (Mcever et al. 1989) (Table 1).

Both selectins are calcium-dependent members of the lectin family (Fig. 1). They display an extracellular amino-terminal C-type carbohydrate-recognition region (CRD; six motives for E-selectin and nine motives for P-selectin) that mediates binding to extracellular ligands, an epidermal growth factor (EGF)-like domain, and a variable number of complement regulatory modules (Vestweber and Blanks 1999, Ehrhardt et al. 2004). This extracellular region is followed by a transmembrane domain and a cytosolic tail that interacts with intracellular partners. Specifically, selectins do not recognize and bind proteins *per se*, but oligosaccharides that decorate these macromolecules. In particular, these are sialyl Lewis x (sLe^x) and sialyl Lewis a (sLe^a) sugars, which can be found on a variety of proteins upon post-translational modification (Phillips et al. 1990). Ligands of E-selectin are E-selectin ligand-1 (ESL-1), P-selectin glycoprotein ligand-1 (PSGL-1), L-selectin, CD43, CD44, β_2 integrins, and DR3 (Sako et al. 1993, Ehrhardt et al. 2004), while P-selectin ligands are PSGL-1 and CD24 (Ehrhardt et al. 2004).

Table 1. Leukocyte-interacting endothelial cell adhesion molecules.

	Molecule			Structure				Expression		Activity	
Full name	Cell determinant	MW		Extracellular	Transmembrane	Cytosolic	Cell types	Quiescent endothelium*	Activated endothelium*	Ligands	Function
E-selectin	CD62E	65–115 kDa		Lectin-like domain, EGF domain, 2–9 CR domains (535 aa)	21 aa (Type I)	32 aa	ECs	2.5×10^3–3.7×10^5	10^6	ESL-1, PSGL-1, L-selectin, CD43, CD44, β_2 integrins, DR3	Leukocyte rolling adhesion
P-selectin	CD62P	86–140 kDa		Lectin-like domain, EGF domain, 6 CR domains (668 aa)	24 aa (Type I)	35 aa	ECs, platelets	2×10^4	5×10^4	PSGL-1, CD24	Leukocyte rolling adhesion
VCAM-1	CD106	74–110 kDa		7 IgG-like domains (674 aa)	22 aa (Type I)	19 aa	ECs, bone marrow cells, leukocytes, dendritic cells	10^3	10^5	$\alpha_4\beta_1$, $\alpha_4\beta_7$, $\alpha_D\beta_2$, galectin-3	Leukocyte firm adhesion, transmigration (paracellular)
ICAM-1	CD54	70–120 kDa		5 IgG-like domains (453 aa)	24 aa (Type I)	28 aa	ECs, EPs, fibroblasts, leukocytes, dendritic cells, neurons, myocytes, etc.	1.6×10^4	7.2×10^4–1.6×10^6	LFA-1, Mac-1, fibrin(ogen)	Leukocyte firm adhesion, transmigration (paracellular and transcellular)
PECAM-1	CD31	130-140 kDa		6 IgG-like domains (574 aa)	19 aa (Type I)	118 aa	ECs, leukocytes, platelets	10^6	1.7×10^6–10^7	PECAM-1	Leukocyte transmigration (paracellular)

ECs = Endothelial cells; EPs = Epithelial cells; *Molecules per cell

Although similar in structure, E- and P-selectin display different expression patterns. E-selectin is most commonly inducible under inflammation at the transcriptional level, which is mediated by a variety of factors, including TNFα, IL-1β, IL-10, IL-3, oncostatin M, LPS, and lipoteichoic acid, as well as exposure of endothelial cells to monocytes or T-cells (Vestweber and Blanks 1999). Only under certain circumstances E-selectin can be expressed constitutively, for instance in hematopoietic tissue, as well as angiogenic and metastatic tissues (Ehrhardt et al. 2004). Conversely, P-selectin is constitutively expressed by endothelial cells, although it is not constitutively exposed on the plasmalemma but stored in secretory granules known as Weibel-Palade bodies. It is upon exposure to factors such as thrombin, histamine, IL-4, IL-3, and oncostatin M, that P-selectin reaches the plasma membrane (Kneuer et al. 2006). Therefore, surface exposure of this molecule upon stimulation is rapid (minutes versus hours) compared to E-selectin (Vestweber and Blanks 1999). In both cases, in addition to their endothelial surface counterparts, circulating forms of E- and P-selectin can be released from the plasmalemma upon enzymatic cleavage or as a result of endothelial damage, which seems related to downregulation of these molecules at the endothelial surface (Kneuer et al. 2006). In addition, E- and P-selectin are endocytosed via clathrin-mediated uptake (Vestweber and Blanks 1999, Kneuer et al. 2006, Setiadi and Mcever 2008). However, their intracellular trafficking seems to differ: while a fraction of internalized P-selectin traffics back to Weibel-Palade bodies for storage, E-selectin traffics fully to lysosomes (Subramaniam et al. 1993). These time- and cytokine-dependent differential patterns of surface exposure and endocytosis lead to what is known as a tripartite expression pattern, where there is first an increase of P-selectin, then E-selectin, and finally, again, P-selectin on the endothelial surface, which helps modulate leukocyte recruitment at areas of extravasation (Vestweber and Blanks 1999).

Despite these differences in expression patterns, E- and P-selectin share functionalities. Their extended conformation over the endothelial surface, with low affinity binding motifs located at the most membrane-distal regions, correlates well with their function as the first step of endothelial association with leukocytes, particularly via rolling interactions (Lawrence and Springer 1991). This is mediated by rapid association-dissociation kinetics toward leukocyte ligands, coupled with weak mechanical bond strengths (Mehta et al. 1998, Wild et al. 2001). This interaction can be additionally modulated by association of cytoskeletal proteins with the cytosolic tail of selectins, such as vinculin, paxillin, filamin and FAK described in the case of E-selectin (Fig. 1). Hence, the role of the cytosolic domain of these molecules extends beyond simply structural (Fig. 2); it can lead to β_2-integrin activation or stimulation of certain types of leukocytes and it can also serve as an outside-in signal transducer in the endothelium. Although these signaling events remain to be fully explored, selectin engagement has been shown to lead to MAPK and ERK1/2 signaling and increase endothelial permeability in a calcium-dependent manner, associated with expression of inflammatory genes, rearrangement of the actin cytoskeleton and cell-cell junction dissociation (Lorenzon et al. 1998). For instance, in the case of E-selectin, its engagement is associated with PLCγ-mediated cytoskeletal reorganization and ERK1/2-mediated disassembly of cell junctions via Src-homology region 2-domain-containing tyrosine phosphatase 2 (SHP2) (Lorenzon et al. 1998).

Due to their involvement in leukocyte recruitment to inflammatory sites, selectins are closely associated with a variety of vascular pathologies. For instance, P-selectin is linked to adult respiratory distress syndrome, acute lung injury, ischemia-reperfusion injury, septic shock, thrombosis, rheumatoid arthritis, tumor metastasis, systemic sclerosis, and others, while E-selectin has been linked to acute ischemic stroke, malaria, rheumatoid arthritis, dermatitis, and psoriasis (Ehrhardt et al. 2004). As such, these molecules have been proposed as targets for therapeutic applications coping with these pathologies (Jubeli et al. 2011).

Drug delivery

As described above, E- and P-selectin are overexpressed in inflammation and angiogenic vessels, and are involved in leukocyte recruitment via rolling interactions. Consequently, these molecules have been used as targets to block leukocyte-endothelial binding, as well as to deliver imaging or therapeutic agents to activated endothelium in inflammation and cancer (Table 2). Blocking has been pursued by using natural ligands (sLe$^{x/a}$) or high affinity analogs (Todderud et al. 1997, John et al. 2003, Ali et al. 2004). Endothelial anchorage of imaging agents and delivery of therapeutics can also be achieved using these affinity moieties, as high specificity and affinity for E- and/or P-selectin is desirable in both scenarios (Dickerson et al. 2001, Kessner et al. 2001, Stahn et al. 2001, Everts et al. 2002, Jamar et al. 2002, Eniola and Hammer 2005, Ferrante et al. 2009, Scott et al. 2009, Shamay et al. 2009a, Theoharis et al. 2009, Asgeirsdottir et al. 2010, Gunawan and Auguste 2010, Mann et al. 2011). Most efforts regarding imaging and drug delivery have focused on using liposomes (Kessner et al. 2001, Stahn et al. 2001, Scott et al. 2009, Asgeirsdottir et al. 2010, Gunawan and Auguste 2010), yet other drug delivery systems have also been tested, including conjugates, inert or biodegradable particles, and dendrimers (Eniola and Hammer 2005, Theoharis et al. 2009). Of note, it has been reported that selectin inducible expression during pathology is somewhat transient, which narrows down the window of opportunity for intervention using these targeting strategies. Nevertheless, low expression (if any) of these molecules in normal tissues represents an advantage to particularly address affected areas.

Table 2. Therapeutic targeting of leukocyte-interacting endothelial cell adhesion molecules.

Target molecule	Affinity moieties	Carrier types	Endocytosis Physiological	Endocytosis Carrier-induced	Transport	Applications
E-selectin	Antibodies, sLe^x, sLe^x analogues	Polymer nano and microparticles, liposomes, dendrimers	Clathrin-mediated	Clathrin-mediated	Lysosomal	Tumors, thrombosis
P-selectin	Antibodies, sLe^x analogues, peptides	Polymer nano and microparticles, liposomes	Clathrin-mediated	Clathrin-mediated	Lysosomal and Weibel-palade body recycling	Tumors, inflammation
VCAM-1	Antibodies, peptides	Liposomes, fusion proteins, microbubbles, contrast agent particles, polymer particles	Undetermined	Clathrin-mediated	Lysosomal	Tumors, inflammation, atherosclerosis
ICAM-1	Antibodies, peptides	Polymer nano and microparticles, liposomes, conjugates, gold nanorods, microbubbles	Undetermined	CAM-mediated	Lysosomal and transcytosis	Tumors, inflammation, arthritis, atherosclerosis, oxidative stress, thrombosis, lysosomal storage disorders
PECAM-1	Antibodies	Protein conjugates, polymer nano and microparticles	Undetermined	CAM-mediated	Lysosomal, lateral recycling compartment	Thrombosis, endothelial oxidative stress, inflammation

CAM = Cell adhesion molecule

With regard to potential applications, numerous works have tested blocking of E- and/or P-selectin in the context of preventing inflammation, showing positive results in attenuating leukocyte extravasation in various mouse models. Furthermore, selectin targeting for imaging of inflammatory pathologies or drug delivery purposes has been successfully demonstrated in models of arthritis, colitis, ischemia, and others. In the context of cancer, these molecules have been used to image tumors in rodents with techniques that range from radioactive probes, to fluorescence, to MRI (Boutry et al. 2005, Funovics et al. 2005). In the context of drug delivery, several selectin-targeted anti-inflammatory platforms, involving formulation of dexamethasone as immunoconjugates or immunoliposomes, have been also tested *in vivo*, e.g., in the case of a murine model of delayed-type hypersensitivity, where dexamethasone-loaded drug vehicles specifically targeted inflamed areas. Selectins have also been explored for gene and siRNA delivery (Theoharis et al. 2009), and targeting of the bone marrow due to constitutive E-selectin expression (Mann et al. 2011).

At the cellular level, the property of selectins of being able to undergo endocytosis by endothelial cells has been used to achieve intracellular transport and delivery of drugs (Fig. 3). For instance, targeting of E-selectin has been shown to result in endocytic uptake of drug delivery systems, such as the case anti-E-selectin conjugates and liposomes (Kessner et al. 2001). Notably, the level of uptake of these artificial "ligands" is low as compared to natural endocytosis of selectins via clathrin-coated pits; e.g., 25 percent of E-selectin-bound liposomes in 90 min. Also, only a 30–60 percent fraction of liposomal uptake seems to associate with K^+-dependent clathrin-mediated endocytosis, while passive uptake appears to account for the remaining internalizable fraction (Kessner et al. 2001). These results may be due to suboptimal induction of this endocytic pathway by drug delivery systems and/or size restrictions of clathrin-coated pits. Indeed, a comparative study showed higher uptake of E-selectin immunoconjugates versus liposomes (Everts et al. 2003). Nevertheless, this level of endocytosis seems significant from a translational perspective. For instance, drug carriers prepared using a biodegradable co-polymer and coupled to a quinic acid derivative that serves as a high affinity analogue of sLe^x are internalized by human endothelial cells in culture while their non-targeted counterparts are not (Shamay et al. 2009b). Intracellular transport is also documented in the case of E-selectin antibodies conjugated to the anti-inflammatory agent dexamethasone, where fluorescence and transmission electron microscopy showed that these conjugates bind to E-selectin on human endothelial cells in culture, are internalized by cells, and finally traffic to lysosomes (Everts et al. 2002), and a similar transport has been shown in the case of E-selectin-targeted polymer carriers (Shamay et al. 2009a).

Despite the potential translational implications of this evidence that selectin-targeted carriers are endocytosed by endothelial cells, no systematic studies exist on this matter. It will be particularly important to determine the levels of carrier uptake depending on targeting moiety (antibody, peptide, number of targeting molecules, etc.), carrier chemistry, surface modification, and size, as different formulations have shown different results. A direct comparison of targeting E- versus P-selectin is also missing. This matter is of crucial importance because the post-endocytic fate of E-selectin differs compared to that of P-selectin, where the former is exclusively transported to lysosomes, while a fraction of the latter traffics to storage granules. This differential transport could be exploited for advanced modulation of drug delivery at the subcellular level, which could be achieved by using specific antibodies or peptides. For instance, similarly to the immunocarriers described above, E-selectin-binding peptides, which do not bind P- or L-selectin, have been successfully used for imaging in the context cancer, due to the close association between selectins and tumor metastasis (Funovics et al. 2005).

There are far fewer examples of studies at the subcellular level regarding P-selectin-targeting, although this strategy also holds potential in drug delivery. As an example, in a mouse model of myocardial infarction, 100 nm PEGylated liposomes conjugated to anti-P-selectin preferentially bind to infarcted tissue versus non-infarcted myocardium, and when loaded with vascular endothelial growth factor, these immunoliposomes ameliorate symptoms associated with myocardial infarction, improving cardiac function versus blank liposomes or free VEGF (Scott et al. 2009). Yet, at the cellular level, most studies using P-selectin targeting have focused on binding kinetics rather than intracellular transport. This is, nevertheless, important in considering drug delivery applications as well as a tool to study rolling interactions on the endothelial surface. For instance, recombinant PSGL-1-coated spheres of 5, 10, 15 and 20 µm in diameter behave differently on P-selectin binding, particularly in a shear rate-dependent manner (Patil et al. 2001). At 75 s^{-1}, spheres of all sizes attach similarly to P-selectin, but at high shear rates such as 600 s^{-1}, only the 5 µm spheres attach to P-selectin. At 400 s^{-1}, 5, 10 and 15 but not 20 µm particles attach to P-selectin. Under low shear stress, where particles of all sizes adhere, these firmly bind to P-selectin and do not roll in the direction of flow. However, when flow is increased at this point, the particles begin rolling, with smaller particles requiring higher shear rates to begin rolling, and the rolling velocity being higher for larger particles at any given shear stress (Patil et al. 2001). This notion is supported by the use of 5 versus ~8 µm diameter spherical drug carriers coated with an antibody that binds both E- and P-selectin. Whereas the 8 µm particles bind to activated endothelial cells at a maximum shear of

0.3 dyn/cm², the 5 μm spheres exhibit binding up to 1.5 dyn/cm² (Dickerson et al. 2001). Although these studies were performed in an *in vitro* model with no blood components present, similar physical properties of carriers have been shown to affect targeting performance of carriers *in vivo* (Muro et al. 2008), and hence are likely to represent important clues to take into account in the development of selectin-targeted therapies. These considerations will be important for furthering endothelial targeting via E- and P-selectin, a currently prolific field with considerable potential.

VCAM-1

Biology

Following rolling interactions between leukocytes and selectins on the endothelial surface, the former firmly adhere via integrins to Ig CAMs on the endothelial lining. One of these Ig CAMs is vascular cell adhesion molecule 1 (VCAM-1; CD106), which was described upon the observation that lymphocyte adhesion could not be fully inhibited using antibodies that block ICAM-1/LFA-1 interaction. This led to a cDNA screening which, after subtraction of ICAM-1 and E-selectin, yielded the clone designated as VCAM-1 (Osborn et al. 1989). This 74-110 kDa molecule is expressed at very low levels (if any) by quiescent endothelial cells, while upregulation of its expression is induced by TNFα, IL-1β, LPS, vistatin, homocysteine, proatherogenic glycemia, and several other stimuli, mediated through the NFκB pathway in a reactive oxygen species (ROS)-dependent manner (Henninger et al. 1997). VCAM-1 is also expressed by bone marrow cells, B cells, follicular dendritic cells, and macrophages (Freedman et al. 1990).

Regarding structural features of VCAM-1 (Fig. 1), this molecule displays 7 extended Ig-like domains (D1 to D7) on its extracellular region, with six N-glycosylation sites in domains D3 through D6, followed by a type I transmembrane domain, and a short cytosolic tail. In addition, a human isoform exists which lacks extracellular domain D4, while in mice there is a GPI-anchored splice variant which only exhibits D1 through D3 (Moy et al. 1993). As observed with most other endothelial CAMs, the extracellular domain of VCAM-1 can be cleaved off the endothelial plasma membrane by metalloproteases. This results in the release of a soluble form of the molecule into the circulation, which is thought to regulate leukocyte-endothelial interaction by competing with the non-soluble form of VCAM-1 expressed on the endothelial surface (Lobb et al. 1991).

Indeed, the VCAM-1 extracellular region mediates binding to leukocyte integrins, mainly $\alpha_4\beta_1$ (VLA-4), which is found on eosinophils, basophils, lymphocytes, mast cells, and monocytes (Table 1). Apart from this ligand, VCAM-1 can also bind to integrins $\alpha_4\beta_7$ and $\alpha_D\beta_2$ (Chan et al. 1992, Grayson et al. 1998, Rao et al. 2007). Interaction of endothelial VCAM-1 with these partners supports rolling of leukocytes when their integrins are not activated, but firm adhesion upon activation (Alon et al. 1995). Integrin $\alpha_4\beta_1$ and $\alpha_4\beta_7$ have different binding properties. For instance, $\alpha_4\beta_1$ binds to domains D1 and D4 of the VCAM-1 extracellular region, while $\alpha_4\beta_7$ binds to D1 only (Osborn et al. 1992). Further, these integrins have different activation requirements for maximal binding, which might provide additional means for regulating leukocyte-endothelial binding affinities under different patho-physiological circumstances.

Through these interactions, VCAM-1 exerts its main functional role which, as mentioned, pertains to firm adhesion of leukocytes to the endothelium (Springer 1994). This function is shared with ICAM-1 (as discussed in the following section), yet ICAM-1 and VCAM-1 have different roles beyond firm adhesion. For instance, in a model of transmigration where ICAM-1, PECAM-1 and selectin interactions are absent, VCAM-1 proves sufficient for transmigration of lymphocytes across an endothelial monolayer, which appears to be independent from wortmannin-sensitive, endothelial PI3 kinase signaling (Matheny et al. 2000). Yet, in other contexts, such as leukocyte transmigration across the blood-brain barrier, VCAM-1 has been shown to be non-crucial for diapedesis, but important only in firm adhesion (Laschinger and Engelhardt 2000).

Although not entirely understood, these functions have been associated with signaling events mediated by the cytosolic domain of VCAM-1 upon engagement by its natural ligands (Fig. 2). As an example, binding to the extracellular region of VCAM-1 induces calcium and G protein (e.g., Rac1) signaling, which in turn activates NADPH oxidase NOX2, leading to formation of low levels reactive oxygen species (ROS). Said ROS production is known to be involved in further signaling events that reinforce the inflammatory cascade. For instance, ROS activates transcriptional expression of genes encoding for inflammatory mediators, which includes endothelial CAMs. This represents a feed-forward expression loop, since VCAM-1 expression is also induced by ROS (Cook-Mills et al. 2011). VCAM-1-mediated ROS production also leads to activation of matrix metalloproteinases, which might help degrade the extracellular matrix during leukocyte transmigration. In addition, low ROS concentrations induce PKCα activity, which in turn activates protein tyrosine phosphatase 1. This leads to downstream opening of cell junctions, although this pathway remains to be fully mapped (Cook-Mills et al. 2011). Concomitant to these events, a ROS-dependent modification of the endothelial actin cytoskeleton has been reported, which appears to favor endothelial cell retraction during passage of leukocytes between endothelial junctions (Matheny et al. 2000). Said cytoskeletal changes seem to be mediated by Rac1, likely via PKC and MAPK activity.

However, whether the cytosolic domain of VCAM-1 is directly involved in such signaling events remains to be fully understood. For instance, this short region of the molecule (19 residues) contains a type I PDZ-binding motif, but it has not been reported to interact with molecules other than cytoskeletal partners ezrin and moesin (Barreiro et al. 2002). Indeed, the VCAM-1 cytosolic tail is not needed for recruitment of VCAM-1 to, nor formation of, specialized cup-like structures that appear on the apical surface of activated endothelial cells, which have been associated with firm adhesion and transmigration of leukocytes (Barreiro et al. 2008). This suggests that the cytosolic domain of VCAM-1 is not absolutely necessary in supporting VCAM-1 behavior on the endothelial plasmalemma. Perhaps other regions of the molecule or association with other membrane components are responsible for this. As an example, in activated endothelium, VCAM-1 resides in lipid raft-like domains which also house ICAM-1 and tetraspanins CD9, CD81, and CD151 (Barreiro et al. 2002, Barreiro et al. 2008). These tetraspanin domains are thought to support effective VCAM-1 and ICAM-1 surface expression, along with formation of signaling platforms, under shear stress.

As a consequence of these biological features and function, VCAM-1 is linked to many disorders, including various inflammatory conditions, atherosclerosis, asthma, and tumor metastasis (Sagara et al. 1997). Its overexpression pattern during these conditions appears to be more distinct and specific than that of other endothelial CAMs and, thus, targeting of endothelial VCAM-1 has been largely explored.

Drug delivery

As in the case of other endothelial CAMs, VCAM-1 overexpression during inflammation and other pathological events makes this molecule suitable for diagnostic and/or therapeutic intervention (Table 2), which has been explored in the context of blocking this molecule, as well as using it as a target for endothelial addressing of imaging agents or therapeutics (Chiu et al. 2003, Dienst et al. 2005, Tsourkas et al. 2005, Deosarkar et al. 2008, Ferrante et al. 2009, Serres et al. 2011). Blocking of VCAM-1 during inflammation has been explored using antibodies to this molecule. Yet, as in other cases, this approach renders suboptimal results, likely due to functional redundancy of endothelial CAMs regarding the recruitment and extravasation of leukocytes, which prevents VCAM-1 blocking from being efficient. This has been shown in the context of preventing ischemic damage (Justicia et al. 2006). Conversely, this might prove beneficial when the intention is to target sites of inflammation without eliciting secondary blocking effects, such as in the case of imaging and therapeutic applications. In this case, VCAM-1 can be targeted with probes or drugs, also using antibodies and affinity peptides (Chiu et al. 2003, Dienst et al. 2005, Tsourkas et al. 2005, Deosarkar et al. 2008, Burtea et al. 2009, Ferrante et al. 2009, Serres et al. 2011). In this context, several design modalities have been studied, including conjugates and fusion proteins, microbubbles and perfluorocarbon nanoparticles, liposomes and biodegradable particles, ferromagnetic particles, and others (Tsourkas et al. 2005, Serres et al. 2011). Similar to the case of selectins, the transient nature of VCAM-1 over-expression in pathology reduces the time frame for intervention using targeting strategies toward this molecule, yet its near absence in normal tissues represents an advantage to selective drug delivery to diseased tissues.

Imaging and/or delivery of therapeutics have been pursued in the context of cancer, atherosclerosis, multiple sclerosis, renal inflammation, and colitis, among others pathologies (Dienst et al. 2005, Ferrante et al. 2009, Serres et al. 2011). Most commonly, VCAM-1 targeting has been used with the purpose of non-invasive imaging. For instance, in some imaging settings, such as the case of magnetic resonance imaging (MRI) in multiple sclerosis, conventional techniques cannot detect tissue damage. However, high levels of expression of endothelial CAMs (e.g., VCAM-1) have been reported to occur prior to and after the existence of obvious symptoms. Based on this premise, coating of 1 μm iron oxide particles with anti-VCAM-1 has been shown to allow for detection of early asymptomatic stages of experimental autoimmune encephalomyelitis, a model of multiple sclerosis (Serres et al. 2011). Another example is that of ^{19}F-coated perfluorocarbon nanoparticles displaying VCAM-1-targeting peptide, which specifically bind VCAM-1 in cell culture, as well as in kidneys after administration in an ApoE$^{-/-}$ mouse model (compared to control mice or non-targeted particles), allowing MRI detection of affected areas undergoing inflammatory VCAM-1 overexpression (Serres et al. 2011). Several other studies additionally show the usefulness of VCAM-1 targeting in imaging of inflammatory tissue (Tsourkas et al. 2005, Jefferson et al. 2011).

Because VCAM-1 expression is also increased in tumors and vicinal vasculature, targeting via this molecule has also been explored for potential cancer treatment strategies. Recombinant fusion proteins using the single-chain variable fragment (scFv) of anti-VCAM-1 antibodies and soluble tissue factor have been made with the purpose of inducing thrombosis on-site to cause intratumoral vessel occlusion and necrosis. This has been tested in mice bearing human Hodgkni lymphoma and SCLC xenograft tumors, leading to tumor necrosis after 3 d and slower tumor growth or tumor regression (Dienst et al. 2005). A similar strategy has been tested using thrombogenic phosphatydilserine-containing liposomes targeted to VCAM-1 with an antibody, showing binding and thrombogenic activity *in vitro* and in cell cultures (Chiu et al. 2003). Interestingly, VCAM-1 targeting seems to be improved when combined with targeting to other endothelial

CAMs. For instance, it has been shown that anti-VCAM-1 displayed on perfluorocarbon-filled lipid-shelled microbubbles can bind to its endothelial target under shear stress conditions (around 1.5 dyn/cm^2). Yet, targeting is improved when the microbubbles carry additionally P-selectin-targeting motifs, particularly under higher sheer stress levels of 3–6 dyn/cm^2 (Ferrante et al. 2009). This may be due to the fact that this combined targeting strategy mimics more closely multimolecular interactions of leukocytes with the endothelium.

Much less is known about the behavior of VCAM-1-targeting systems in terms of the cellular fate of these entities. Cell culture experiments studying in detail the mechanisms of targeting via VCAM-1 are less prevalent than those conducted for other endothelial CAMs, such as selectins, ICAM-1, or PECAM-1. VCAM-1 cross-linking using primary and, subsequently, secondary antibodies induces endocytosis via clathrin-coated pits, with 70–90 percent of all engaged VCAM-1 being internalized by 60–90 min (Ricard et al. 1998). This route of endocytosis is similar to that seen with endothelial selectins and is hypothesized to contribute to endothelial VCAM-1 turnover, possibly helping in detachment from integrin $\alpha_4\beta_1$ during leukocyte lateral migration on the endothelium and extravasation across the endothelial lining. As evidence of this endocytosis, several drug carriers targeted to VCAM-1 have also been shown to be internalized by endothelial cells (Voinea et al. 2005). An example is that of 90 nm diameter liposomes targeted to VCAM-1, which show specific endothelial accumulation near tumors in a COLO-677 xenograft model and are endocytosed (although somewhat poorly) by endothelial cells in culture. This may be due to size restrictions of clathrin pits, as discussed in the case of selectin-targeted drug delivery systems. Indeed, it has been shown that endothelial cells do not internalize micron-sized objects targeted to VCAM-1 (Matheny et al. 2000). However, to date, there has been no systematic study on these limitations, intracellular fate, and other parameters associated with VCAM-1 endocytosis of drug carriers.

Yet, in perspective, VCAM-1 targeting represents an interesting tool in imaging modalities and the delivery of therapeutics, since it exhibits almost null levels of constitutive expression in physiological situations, but it is highly overexpressed during inflammation and associated pathologies.

ICAM-1

Biology

Another endothelial CAM involved in firm adhesion and transmigration of leukocytes is intercellular adhesion molecule-1 (ICAM-1; CD54), a type I transmembrane glycoprotein of the Ig superfamily (Table 1). ICAM-1 was first characterized in 1986 as a 90 kDa molecule necessary for leukocyte homotypic adhesion, but was later found to be expressed by endothelial cells as well (Rothlein et al. 1986). Since then, ICAM-1 has been described as a molecule crucial to the immune response and, as such, is associated with a variety of vascular pathologies (Muro 2007).

The extracellular region of ICAM-1 (Fig. 1) consists of 5 Ig-like domains arranged linearly in a rod-like conformation that extends from the endothelial surface and, thus, positions this molecule in a manner suitable for its function as an adhesive co-receptor for leukocytes. This region is connected via a single transmembrane domain to the cytosolic tail of the molecule. The ICAM-1 cytosolic region is relatively short and lacks conventional protein-protein interaction domains. However, it contains several positively-charged residues and a tyrosine, which seems to account for its protein-binding and signaling activities (Federici et al. 1996, Pluskota et al. 2000, Sans et al. 2001). Natural ligands that bind to ICAM-1 include leukocyte integrins $\alpha_L\beta_2$ (LFA-1) and $\alpha_M\beta_2$ (Mac-1) which bind, respectively, to the extracellular domains D1 and D3 of the molecule, and fibrinogen and fibrin, which bind to domain D1 and serve as an additional bridge between macrophages and endothelial cells (D'souza et al. 1996).

With regard to its expression, endothelial ICAM-1 is constitutive, albeit low, yet is upregulated during inflammatory and thrombogenic conditions by cytokines and other factors, such as TNFα, IFNγ, TGFβ, thrombin, fibrinogen or fibrin, and ROS (Qi et al. 1997, Rahman et al. 1999). In addition, contrarily to other CAMs discussed in this chapter, ICAM-1 is also found on a wide range of cell types, including not only endothelium and leukocytes, but also dendritic cells, epithelial cells, fibroblasts, neurons, myocytes, etc. (Table 1). Several isoforms of this protein have been identified, including splicing and post-translational glycosylation variants, along with a soluble form of ICAM-1 that can appear in circulation and corresponds to the extracellular domain of this protein, which might compete against endothelial ICAM-1 for integrin binding as a means of regulating inflammatory events (Kusterer et al. 1998).

As mentioned above, ICAM-1 function pertains to firm adhesion and extravasation of leukocytes, which is favored by the particular distribution and related changes associated with this molecule (Fig. 1). For instance, interactions of cytoskeletal partners with the cytosolic tail of ICAM-1 support enrichment of this molecule in microvillar structures, as deletion of ICAM-1 cytoplasmic domain shifts its distribution to more homogeneous on the endothelial surface (Heiska et al. 1998, Oh et al. 2007). This association with the actin cytoskeleton and specialized membrane domains helps regulate leukocyte transmigration. As described above, when leukocytes crawl laterally on the endothelial surface, endothelial cells extend protrusions around leukocytes, forming a "docking" structure known as the transmigratory cup, which associates with both paracellular and transcellular transmigration (Barreiro et al. 2002). This structure is enriched in ICAM-1, dependent on its cytosolic tail, and is associated with formation of actin stress fibers (Barreiro et al. 2002, Oh et al. 2007). Maintenance of these docking structures is associated with ICAM-1-dependent signaling at specialized protein and lipid domains, manifested by enrichment in tetraspanins and redistribution of ICAM-1 to detergent-resistant membrane fractions (Barreiro et al. 2002).

In the case of transcellular transmigration, along with formation of this transmigratory cup, there is generation of 0.2–1 μm invaginations and vesicles at the endothelial plasmalemma. These vesicles coalesce into larger structures forming a pore (several μm in diameter) that transverses the entire endothelial cell body, through which leukocytes can transmigrate (Millán et al. 2006, Carman et al. 2007). There is clear evidence that ICAM-1 mediates this process in the endothelium. Indeed, ICAM-1 engagement with inert ligands (e.g., micron-sized anti-ICAM-coated particles) seems to be sufficient to elicit formation of endothelial engulfment structures and endocytic-type membrane invaginations and vesicles (Barreiro et al. 2002, Muro et al. 2003, Serrano et al. 2012). However, the mechanistics of these ICAM-1-mediated events are still obscure; some data suggest the involvement of caveolar endocytosis and/or the related vesiculo-vacuolar organelle (Millán et al. 2006), while other reports show no or partial association with caveolin-1 structures (Carman et al. 2007). Indeed, ICAM-1 engagement with anti-ICAM-coated particles results in formation of docking-like structures and endocytic vesicles by a pathway that depends on specialized lipid domains at the plasma membrane, but differs from the caveolar route. Lipid domains, plasmalemma deformability, molecular signaling, and cytoskeletal reorganization associated with these events seem rather similar to those related to a non-conventional endocytic pathway, termed cell adhesion molecule-(CAM)-mediated endocytosis (Muro et al. 2003, Muro et al. 2008, Serrano et al. 2012). Interestingly, ICAM-1 becomes internalized and undergoes transcytosis to the basal membrane of endothelial cells during transcellular extravasation of leukocytes (Millán et al. 2006), and vesicular transcytosis across cellular layers, via the CAM pathway, can also be induced by engaging ICAM-1 with anti-ICAM-coated particles (Ghaffarian et al. 2012).

These different events have been related to outside-in signal transduction mediated through the cytosolic domain of ICAM-1 (Fig. 2). ICAM-1 engagement generates ROS via xanthine oxidase activity and activates Src-kinases, small GTPase Rho and related partners, and MAPK, resulting in actin cytoskeletal rearrangement into stress fibers (Etienne et al. 1998, Wang and Doerschuk 2001, Yang et al. 2006, Serrano et al. 2012). Along with this, the cytoplasmic tail of ICAM-1 interacts with cytoskeleton adaptors, e.g., α-actinin, β-tubulin, ezrin, radixin and moesin family proteins, and others (Heiska et al. 1998). This occurs in parallel to ICAM-1 redistribution to detergent-resistant membrane fractions, which seems associated with SHP-2-dependent Src activation and ERK1/2, leading to increased expression of inflammation-related genes like IL-1β, RANTES, VCAM-1, and ICAM-1 itself (Pluskota et al. 2000). Specialized lipid domains serve as low-diffusion platforms to allow localized ICAM-1-dependent signaling and cytoskeletal remodeling (Van Buul et al. 2010). Indeed, the sole engagement of ICAM-1 using antibody-coated particles leads to localized formation of ceramide via the action of acid sphingomyelinase (Serrano et al. 2012). Ceramide is closely linked to membrane and cytoskeletal dynamics due to its ability to serve as a signaling messenger and to modify the biophysical properties of membranes. Hence, these events seem supportive of endothelial structures that contribute to leukocyte transmigration.

The pattern of expression and role of ICAM-1 positions this molecule as an interesting target for clinical intervention in many pathologies, including inflammatory and autoimmune conditions, infections and septic shock, atherosclerosis and metabolic disorders, asthma, and cancer, among others.

Drug delivery

Similarly to the other endothelial CAMs discussed, the large differences in expression levels of ICAM-1 between resting and inflammatory or pathological states, along with its physiological role, have been explored in the context of blocking or attenuating inflammation, as well as a means for delivery of drugs and imaging agents for a variety of applications (Table 2). ICAM-1 blocking and targeting have both been pursued using antibodies or peptides in different modalities, such as full monoclonal antibodies, humanized counterparts, different types of antibody fragments, and peptides derived from natural ligands or obtained by other technologies such as phage display (Isobe et al. 1992, Kavanaugh et al. 1997, Burns et al. 2001, Kuwahara et al. 2003). In the case of ICAM-1 targeting for imaging or drug delivery purposes, a variety of systems have also been explored, including conjugates, liposomes, polymer particles, gold nanorods, and iron oxide nanoparticles (Murciano et al. 2003, Hamilton et al. 2004, Weller et al. 2005, Choi et al. 2007, Kim et al. 2007, Zhang et al. 2008, Gunawan and Auguste 2010, Park et al. 2010, Shao et al. 2011, Garnacho et al. 2012, Hsu et al. 2012).

Numerous studies in animal models and/or human clinical trials have shown some potential regarding ICAM-1 blocking in the context of diseases characterized by an important inflammatory component, such as graft-versus-host disease, ischemia-reperfusion injury, Crohn's disease, myocardial hypertrophy, and rheumatoid arthritis (Isobe et al. 1992, Kavanaugh et al. 1997, Burns et al. 2001, Kuwahara et al. 2003, Merchant et al. 2003). In these cases, the importance of ICAM-1 in pathology is evident, as the therapeutic effect achieved by targeting this molecule is quite efficient. For instance, an ICAM-1-binding cyclic peptide can reduce neutrophil infiltration by 56% and infarct size by 40% in a mouse model of ischemia-reperfusion injury (Merchant et al. 2003). Nonetheless, in many cases, ICAM-1 therapy must be complemented with targeting of other molecules, such as the case of transplantation, where antibodies to ICAM-1 are less effective at prolonging graft survival than a combination of anti-ICAM-1 and anti-CD11a (Isobe et al. 1992). This, as seen with VCAM-1, shows that functional redundancy might ameliorate the effects of CAM blocking for therapeutic effects. Yet, these phenomena do not hinder the potential therapeutic effect of targeting these molecules, as seen with clinical trials of antibodies to ICAM-1 for treatment of rheumatoid arthritis, where benefits extend to 2 months or more past therapeutic intervention (Kavanaugh et al. 1996, Kavanaugh et al. 1997).

Imaging and therapeutic delivery have also been studied for numerous applications, including inflammation caused by pathogen infections, autoimmune inflammatory disorders such as arthritis, oxidative stress and thrombosis, atherosclerosis, genetic lysosomal storage disorders, and cancer models (Murciano et al. 2003, Hamilton et al. 2004, Weller et al. 2005, Choi et al. 2007, Kim et al. 2007, Zhang et al. 2008, Gunawan and Auguste 2010, Park et al. 2010, Shao et al. 2011, Garnacho et al. 2012, Hsu et al. 2012). An important aspect to take into account in the case of ICAM-1 targeting, is that this molecule is not completely void from tissues in the absence of pathological stimuli, at least not to the extent of selectins and VCAM-1. Indeed, intravenous injection of ICAM-1-targeting systems in normal laboratory animals leads to effective accumulation of carriers and their cargoes in multiple organs through the body (Murciano et al. 2003, Hsu et al. 2012). However, this can be modulated by modifying the avidity of the targeting system so that accumulation is specific to areas in the body displaying high ICAM-1 expression (Calderon et al. 2011). Indeed, local upregulation of ICAM-1 expression correlates well with enhanced targeting for imaging and therapeutic applications (Garnacho et al. 2008b). Importantly, the inducible over-expression of this molecule during pathology is more sustained than that of the selectins and VCAM-1, supporting a wider window time for intervention. In addition, this property can be used as an advantage when no toxic effects are expected for targeting normal tissues or when profuse delivery throughout the body is intended, such as the case of delivery of enzyme replacement therapies for genetic conditions (Hsu et al. 2012). Combinatorial approaches pursuing ICAM-1 targeting along with targeting to other endothelial CAMs are also being explored to improve specificity and efficiency. For instance, combination of particular molar ratios of (a) antibodies to ICAM-1 and (b) sLe[x] or antibodies to selectins on the surface of liposomes or particles, seems to provide enhanced targeting properties (Gunawan and Auguste 2010, Robbins et al. 2010).

Regarding intracellular transport, there is considerably extensive knowledge on the behavior of ICAM-1-targeting systems at the subcellular level (Fig. 3), showing great potential not only in the context of intracellular drug delivery but, more importantly, for delivery across cellular barriers in the body. Although ICAM-1 had not been associated with endocytic uptake, studies focusing on carrier or conjugate targeting led to the discovery that ICAM-1 engagement induces endocytosis by cells (Muro et al. 2003). The mechanism of uptake, called CAM-mediated endocytosis, is unrelated to clathrin- and caveolae-dependent endocytosis and, although holding some similarities, differs from macropinocytosis and phagocytosis. CAM endocytosis involves signaling through Src, PKC, and Rho-related molecules, and associates with enzyme-mediated formation of ceramide-rich domains at the plasmalemma and appearance of actin stress fibers, likely reminiscent of ICAM-1 signaling during leukocyte adhesion and transmigration (Muro et al. 2003, Serrano et al. 2012). During this process, ICAM-1 interacts with the sodium/proton exchanger 1 (NHE1) protein at the endothelial plasma membrane, which seems to contribute to uptake via regulating local pH, osmolarity, and crosslinking with cytoskeletal elements (Muro et al. 2006, Serrano et al. 2012). These properties have shown to be particularly amenable for cell internalization of ICAM-1-targeting systems displaying not only different chemistries, but also sizes and shapes: 200 nm to 10 μm, spheres, discs, or amorphous conjugates, providing flexibility in the design of drug carriers targeted to this molecule (Muro et al. 2003, Muro et al. 2008). Following CAM endocytosis, ICAM-1 recycles back to the plasma membrane, while carriers can follow two transport routes. In one instance, they can travel to endosomes and lysosomes in cells, which could be exploited for delivery of endosome-permeable drugs, pH-sensitive drug delivery systems, or for delivery of lysosomal enzymes in genetic lysosomal storage disorders (Muro et al. 2005, Hsu et al. 2012). In a second instance, ICAM-1-targeted carriers are transported via transcytosis across cellular barriers in cell culture and animal models, such as the case of the gastrointestinal epithelium or the blood-brain barrier, without opening of the cell junctions (Ghaffarian et al. 2012, Hsu et al. 2012).

Hence, targeting ICAM-1 provides means to address imaging and therapeutic agents to sites affected by disease, with subsequent transport both across cell barriers and into cells. Said transport, contrarily to clathrin-

or caveolar-pathways that support endocytosis and transcytosis only of small conjugates and carriers below 100 nm in diameter, is mediated by the CAM pathway, which holds an unprecedented potential in supporting transport of carriers with more varied designs.

PECAM-1

Biology

Platelet endothelial cell adhesion molecule 1 (PECAM-1; CD31) was initially identified in the context of the coagulation system. This molecule was first obtained in a study that attempted to improve the isolation of platelet membrane proteins (Phillips and Agin 1977) and it remained uncharacterized, known as platelet glycoprotein IIa, for almost a decade. In 1985, another study focused on generating a series of monoclonal antibodies against endothelial cells (Van Mourik et al. 1985). From these, an antibody that also bound platelets was further characterized and found to bind to an endothelial antigen. This antigen migrated by electrophoresis similarly to platelet glycoprotein IIa, under both reduced versus non-reduced conditions. Cloning of this endothelial antigen led to the identification and re-naming of PECAM-1 (Newman et al. 1990).

Regarding its molecular organization, as in the case of VCAM-1 and ICAM-1, PECAM-1 is also an Ig-like type I transmembrane glycoprotein (Fig. 1). It has a molecular weight of 130–140 kDa (depending on glycosylation states), comprising an extracellular region with 6 Ig-like domains (D1–D6) that offers an affinity surface for its ligands, a single-pass transmembrane domain, and a cytosolic domain, larger than that of the other CAMs discussed, which has been associated to various signaling events (Newman et al. 1990). Alternative splicing of the PECAM-1 transcript can also lead to diversity in its cytosolic domain, suggesting the functional importance of this region of the molecule (Privratsky et al. 2010). In addition, an isoform exists which lacks the transmembrane and cytosolic domains, and represents the soluble fraction of PECAM-1 in circulation (Goldberger et al. 1994), although soluble PECAM-1 might also arise from cleavage of its extracellular domain after expression on the plasmalemma (Ilan et al. 2001). The most well characterized binding interaction of PECAM-1 is that of homophilic binding, or PECAM-1:PECAM-1 interactions. However, PECAM-1 can also bind $\alpha_v\beta_3$ on the surface of leukocytes, hypothetically supporting at some level leukocyte transendothelial migration (Piali et al. 1995). Although still unclear, other cases of heterophilic binding interactions associated with PECAM-1 have been documented, namely with CD177, CD38, and glycosaminoglycans (Privratsky et al. 2010).

As mentioned, PECAM-1 is mainly expressed by endothelial cells and platelets, but also other circulating cells, such as monocytes, neutrophils, and T cells. In endothelial cells specifically, this molecule localizes at the cell-cell border of confluent monolayers and has a molecular weight of 130 kDa, with low expression on the apical surface (Albelda et al. 1990, Newman et al. 1990, Newman and Newman 2003). As opposed to selectins, VCAM-1 and ICAM-1, whose display on the endothelial plasmalemma is mostly inducible, PECAM-1 is constitutively expressed on endothelial cells at a relatively high level and it appears that its expression is not further upregulated by inflammatory factors such as TNFα and IFNγ. However, exposure to these cytokines causes PECAM-1 to distribute away from the cell-cell borders, yet it seems unclear whether this is fully due to redistribution to other cell compartments or whether there is some inhibition of PECAM-1 transcription or translation (Romer et al. 1995). Nonetheless, it is possible that other non-conventional stimuli might induce increased expression of PECAM-1, as it appears to be the case with lactosylceramide (Gong et al. 2004).

PECAM-1 appears to be involved in formation and/or maintenance of endothelial cell-cell junctions, as it has been shown both in cell culture and *in vivo* that disruption of the endothelial monolayer integrity can occur upon exposure to certain antibodies that bind PECAM-1 (Ferrero et al. 1995). This role has been associated with its capacity to interact with a homologous molecule displayed by an adjacent endothelial cell, via homophilic PECAM-1:PECAM-1 interactions. Yet, in addition to the contribution of this homophilic binding, it is possible that PECAM-1 signaling also mediates this function. For instance, some evidence indicates association between PECAM-1 and sphingosine 1-phosphate (S1P), a known regulator of endothelial barrier integrity (Huang et al. 2008). In any instance, this role also correlates with PECAM-1 contribution to leukocyte transmigration across the endothelium. Due to its preferential localization at the cell border, it was originally thought that PECAM-1 solely mediated paracellular extravasation, particularly via the interaction between endothelial and leukocyte PECAM-1 molecules (Liao et al. 1995, Ley et al. 2007). Yet, it appears this molecule is transported to the cell surface from a lateral-border recycling compartment, rich in PECAM-1, by a complex recycling system which consists of a series of interconnected vesicles that traffic from cell junction regions to the apical membrane of endothelial cells. This is induced by leukocyte adhesion to the endothelium, and hence, it is thought to provide non-bound endothelial PECAM-1 to interact with leukocyte PECAM-1, facilitating transmigration in the case of paracellular as well as transcellular extravasation (Mamdouh et al. 2009). Despite this, it is unclear whether PECAM-1 actively mediates transcellular leukocyte transmigration or simply participates downstream of another initiator, as apparently this function is rather associated with ICAM-1 (Carman 2009).

In addition, PECAM-1 has been implicated in endothelial cell motility by regulating formation of filopodia (Delisser et al. 1993). This is in agreement with the association of PECAM-1 with the cytoskeleton (e.g., F-actin, filamin and spectrin), which increases upon cross-linking of PECAM-1 with antibodies or during cell migration (Newman and Newman 2003). In support of this, antibody crosslinking of PECAM-1 leads to cell spreading and rearrangement of actin into foci, and PECAM-1 co-precipitates with β-catenin and γ-catenin, adaptors of the actin cytoskeleton (Newman and Newman 2003).

Signaling mediated through the cytosolic domain of PECAM-1 seems crucial in the context of regulating all of these functions. Indeed, PECAM-1 signaling is well documented (Fig. 2). Its cytosolic domain houses potential phosphorylation sites, some of which are phosphorylated in response to several forms of stimulation, such as shear stress or exposure to fibronectin or collagen. Among these, tyrosine 663 and 686 are found within immunoreceptor tyrosine-based inhibitor motifs (ITIMS), whose phosphorylation leads to recruitment of SHP-2, PLCγ, and phosphatase SHIP. PECAM-1 ITIMs are not necessary for cell junction-related functions, yet the cytosolic domain of this molecule seems to be required for homophilic interaction (Privratsky et al. 2011). Further, PECAM-1 tyrosine residues are phosphorylated in the presence of S1P via Gi and SFK, cSrc and Fyn, resulting in decreased monolayer permeability (Huang et al. 2008). However, it appears that PECAM-1 does not only function to maintain junctional integrity, as it seemingly has divergent roles depending on context, where it can act as a pro-inflammatory molecule (via PI3K and NFκB) or as an anti-inflammatory molecule (via STAT3). Indeed, PECAM-1 is involved in maintenance endothelial barriers but also leukocyte transmigration (as described above), as well activation of leukocyte integrins and regulation of cytokine production but also inhibition of leukocyte activation (Carrithers et al. 2005, Vernon-Wilson et al. 2006, Woodfin et al. 2007, Privratsky et al. 2010). Although apparently conflicting, these functions of PECAM-1 are context-dependent. For instance, differential PECAM-1 functions have been observed in responses to shear stress at sites of laminar versus turbulent flow (Privratsky et al. 2010).

The functional role of PECAM-1, as well as its expression pattern, confers this molecule high interest from the perspective of therapeutic applications, as described below.

Drug delivery

As in the case of the other endothelial CAMs discussed above, PECAM-1 has also been explored regarding blocking leukocyte-endothelial interaction during inflammation, as well as for drug targeting to the endothelium, which has proven useful in the context of systemic vascular pathologies. Blocking of PECAM-1 function has been attempted using antibodies as a means for therapeutic intervention for inflammatory pathologies and also cancer, as this molecule has been associated to angiogenesis (Bogen et al. 1994, Murohara et al. 1996, Zhou et al. 1999). PECAM-1 targeting of drugs and imaging agents have been shown to provide systemic endothelial therapy or labeling, respectively, and this has been mostly tested using full antibodies or their fragments (Muzykantov et al. 1999, Muro et al. 2003, Ding et al. 2008, Garnacho et al. 2008a, Ding et al. 2009, Shuvaev et al. 2011, Chacko et al. 2012, Xiao et al. 2012). Targeting modalities designed for this purpose include fusion proteins, protein conjugates, and polymeric drug carriers (Ding et al. 2008, Ding et al. 2009, Shuvaev et al. 2011). Importantly, it must be noted that PECAM-1 (a) is constitutively expressed by endothelial cells, (b) its expression is not upregulated by conventional inflammatory cytokines and, hence, (c) it is distributed through the body (a pan-endothelial marker). Therefore, applications of targeting to this molecule are mostly relevant where broad intervention is needed, no major side effects are expected from the delivered drugs, or for prophylactic applications.

Blocking of PECAM-1 has been tested in the context of acute peritonitis, tumor angiogenesis, ischemia-reperfusion injury, and arthritis, in several animal models (Bogen et al. 1994, Zhou et al. 1999). In the case of imaging and therapeutic delivery, PECAM-1 targeting has been also explored in a variety of settings, including LPS-induced inflammation, graft damage upon lung transplantation, myocardial infarction, thrombosis prevention during ischemia-reperfusion, and other injuries that are mediated through signaling associated with oxidative stress (Kozower et al. 2003, Dziubla et al. 2008, Ding et al. 2009). For instance, numerous efforts have been focused on delivery of antioxidant enzymes to the vasculature in these settings, using catalase or superoxide dismutase (SOD) complexed to antibodies to PECAM-1 or carried by polymeric drug delivery systems which display anti-PECAM-1 on their surface. An example is that of anti-PECAM-1/SOD complexes that, after intravenous injection in mice, accumulate in lungs at 20- to 40-fold higher levels compared to non-targeted counterparts and, as a consequence, ameliorate VCAM-1 overexpression induced by LPS (Shuvaev et al. 2011). It should be noted, however, that enzymes simply conjugated to targeting antibodies, are exposed to blood components and might be vulnerable to degradation. Thus, other efforts have used nanoparticles made of biodegradable polymers to encapsulate these enzymes, e.g., catalase. These PECAM-1-targeted nanoparticle formulations have been shown to bind to endothelial cells 40-fold higher compared to non-targeted carriers and, when injected in mice, bind preferentially to the pulmonary endothelium, protecting lung tissue from oxidative stress (Dziubla et al. 2008).

PECAM-1 targeting has also been explored in the context of thrombosis, particularly for prevention of thrombosis caused by ischemia-reperfusion. This is due to the role of PECAM-1 in blood hemostasis via homophilic interaction, which contributes to platelet aggregation and also binding of platelets to endothelial cells at sites of endothelial damage. In this regard, PECAM-1 has been used for targeting plasminogen activators, molecular proteases that cleave plasminogen into plasmin leading to fibrinolysis of blood clots (Ding et al. 2008). Similarly, PECAM-1 targeting has been tested for endothelial anchorage of thrombomodulin, an endothelial membrane molecule that modulates the ability of thrombin to activate protein C, leading to anti-inflammatory activity (Ding et al. 2009). Since during inflammation thrombomodulin expression is downregulated, supplying this molecule at sites of inflammation might ameliorate this feed-forward series of events that increase inflammation.

With regard to the subcellular transport of PECAM-1-targeting systems, early work showed binding of PECAM-1 antibodies to the endothelial surface with no internalization (Muzykantov et al. 1999, Muro et al. 2003). Yet, multivalent display of these antibodies, e.g., through formation of biotin-streptavidin linkages or by coupling said antibodies on the surface of polymer nanoparticles, was observed to induce endothelial endocytosis (Muzykantov et al. 1999). The pathway responsible for this uptake is accepted to be similar to that described above for ICAM-1, CAM-mediated endocytosis, although it is unknown whether differences exist between CAM-mediated uptake driven by ICAM-1 versus PECAM-1. However, dependency on NHE1 activity has also been documented, as endocytosis via PECAM-1 can be inhibited by drugs against this molecule, and actin reorganization into stress fibers have also been observed in this case (Muro et al. 2003). This leads to uptake of polymer particles 100–300 nm in diameter, which subsequently follow endo-lysosomal transport (Kozower et al. 2003, Muro et al. 2003, Garnacho et al. 2008b). This has been proven beneficial for delivery of therapeutic agents into cells, such as the case of PECAM-1-targeted polyplexes of polyethylimine (PEI) and plasmid DNA, capable of accumulating in lung tissue and transfecting endothelial cells (Li et al. 2000). Another example is that of PECAM-1-targeted conjugates of enzymes such as catalase or SOD, which provide antioxidant protection both when anchored to the endothelial plasmalemma and after internalization within endosomes in these cells. These strategies additionally prevent over-expression of other endothelial CAMs upon induction by inflammatory molecules, such as TNFα, IL-1β and LPS (Muzykantov 2001, Shuvaev et al. 2011).

Interestingly, a somewhat new concept in drug delivery has arisen specifically from studies on PECAM-1 targeting. This refers to the idea of differential drug delivery outcomes by targeting different epitopes within the same target molecule. When PECAM-1 is targeted using antibodies that bind to different regions of the protein, targeting outcomes can be radically different (Garnacho et al. 2008a). For instance, among a collection of five different antibodies that bind similarly to endothelial PECAM-1, an antibody directed to a PECAM-1 domain proximal to the plasma membrane fails to target 200 nm nanoparticles to the endothelium, likely due to steric hindrances. More surprisingly, among the remaining antibodies, all of which target 200 nm nanoparticles similarly to endothelial PECAM-1, one does not induce endocytosis of drug carriers and remains anchored to the cell surface, while the others elicit uptake within cells. Furthermore, when two of these antibodies capable of inducing nanoparticle uptake were tested in the context of intracellular trafficking, one of them induced nanoparticle transport to lysosomal compartments while carriers coated with the other antibody remained in pre-lysosomal compartments (Garnacho et al. 2008a). In another scenario, the use of a combination of two antibodies directed to extracellular domain D1 of PECAM-1, one as a part of a fusion protein consisting of anti-PECAM scFv plus thrombomodulin, and the other one administered free in solution, resulted in an increased binding capacity toward PECAM-1, leading to prevention of thrombotic activity (Chacko et al. 2012).

This series of unpredictable and diverse outcomes reflects the complex nature of PECAM-1 signaling and function. Thus, this versatility might be exploited in the future to expand these targeting systems into other applications.

Other Adhesion Molecules

Above, we provided a review of the most widely studied and used endothelial CAMs for targeted drug delivery. For these molecules, a vast pool of studies exists detailing their structure, function, signaling, expression patterns, and behavior when presented with a variety of drug targeting or carrier strategies under different settings. Yet, several other endothelial adhesion molecules have been described which may be exploited for similar purposes. In particular, these molecules might not be expressed as ubiquitously as ICAM-1 or PECAM-1, for instance, making them suitable for more tissue-specific drug delivery; or they might exhibit different expression patterns compared to the relatively similar inflammation-dependent expression of the selectins and Ig-like CAMs. We will focus this section on briefly describing a couple of these molecules, namely integrin $\alpha_v\beta_3$ and mucosal vascular addressin cell adhesion molecule 1 (MAdCAM-1), due to the more extensive body of work completed on these, compared to the remaining endothelial adhesion molecules potentially suitable for drug delivery. Although a more detailed understanding of the features and roles of these molecules is still needed, it is becoming apparent these markers also hold potential in the context of targeting therapeutic applications.

Similar to other integrins, $\alpha_v\beta_3$ is a heterodimer expressed on the cell surface (Wilder 2002), while MAdCAM-1 displays three extracellular regions, two N-terminal Ig-like domains and a mucin-like domain, which differentiates it from other Ig-like CAMs (Shyjan et al. 1996). Integrin $\alpha_v\beta_3$ binds several extracellular matrix proteins and vitronectin, along with other cell surface molecules (Wilder 2002), while MAdCAM-1, similar to the other molecules described here, binds to leukocyte integrins, particularly $\alpha_4\beta_7$ (Berlin et al. 1993).

The expression patterns of these molecules are not as clearly defined as those of the selectins and main Ig-like CAMs, but it is known that $\alpha_v\beta_3$ is expressed at low levels on most tissues under normal conditions and is highly expressed at inflammatory regions and tumors, as well as endothelial cells in angiogenic vessels (Scatena and Giachelli 2002). Expression of MAdCAM-1 also seems more restricted than that of the endothelial selectins , VCAM-1, ICAM, and PECAM-1, as it is found only in small and large intestine and spleen endothelium, with lower expression in the brain endothelium (Oshima et al. 2001). However, like the other CAMs, MAdCAM-1 becomes overexpressed in the presence of inflammatory cytokines (Sikorski et al. 1993). In both cases, these limited expression patterns appear to associate with the function of these molecules. Also known as the vitronectin receptor, $\alpha_v\beta_3$ has been implicated in angiogenesis, yet it is unclear what its specific role is in this phenomenon (Hodivala-Dilke 2008). For instance, inhibition and knock-out studies have shown that $\alpha_v\beta_3$ can have angiogenic and anti-angiogenic activity depending on context. Thus, the detailed function of this molecule remains elusive. Nonetheless, certain properties of $\alpha_v\beta_3$ are known, such as its expression patterns and its ability to bind given growth factors or anti-angiogenic molecules (Hodivala-Dilke 2008). Functionally, MAdCAM-1 appears to be similar to ICAM-1 and VCAM-1 regarding recruitment of leukocytes, yet due to its limited expression patterns, MAdCAM-1 seems to be responsible for leukocyte homing particularly at the gastrointestinal tract (Briskin et al. 1997). Whether this molecule shares similar function in leukocyte recruitment to ICAM-1 beyond a role in firm adhesion (i.e., signaling leading to cytoskeletal rearrangement or formation of transmigratory cups or pores) is still unknown.

As such, use of these molecules for therapeutic intervention has been proven useful in a number of scenarios. Blocking of $\alpha_v\beta_3$ with peptides or short proteins can ameliorate pathologies associated with angiogenesis, such as early glomerulonephritis and melanoma metastasis (Ramos et al. 2008). MAdCAM-1 blocking has been observed to decrease rejection of intestinal grafts in mice, due to the reduced infiltration of leukocytes, and also results in a positive therapeutic effect in mouse colitis model (Farkas et al. 2006). In addition, targeting of therapeutics and imaging strategies have been also pursued in the case of $\alpha_v\beta_3$, precisely in the realm of its function in angiogenesis. In particular, since tumor growth is associated with and depends largely on increasing blood supply via angiogenesis, a considerable effort has been dedicated to imaging these regions via $\alpha_v\beta_3$ (Schmieder et al. 2008). Lipid-based nanoparticles coated with an $\alpha_v\beta_3$ ligand can be used to deliver anti-cancer agents via specific targeting of angiogenic endothelium at sites of tumor growth, which has rendered positive therapeutic effects in mice for both implanted and established metastatic tissues (Hood et al. 2002). Thus, it appears that $\alpha_v\beta_3$ targeting has potential for treatment of cancer, yet any example of pathological neovascularization could also be an appropriate context for this type of targeting approach, as is the case of atherosclerosis or arthritis (Winter et al. 2006). In addition, using $\alpha_v\beta_3$ targeting in choroidal membrane neovascularization, which is linked to macular degeneration and blindness, dominant negative kinase Raf can be delivered *in vivo*, reducing the size of neovascular membranes by promoting site-specific apoptosis of these cells (Salehi-Had et al. 2011). Other potential applications for targeting of $\alpha_v\beta_3$ might include modulation of aberrant vessel remodeling or building which, although yet unexplored, holds potential in the context of several vascular conditions. In the case of MAdCAM-1, it has been shown that targeting of lipid shell microbubbles to this molecule supports imaging of inflammatory bowel disease. These formulations bind to endothelial cells in culture under physiological shear stress and, when injected in mouse model of Crohn's-like ileitis, MAdCAM-1-targeted microbubbles accumulate at areas of high expression, allowing for specific imaging of intestinal inflammation (Bachmann et al. 2006).

Despite a lack of extensive studies, there is also potential for the use of several other endothelial adhesion molecules in the context of drug delivery. For instance ICAM-2 and ICAM-3 display two and five extracellular Ig-like domains respectively, through which they bind LFA-1, like ICAM-1. Indeed, ICAM-2 has been implicated in leukocyte transmigration, yet only under specific circumstances (Huang et al. 2006). Although the signaling pathways associated with ICAM-2 and ICAM-3 are not as well described as those linked to ICAM-1, similar events have been observed, such as modulation of cell-cell junction integrity, and association with the actin cytoskeleton or actin adaptors (Serrador et al. 1997). More is known about ICAM-2 and ICAM-3 expressed on circulating cells, and, thus, better knowledge about the behavior of these molecules will be required regarding endothelial cells in order to develop targeting strategies in this direction. Current studies only have explored targeting of dendritic cells via these molecules (Cruz et al. 2011), but their findings could promote further exploration of their use in endothelial targeting. Finally, CD34, CD99, JAM-A and JAM-C, and VAP-1 are surface adhesion molecules that have been targeted in other cell types, for either imaging purposes or antibody-based blocking therapies (Scotlandi et al. 2006, Ujula et al. 2009, Goetsch et al. 2012). Further exploration of this variety of molecules will reveal their potential use for diversifying future endothelial targeting technologies.

Acknowledgements

The authors thank the National Science Foundation Graduate Research Fellowship Program (D.S.), and the National Institutes of Health (R01-HL98416) and the American Heart Association (09BGIA2450014) (S.M.).

References

Aird, W.C. 2008. Endothelium in health and disease. *Pharmacol. Rep.* 60: 139–143.

Albelda, S.M., P.D. Oliver, L.H. Romer and C.A. Buck. 1990. EndoCAM: a novel endothelial-cell cell-adhesion molecule. *J. Cell Biol.* 110: 1227–1237.

Ali, M., A.E. Hicks, P.G. Hellewell, G. Thoma and K.E. Norman. 2004. Polymers bearing sLe^x-mimetics are superior inhibitors of E-selectin-dependent leukocyte rolling *in vivo*. *FASEB J.* 18: 152–154.

Alon, R., P.D. Kassner, M.W. Carr, E.B. Finger, M.E. Hemler and T.A. Springer. 1995. The Integrin VLA-4 Supports Tethering and Rolling in Flow on VCAM-1. *J. Cell Biol.* 128: 1243–1253.

Asgeirsdottir, S.A., E.G. Talman, I.A. De Graaf, J.A.A.M. Kamps, S.C. Satchell, P.W. Mathieson, M.H.J. Ruiters and G. Molema. 2010. Targeted transfection increases siRNA uptake and gene silencing of primary endothelial cells *in vitro*—A quantitative study. *J. Control. Release* 141: 241–251.

Bachmann, C., A.L. Klibanov, T.S. Olson, J.R. Sonnenschein, J. Rivera-Nieves, F. Cominelli, K.F. Ley, J.R. Lindner and T.T. Pizarro. 2006. Targeting mucosal addressin cellular adhesion molecule (MAdCAM)-1 to noninvasively image experimental Crohn's disease. *Gastroenterology* 130: 8–16.

Barreiro, O., M. Yanez-Mo, J.M. Serrador, M.C. Montoya, M. Vicente-Manzanares, R. Tejedor, H. Furthmayr and F. Sanchez-Madrid. 2002. Dynamic interaction of VCAM-1 and ICAM-1 with moesin and ezrin in a novel endothelial docking structure for adherent leukocytes. *J. Cell Biol.* 157: 1233–1245.

Barreiro, O., M. Zamai, M. Yáñez-Mó, E. Tejera, P. López-Romero, P.N. Monk, E. Gratton, V.R. Caiolfa and F. Sánchez-Madrid. 2008. Endothelial adhesion receptors are recruited to adherent leukocytes by inclusion in preformed tetraspanin nanoplatforms. *J. Cell Biol.* 183: 527–542.

Berlin, C., E.L. Berg, M.J. Briskin, D.P. Andrew, P.J. Kilshaw, B. Holzmann, I.L. Weissman, A. Hamann and E.C. Butcher. 1993. $\alpha_4\beta_7$ integrin mediates lymphocyte binding to the mucosal vascular addressin MAdCAM-1. *Cell* 74: 185–195.

Bevilacqua, M.P., J.S. Pober, D.L. Mendrick, R.S. Cotran and M.A. Gimbrone. 1987. Identification of an inducible endothelial leukocyte adhesion molecule. *Proc. Natl. Acad. Sci. USA* 84: 9238–9242.

Bogen, S., J. Pak, M. Garifallou, X. Deng and W.A. Muller. 1994. Monoclonal antibody to murine PECAM-1 (CD31) blocks acute inflammation *in vivo*. *J. Exp. Med.* 179: 1059–1064.

Boutry, S., C. Burtea, S. Laurent, G. Toubeau, L. Vander Elst and R.N. Muller. 2005. Magnetic resonance imaging of inflammation with a specific selectin-targeted contrast agent. *Magn. Reson. Med.* 53: 800–807.

Briskin, M., D. Winsor-Hines, A. Shyjan, N. Cochran, S. Bloom, J. Wilson, L.M. Mcevoy, E.C. Butcher, N. Kassam, C.R. Mackay, W. Newman and D.J. Ringler. 1997. Human mucosal addressin cell adhesion molecule-1 is preferentially expressed in intestinal tract and associated lymphoid tissue. *Am. J. Pathol.* 151: 97–110.

Burns, R.C., J. Rivera-Nieves, C.A. Moskaluk, S. Matsumoto, F. Cominelli and K. Ley. 2001. Antibody blockade of ICAM-1 and VCAM-1 ameliorates inflammation in the SAMP-1/Yit adoptive transfer model of Crohn's disease in mice. *Gastroenterology* 121: 1428–1436.

Burtea, C., S. Laurent, M. Port, E. Lancelot, S. Ballet, O. Rousseaux, G. Toubeau, L.V. Elst, C. Corot and R.N. Muller. 2009. Magnetic resonance molecular imaging of vascular cell adhesion molecule-1 expression in inflammatory lesions using a peptide-vectorized paramagnetic imaging probe. *J. Med. Chem.* 52: 4725–4742.

Calderon, A.J., T. Bhowmick, J. Leferovich, B. Burman, B. Pichette, V. Muzykantov, D.M. Eckmann and S. Muro. 2011. Optimizing endothelial targeting by modulating the antibody density and particle concentration of anti-ICAM coated carriers. *J. Control. Release* 150: 37–44.

Carman, C.V. 2009. Mechanisms for transcellular diapedesis: probing and pathfinding by 'invadosome-like protrusions'. *J. Cell Sci.* 122: 3025–3035.

Carman, C.V., P.T. Sage, T.E. Sciuto, M.A. De La Fuente, R.S. Geha, H.D. Ochs, H.F. Dvorak, A.M. Dvorak and T.A. Springer. 2007. Transcellular diapedesis is initiated by invasive podosomes. *Immunity* 26: 784–797.

Carrithers, M., S. Tandon, S. Canosa, M. Michaud, D. Graesser and J.A. Madri. 2005. Enhanced susceptibility to endotoxic shock and impaired STAT3 signaling in CD31-deficient mice. *Am. J. Pathol.* 166: 185–196.

Chacko, A.-M., M. Nayak, C.F. Greineder, H.M. Delisser and V.R. Muzykantov. 2012. Collaborative enhancement of antibody binding to distinct PECAM-1 epitopes modulates endothelial targeting. *PloS one* 7: e34958.

Chan, B.M.C., M.J. Elices, E. Murphy and M.E. Hemler. 1992. Adhesion to vascular cell adhesion molecule 1 and fibronectin. Comparison of $\alpha^4\beta_1$alpha 4 beta 1 (VLA-4) and $\alpha^4\beta_7$ on the human B cell line JY. *J. Biol. Chem.* 267: 8366–8370.

Chiu, G.N.C., M.B. Bally and L.D. Mayer. 2003. Targeting of antibody conjugated, phosphatidylserine-containing liposomes to vascular cell adhesion molecule 1 for controlled thrombogenesis. *Biochim. Biophys. Acta-Biomembranes* 1613: 115–121.

Choi, K.S., S.H. Kim, Q.Y. Cai, S.Y. Kim, H.O. Kim, H.J. Lee, E.A. Kim, S.E. Yoon, K.J. Yun and K.H. Yoon. 2007. Inflammation-specific T1 imaging using anti-intercellular adhesion molecule 1 antibody-conjugated gadolinium diethylenetriaminepentaacetic acid. *Mol. Imaging* 6: 75–84.

Cook-Mills, J.M., M.E. Marchese and H. Abdala-Valencia. 2011. Vascular cell adhesion molecule-1 expression and signaling during disease: regulation by reactive oxygen species and antioxidants. *Antioxid. Redox Signal.* 15: 1607–1638.

Cruz, L.J., P.J. Tacken, F. Bonetto, S.I. Buschow, H.J. Croes, M. Wijers, I.J. De Vries and C.G. Figdor. 2011. Multimodal imaging of nanovaccine carriers targeted to human dendritic cells. *Mol. Pharm.* 8: 520–531.

D'souza, S.E., V.J. Byers-Ward, E.E. Gardiner, H. Wang and S.S. Sung. 1996. Identification of an active sequence within the first immunoglobulin domain of intercellular cell adhesion molecule-1 (ICAM-1) that interacts with fibrinogen. *J. Biol. Chem.* 271: 24270–24277.

Delisser, H.M., H.C. Yan, P.J. Newman, W.A. Muller, C.A. Buck and S.M. Albelda. 1993. Platelet/endothelial cell adhesion molecule-1 (CD31)-mediated cellular aggregation involves cell surface glycosaminoglycans. *J. Biol. Chem.* 268: 16037–16046.

Deosarkar, S.P., R. Malgor, J. Fu, L.D. Kohn, J. Hanes and D.J. Goetz. 2008. Polymeric particles conjugated with a ligand to VCAM-1 exhibit selective, avid, and focal adhesion to sites of atherosclerosis. *Biotechnol. Bioeng.* 101: 400–407.

Dickerson, J.B., J.E. Blackwell, J.J. Ou, V.R.S. Patil and D.J. Goetz. 2001. Limited adhesion of biodegradable microspheres to E- and P-selectin under flow. *Biotechnol. Bioeng.* 73: 500–509.

Dienst, A., A. Grunow, M. Unruh, B. Rabausch, J.E. Nor, J.W. Fries and C. Gottstein. 2005. Specific occlusion of murine and human tumor vasculature by VCAM-1-targeted recombinant fusion proteins. *J. Natl. Cancer Inst.* 97: 733–747.

Ding, B.-S., N. Hong, J.-C. Murciano, K. Ganguly, C. Gottstein, M. Christofidou-Solomidou, S.M. Albelda, A.B. Fisher, D.B. Cines and V.R. Muzykantov. 2008. Prophylactic thrombolysis by thrombin-activated latent prourokinase targeted to PECAM-1 in the pulmonary vasculature. *Blood* 111: 1999–2006.

Ding, B.S., N. Hong, M. Christofidou-Solomidou, C. Gottstein, S.M. Albelda, D.B. Cines, A.B. Fisher and V.R. Muzykantov. 2009. Anchoring fusion thrombomodulin to the endothelial lumen protects against injury-induced lung thrombosis and inflammation. *Am. J. Respir. Crit. Care Med.* 180: 247–256.

Discher, D.E. and A. Eisenberg. 2002. Polymer vesicles. *Science* 297: 967–973.

Duncan, R. 2003. The dawning era of polymer therapeutics. *Nat. Rev. Drug Discov.* 2: 347–360.

Dziubla, T.D., V.V. Shuvaev, N.K. Hong, B.J. Hawkins, M. Madesh, H. Takano, E. Simone, M.T. Nakada, A. Fisher, S.M. Albelda and V.R. Muzykantov. 2008. Endothelial targeting of semi-permeable polymer nanocarriers for enzyme therapies. *Biomaterials* 29: 215–227.

Ehrhardt, C., C. Kneuer and U. Bakowsky. 2004. Selectins—An emerging target for drug delivery. *Adv. Drug Delivery Rev.* 56: 527–549.

El-Sayed, M.E., C.A. Rhodes, M. Ginski and H. Ghandehari. 2003. Transport mechanism(s) of poly (amidoamine) dendrimes across Caco-2 cell monolayers. *Int. J. Pharm.* 265: 151–157.

Eniola, A.O. and D.A. Hammer. 2005. *In vitro* characterization of leukocyte mimetic for targeting therapeutics to the endothelium using two receptors. *Biomaterials* 26: 7136–7144.

Etienne, S., P. Adamson, J. Greenwood, A.D. Strosberg, S. Cazaubon and P.O. Couraud. 1998. ICAM-1 signaling pathways associated with Rho activation in microvascular brain endothelial cells. *J. Immunol.* 161: 5755–5761.

Everts, M., R.J. Kok, S.A. Asgeirsdottir, B.N. Melgert, T.J.M. Moolenaar, G.A. Koning, M.J.A. Van Luyn, D.K.F. Meijer and G. Molema. 2002. Selective intracellular delivery of dexamethasone into activated endothelial cells using an E-selectin-directed immunoconjugate. *J. Immunol.* 168: 883–889.

Everts, M., G.A. Koning, R.J. Kok, S.A. Asgeirsdottir, D. Vestweber, D.K. Meijer, G. Storm and G. Molema. 2003. *In vitro* cellular handling and *in vivo* targeting of E-selectin-directed immunoconjugates and immunoliposomes used for drug delivery to inflamed endothelium. *Pharm. Res.* 20: 64–72.

Farkas, S., M. Hornung, C. Sattler, K. Edtinger, M. Steinbauer, M. Anthuber, H.J. Schlitt, H. Herfarth and E.K. Geissler. 2006. Blocking MAdCAM-1 *in vivo* reduces leukocyte extravasation and reverses chronic inflammation in experimental colitis. *Int. J. Colorectal Dis.* 21: 71–78.

Federici, C., L. Camoin, M. Hattab, A.D. Strosberg and P.O. Couraud. 1996. Association of the cytoplasmic domain of intercellular-adhesion molecule-1 with glyceraldehyde-3-phosphate dehydrogenase and beta-tubulin. *Eur. J. Biochem.* 238: 173–180.

Ferrante, E.A., J.E. Pickard, J. Rychak, A. Klibanov and K. Ley. 2009. Dual targeting improves microbubble contrast agent adhesion to VCAM-1 and P-selectin under flow. *J. Control. Release* 140: 100–107.

Ferrero, E., M.E. Ferrero, R. Pardi and M.R. Zocchi. 1995. The platelet endothelial cell adhesion molecule-1 (PECAM1) contributes to endothelial barrier function. *FEBS Lett.* 374: 323–326.

Freedman, A.S., J.M. Munro, G.E. Rice, M.P. Bevilacqua, C. Morimoto, B.W. Mcintyre, K. Rhynhart, J.S. Pober and L.M. Nadler. 1990. Adhesion of human B cells to germinal centers *in vitro* involves VLA-4 and INCAM-110. *Science* 249: 1030–1033.

Funovics, M., X. Montet, F. Reynolds, R. Weissleder and L. Josephson. 2005. Nanoparticles for the optical imaging of tumor E-selectin. *Neoplasia* 7: 904–911.

Garnacho, C., S.M. Albelda, V.R. Muzykantov and S. Muro. 2008a. Differential intra-endothelial delivery of polymer nanocarriers targeted to distinct PECAM-1 epitopes. *J. Control. Release* 130: 226–233.

Garnacho, C., R. Dhami, E. Simone, T. Dziubla, J. Leferovich, E.H. Schuchman, V. Muzykantov and S. Muro. 2008b. Delivery of acid sphingomyelinase in normal and niemann-pick disease mice using intercellular adhesion molecule-1-targeted polymer nanocarriers. *J. Pharmacol. Exp. Ther.* 325: 400–408.

Garnacho, C., D. Serrano and S. Muro. 2012. A fibrinogen-derived peptide provides intercellular adhesion molecule-1-specific targeting and intraendothelial transport of polymer nanocarriers in human cell cultures and mice. *J. Pharmacol. Exp. Ther.* 340: 638–647.

Ghaffarian, R., T. Bhowmick and S. Muro. 2012. Transport of nanocarriers across gastrointestinal epithelial cells by a new transcellular route induced by targeting ICAM-1. *J. Control Release* 163: 25–33.

Goetsch, L., J.F. Haeuw, C. Beau-Larvor, A. Gonzalez, L. Zanna, M. Malissard, A.M. Lepecquet, A. Robert, C. Bailly, M. Broussas and N. Corvaia. 2012. A novel role for junctional adhesion molecule-A (JAM-A) in tumor proliferation: Modulation by an anti-JAM-A monoclonal antibody. *Int. J. Cancer.* 132: 1463–1474.

Goldberger, A., K.A. Middleton, J.A. Oliver, C. Paddock, H.C. Yan, H.M. Delisser, S.M. Albelda and P.J. Newman. 1994. Biosynthesis and processing of the cell adhesion molecule PECAM-1 includes production of a soluble form. *J. Biol. Chem.* 269: 17183–17191.

Gong, N., H. Wei, S.H. Chowdhury and S. Chatterjee. 2004. Lactosylceramide recruits PKCα/ε and phospholipase A$_2$ to stimulate PECAM-1 expression in human monocytes and adhesion to endothelial cells. *Proc. Natl. Acad. Sci. USA* 101: 6490–6495.

Grayson, M.H., M. Van Der Vieren, S.A. Sterbinsky, W.M. Gallatin, P.A. Hoffman, D.E. Staunton and B.S. Bochner. 1998. αdβ2 integrin is expressed on human eosinophils and functions as an alternative ligand for vascular cell adhesion molecule 1 (VCAM-1). *J. Exp. Med.* 188: 2187–2191.

Gunawan, R.C. and D.T. Auguste. 2010. Immunoliposomes that target endothelium *in vitro* are dependent on lipid raft formation. *Mol. Pharm.* 7: 1569–1575.

Hamilton, A.J., S.L. Huang, D. Warnick, M. Rabbat, B. Kane, A. Nagaraj, M. Klegerman and D.D. Mcpherson. 2004. Intravascular ultrasound molecular imaging of atheroma components *in vivo*. *J. Am. Coll. Cardiol.* 43: 453–460.

Heiska, L., K. Alfthan, M. Grönholm, P. Vilja, A. Vaheri and O. Carpén. 1998. Association of ezrin with intercellular adhesion molecule-1 and -2 (ICAM-1 and ICAM-2). Regulation by phosphatidylinositol 4, 5-bisphosphate. *J. Biol. Chem.* 273: 21893–21900.

Henninger, D.D., J. Panes, M. Eppihimer, J. Russell, M. Gerritsen, D.C. Anderson and D.N. Granger. 1997. Cytokine-induced VCAM-1 and ICAM-1 expression in different organs of the mouse. *J. Immunol.* 158: 1825–1832.

Hodivala-Dilke, K. 2008. α$_v$β$_3$ integrin and angiogenesis: a moody integrin in a changing environment. *Curr. Opin. Cell Biol.* 20: 514–519.

Hood, J.D., M. Bednarski, R. Frausto, S. Guccione, R.A. Reisfeld, R. Xiang and D.A. Cheresh. 2002. Tumor regression by targeted gene delivery to the neovasculature. *Science* 296: 2404–2407.

Hsu, J., L. Northrup, T. Bhowmick and S. Muro. 2012. Enhanced delivery of alpha-glucosidase for Pompe disease by ICAM-1-targeted nanocarriers: comparative performance of a strategy for three distinct lysosomal storage disorders. *Nanomedicine* 8: 731–739.

Huang, M.T., K.Y. Larbi, C. Scheiermann, A. Woodfin, N. Gerwin, D.O. Haskard and S. Nourshargh. 2006. ICAM-2 mediates neutrophil transmigration *in vivo*: evidence for stimulus specificity and a role in PECAM-1-independent transmigration. *Blood* 107: 4721–4727.

Huang, Y.-T., S.-U. Chen, C.-H. Chou and H. Lee. 2008. Sphingosine 1-phosphate induces platelet/endothelial cell adhesion molecule-1 phosphorylation in human endothelial cells through cSrc and Fyn. *Cell Signal.* 20: 1521–1527.

Ilan, N., A. Mohsenin, L. Cheung and J.A. Madri. 2001. PECAM-1 shedding during apoptosis generates a membrane-anchored truncated molecule with unique signaling characteristics. *FASEB J.* 15: 362–372.

Isobe, M., H. Yagita, K. Okumura and A. Ihara. 1992. Specific acceptance of cardiac allograft after treatment with antibodies to ICAM-1 and LFA-1. *Science* 255: 1125–1127.

Jamar, F., F.A. Houssiau, J.P. Devogelaer, P.T. Chapman, D.O. Haskard, V. Beaujean, C. Beckers, D.H. Manicourt and A.M. Peters. 2002. Scintigraphy using a technetium 99m-labelled anti-E-selectin Fab fragment in rheumatoid arthritis. *Rheumatology* 41: 53–61.

Jefferson, A., R.S. Wijesurendra, M.A. Mcateer, J.E. Digby, G. Douglas, T. Bannister, F. Perez-Balderas, Z. Bagi, A.C. Lindsay and R.P. Choudhury. 2011. Molecular imaging with optical coherence tomography using ligand-conjugated microparticles that detect activated endothelial cells: Rational design through target quantification. *Atherosclerosis* 219: 579–587.

John, A.E., N.W. Lukacs, A.A. Berlin, A. Palecanda, R.F. Bargatze, L.M. Stoolman and J.O. Nagy. 2003. Discovery of a potent nanoparticle P-selectin antagonist with anti-inflammatory effects in allergic airway disease. *FASEB J.* 17: 2296–2298.

Jubeli, E., L. Moine, J. Vergnaud-Gauduchon and G. Barratt. 2011. E-selectin as a target for drug delivery and molecular imaging. *J. Control Release* 158: 194–206.

Justicia, C., A. Martin, S. Rojas, M. Gironella, A. Cervera, J. Panes, A. Chamorro and A.M. Planas. 2006. Anti-VCAM-1 antibodies did not protect against ischemic damage either in rats or in mice. *J. Cereb. Blood Flow Metab.* 26: 421–432.

Kavanaugh, A.F., L.S. Davis, R.I. Jain, L.A. Nichols, S.H. Norris and P.E. Lipsky. 1996. A phase I/II open label study of the safety and efficacy of an anti-ICAM-1 (intercellular adhesion molecule-1; CD54) monoclonal antibody in early rheumatoid arthritis. *J. Rheumatol.* 23: 1338–1344.

Kavanaugh, A.F., H. Schulze-Koops, L.S. Davis and P.E. Lipsky. 1997. Repeat treatment of rheumatoid arthritis patients with a murine anti-intercellular adhesion molecule 1 monoclonal antibody. *Arthritis Rheum.* 40: 849–853.

Kessner, S., A. Krause, U. Rothe and G. Bendas. 2001. Investigation of the cellular uptake of E-Selectin-targeted immunoliposomes by activated human endothelial cells. *Biochim. Biophys. Acta=Biomembranes* 1514: 177–190.

Kim, K., S.W. Huang, S. Ashkenazi, M. O'donnell, A. Agarwal, N.A. Kotov, M.F. Denny and M.J. Kaplan. 2007. Photoacoustic imaging of early inflammatory response using gold nanorods. *Appl. Phys. Lett.* 90: 223901.

Kneuer, C., C. Ehrhardt, M.W. Radomski and U. Bakowsky. 2006. Selectins—Potential pharmacological targets? *Drug Discov. Today* 11: 1034–1040.

Kozower, B.D., M. Christofidou-Solomidou, T.D. Sweitzer, S. Muro, D.G. Buerk, C.C. Solomides, S.M. Albelda, G.A. Patterson and V.R. Muzykantov. 2003. Immunotargeting of catalase to the pulmonary endothelium alleviates oxidative stress and reduces acute lung transplantation injury. *Nat. Biotechnol.* 21: 392–398.

Kusterer, K., J. Bojunga, M. Enghofer, E. Heidenthal, K.H. Usadel, H. Kolb and S. Martin. 1998. Soluble ICAM-1 reduces leukocyte adhesion to vascular endothelium in ischemia-reperfusion injury in mice. *Am. J. Physiol.* 275: G377–380.

Kuwahara, F., H. Kai, K. Tokuda, H. Niiyama, N. Tahara, K. Kusaba, K. Takemiya, A. Jalalidin, M. Koga, T. Nagata, R. Shibata and T. Imaizumi. 2003. Roles of intercellular adhesion molecule-1 in hypertensive cardiac remodeling. *Hypertension* 41: 819–823.

Langer, R. 1998. Drug delivery and targeting. *Nature* 392: 5–10.

Laschinger, M. and B. Engelhardt. 2000. Interaction of α4-integrin with VCAM-1 is involved in adhesion of encephalitogenic T cell blasts to brain endothelium but not in their transendothelial migration *in vitro*. *J. Neuroimmunol.* 102: 32–43.

Lawrence, M.B. and T.A. Springer. 1991. Leukocytes roll on a selectin at physiologic flow rates: distinction from and prerequisite for adhesion through integrins. *Cell* 65: 859–873.

Ley, K., C. Laudanna, M.I. Cybulsky and S. Nourshargh. 2007. Getting to the site of inflammation: the leukocyte adhesion cascade updated. *Nat. Rev. Immunol.* 7: 678–689.

Li, S., Y. Tan, E. Viroonchatapan, B.R. Pitt and L. Huang. 2000. Targeted gene delivery to pulmonary endothelium by anti-PECAM antibody. *Am. J. Physiol. Lung Cell Mol. Physiol.* 278: L504–511.

Liao, F., H.K. Huynh, A. Eiroa, T. Greene, E. Polizzi and W.A. Muller. 1995. Migration of monocytes across endothelium and passage through extracellular matrix involve separate molecular domains of PECAM-1. *J. Exp. Med.* 182: 1337–1343.

Lobb, R., G. Chi-Rosso, D. Leone, M. Rosa, B. Newman, S. Luhowskyj, L. Osborn, S. Schiffer, C. Benjamin, I. Dougas, C. Hession and P. Chow. 1991. Expression and functional characterization of a soluble form of vascular cell adhesion molecule 1. *Biochem. Biophys. Res. Commun.* 178: 1498–1504.

Lorenzon, P., E. Vecile, E. Nardon, E. Ferrero, J.M. Harlan, F. Tedesco and A. Dobrina. 1998. Endothelial cell E- and P-selectin and vascular cell adhesion molecule-1 function as signaling receptors. *J. Cell Biol.* 142: 1381–1391.

Mamdouh, Z., A. Mikhailov and W.A. Muller. 2009. Transcellular migration of leukocytes is mediated by the endothelial lateral border recycling compartment. *J. Exp. Med.* 206: 2795–2808.

Mann, A.P., T. Tanaka, A. Somasunderam, X. Liu, D.G. Gorenstein and M. Ferrari. 2011. E-selectin-targeted porous silicon particle for nanoparticle delivery to the bone marrow. *Adv. Mater.* 23: H278–282.

Matheny, H.E., T.L. Deem and J.M. Cook-Mills. 2000. Lymphocyte migration through monolayers of endothelial cell lines involves VCAM-1 signaling via endothelial cell NADPH oxidase. *J. Immunol.* 164: 6550–6559.

Mcever, R.P., J.H. Beckstead, K.L. Moore, L. Marshallcarlson and D.F. Bainton. 1989. GMP-140, a platelet alpha-granule membrane-protein, is also synthesized by vascular endothelial-cells and is localized in Weibel-Palade bodies. *J. Clin. Investig.* 84: 92–99.

Mehta, P., R.D. Cummings and R.P. Mcever. 1998. Affinity and kinetic analysis of P-selectin binding to P-selectin glycoprotein ligand-1. *J. Biol. Chem.* 273: 32506–32513.

Merchant, S.H., D.M. Gurule and R.S. Larson. 2003. Amelioration of ischemia-reperfusion injury with cyclic peptide blockade of ICAM-1. *Am. J. Physiol. Heart Circ. Physiol.* 284: H1260–1268.

Millán, J., L. Hewlett, M. Glyn, D. Toomre, P. Clark and A.J. Ridley. 2006. Lymphocyte transcellular migration occurs through recruitment of endothelial ICAM-1 to caveola- and F-actin-rich domains. *Nat. Cell Biol.* 8: 113–123.

Moghimi, S.M., A.C. Hunter and J.C. Murray. 2001. Long-circulating and target-specific nanoparticles: theory to practice. *Pharmacol. Rev.* 53: 283–318.

Moy, P., R. Lobb, R. Tizard, D. Olson and C. Hession. 1993. Cloning of an inflammation-specific phosphatidyl inositol-linked form of murine vascular cell adhesion molecule-1. *J. Biol. Chem.* 268: 8835–8841.

Murciano, J.-C., S. Muro, L. Koniaris, M. Christofidou-Solomidou, D.W. Harshaw, S.M. Albelda, D.N. Granger, D.B. Cines and V.R. Muzykantov. 2003. ICAM-directed vascular immunotargeting of antithrombotic agents to the endothelial luminal surface. *Blood* 101: 3977–3984.

Muro, S. 2007. Intercellular adhesion molecule 1 and vascular cell adhesion molecule 1. *In:* W.C. Aird. [ed.]. Endothelial biomedicine. Cambridge University Press. New York, NY, pp. 1058–1070.

Muro, S. 2012. Strategies for delivery of therapeutics into the central nervous system for treatment of lysosomal storage disorders. *Drug Deliv. Transl. Res.* 2: 169–186.

Muro, S., C. Gajewski, M. Koval and V.R. Muzykantov. 2005. ICAM-1 recycling in endothelial cells: a novel pathway for sustained intracellular delivery and prolonged effects of drugs. *Blood* 105: 650–658.

Muro, S., C. Garnacho, J.A. Champion, J. Leferovich, C. Gajewski, E.H. Schuchman, S. Mitragotri and V.R. Muzykantov. 2008. Control of endothelial targeting and intracellular delivery of therapeutic enzymes by modulating the size and shape of ICAM-1-targeted carriers. *Mol. Ther.* 16: 1450–1458.

Muro, S., M. Koval and V. Muzykantov. 2004. Endothelial endocytic pathways: gates for vascular drug delivery. *Curr. Vasc. Pharmacol.* 2: 281–299.

Muro, S., M. Mateescu, C. Gajewski, M. Robinson, V.R. Muzykantov and M. Koval. 2006. Control of intracellular trafficking of ICAM-1-targeted nanocarriers by endothelial Na$^+$/H$^+$ exchanger proteins. *Am. J. Physiol. Lung Cell Mol. Physiol.* 290: L809–817.

Muro, S., R. Wiewrodt, A. Thomas, L. Koniaris, S.M. Albelda, V.R. Muzykantov and M. Koval. 2003. A novel endocytic pathway induced by clustering endothelial ICAM-1 or PECAM-1. *J. Cell Sci.* 116: 1599–1609.

Murohara, T., J.A. Delyani, S.M. Albelda and A.M. Lefer. 1996. Blockade of platelet endothelial cell adhesion molecule-1 protects against myocardial ischemia and reperfusion injury in cats. *J. Immunol.* 156: 3550–3557.

Muzykantov, V.R. 2001. Targeting of superoxide dismutase and catalase to vascular endothelium. *J. Control. Release* 71: 1–21.

Muzykantov, V.R. 2011. Targeted therapeutics and nanodevices for vascular drug delivery: quo vadis? *IUBMB Life* 63: 583–585.

Muzykantov, V.R., M. Christofidou-Solomidou, I. Balyasnikova, D.W. Harshaw, L. Schultz, A.B. Fisher and S.M. Albelda. 1999. Streptavidin facilitates internalization and pulmonary targeting of an anti-endothelial cell antibody (platelet-endothelial cell adhesion molecule 1): a strategy for vascular immunotargeting of drugs. *Proc. Natl. Acad. Sci. USA* 96: 2379–2384.

Newman, P.J., M.C. Berndt, J. Gorski, G.C. White, S. Lyman, C. Paddock and W.A. Muller. 1990. PECAM-1 (CD31) cloning and relation to adhesion molecules of the immunoglobulin gene superfamily. *Science* 247: 1219–1222.

Newman, P.J. and D.K. Newman. 2003. Signal transduction pathways mediated by PECAM-1: new roles for an old molecule in platelet and vascular cell biology. *Arterioscler. Thromb. Vasc. Biol.* 23: 953–964.

Oh, H.-M., S. Lee, B.-R. Na, H. Wee, S.-H. Kim, S.-C. Choi, K.-M. Lee and C.-D. Jun. 2007. RKIKK motif in the intracellular domain is critical for spatial and dynamic organization of ICAM-1: functional implication for the leukocyte adhesion and transmigration. *Mol. Biol. Cell* 18: 2322–2335.

Osborn, L., C. Hession, R. Tizard, C. Vassallo, S. Luhowskyj, G. Chi-Rosso and R. Lobb. 1989. Direct expression cloning of vascular cell adhesion molecule 1, a cytokine-induced endothelial protein that binds to lymphocytes. *Cell* 59: 1203–1211.

Osborn, L., C. Vassallo and C.D. Benjamin. 1992. Activated endothelium binds lymphocytes through a novel binding site in the alternately spliced domain of vascular cell adhesion molecule-1. *J. Exp. Med.* 176: 99–107.

Oshima, T., K.P. Pavlick, F.S. Laroux, S.K. Verma, P. Jordan, M.B. Grisham, L. Williams and J.S. Alexander. 2001. Regulation and distribution of MAdCAM-1 in endothelial cells *in vitro. Am. J. Physiol. Cell Physiol.* 281: C1096–1105.

Panyam, J. and V. Labhasetwar. 2003. Biodegradable nanoparticles for drug and gene delivery to cells and tissue. *Adv. Drug Deliv. Rev.* 55: 329–347.

Park, S., S. Kang, A.J. Veach, Y. Vedvyas, R. Zarnegar, J.Y. Kim and M.M. Jin. 2010. Self-assembled nanoplatform for targeted delivery of chemotherapy agents via affinity-regulated molecular interactions. *Biomaterials* 31: 7766–7775.

Patil, V.R.S., C.J. Campbell, Y.H. Yun, S.M. Slack and D.J. Goetz. 2001. Particle diameter influences adhesion under flow. *Biophys. J.* 80: 1733–1743.

Phillips, D.R. and P.P. Agin. 1977. Platelet plasma-membrane glycoproteins. Evidence for presence of nonequivalent disulfide bonds using nonreduced-reduced two-dimensional gel-electrophoresis. *J. Biol. Chem.* 252: 2121–2126.

Phillips, M.L., E. Nudelman, F.C. Gaeta, M. Perez, A.K. Singhal, S. Hakomori and J.C. Paulson. 1990. ELAM-1 mediates cell adhesion by recognition of a carbohydrate ligand, sialyl-Le^x. *Science* 250: 1130–1132.

Piali, L., P. Hammel, C. Uherek, F. Bachmann, R.H. Gisler, D. Dunon and B.A. Imhof. 1995. CD31/PECAM-1 is a ligand for $\alpha_v\beta_3$ integrin involved in adhesion of leukocytes to endothelium. *J. Cell Biol.* 130: 451–460.

Pluskota, E., Y. Chen and S.E. D'souza. 2000. Src homology domain 2-containing tyrosine phosphatase 2 associates with intercellular adhesion molecule 1 to regulate cell survival. *J. Biol. Chem.* 275: 30029–30036.

Privratsky, J.R., D.K. Newman and P.J. Newman. 2010. PECAM-1: conflicts of interest in inflammation. *Life Sci.* 87: 69–82.

Privratsky, J.R., C.M. Paddock, O. Florey, D.K. Newman, W.A. Muller and P.J. Newman. 2011. Relative contribution of PECAM-1 adhesion and signaling to the maintenance of vascular integrity. *J. Cell Sci.* 124: 1477–1485.

Qi, J., D.L. Kreutzer and T.H. Piela-Smith. 1997. Fibrin induction of ICAM-1 expression in human vascular endothelial cells. *J. Immunol.* 158: 1880–1886.

Rahman, A., K.N. Anwar, A.L. True and A.B. Malik. 1999. Thrombin-induced p65 homodimer binding to downstream NF-κB site of the promoter mediates endothelial ICAM-1 expression and neutrophil adhesion. *J. Immunol.* 162: 5466–5476.

Ramos, O.H., A. Kauskot, M.R. Cominetti, I. Bechyne, C.L. Salla Pontes, F. Chareyre, J. Manent, R. Vassy, M. Giovannini, C. Legrand, H.S. Selistre-De-Araujo, M. Crepin and A. Bonnefoy. 2008. A novel $\alpha_v\beta_3$-blocking disintegrin containing the RGD motive, DisBa-01, inhibits bFGF-induced angiogenesis and melanoma metastasis. *Clin. Exp. Metastasis* 25: 53–64.

Rao, S.P., Z. Wang, R.I. Zuberi, L. Sikora, N.S. Bahaie, B.L. Zuraw, F.-T. Liu and P. Sriramarao. 2007. Galectin-3 functions as an adhesion molecule to support eosinophil rolling and adhesion under conditions of flow. *J. Immunol.* 179: 7800–7807.

Ricard, I., M.D. Payet and G. Dupuis. 1998. VCAM-1 is internalized by a clathrin-related pathway in human endothelial cells but its $\alpha_4\beta_1$ integrin counter-receptor remains associated with the plasma membrane in human T lymphocytes. *Eur. J. Immunol.* 28: 1708–1718.

Robbins, G.P., R.L. Saunders, J.B. Haun, J. Rawson, M.J. Therien and D.A. Hammer. 2010. Tunable leuko-polymersomes that adhere specifically to inflammatory markers. *Langmuir* 26: 14089–14096.

Romer, L.H., N.V. Mclean, H.C. Yan, M. Daise, J. Sun and H.M. Delisser. 1995. IFN-γ and TNF-α induce redistribution of PECAM-1 (CD31) on human endothelial cells. *J. Immunol.* 154: 6582–6592.

Rothlein, R., M.L. Dustin, S.D. Marlin and T.A. Springer. 1986. A human intercellular adhesion molecule (ICAM-1) distinct from LFA-1. *J. Immunol.* 137: 1270–1274.

Sagara, H., H. Matsuda, N. Wada, H. Yagita, T. Fukuda, K. Okumura, S. Makino and C. Ra. 1997. A monoclonal antibody against very late activation antigen-4 inhibits eosinophil accumulation and late asthmatic response in a guinea pig model of asthma. *Int. Arch. Allergy Immunol.* 112: 287–294.

Sako, D., X.J. Chang, K.M. Barone, G. Vachino, H.M. White, G. Shaw, G.M. Veldman, K.M. Bean, T.J. Ahern, B. Furie, D.A. Cumming and G.R. Larsen. 1993. Expression cloning of a functional glycoprotein ligand for P-selectin. *Cell* 75: 1179–1186.

Salehi-Had, H., M.I. Roh, A. Giani, T. Hisatomi, S. Nakao, I.K. Kim, E.S. Gragoudas, D. Vavvas, S. Guccione and J.W. Miller. 2011. Utilizing targeted gene therapy with nanoparticles binding alpha v beta 3 for imaging and treating choroidal neovascularization. *PLoS One* 6: e18864.

Sans, E., E. Delachanal and A. Duperray. 2001. Analysis of the roles of ICAM-1 in neutrophil transmigration using a reconstituted mammalian cell expression model: implication of ICAM-1 cytoplasmic domain and Rho-dependent signaling pathway. *J. Immunol.* 166: 544–551.

Scatena, M. and C. Giachelli. 2002. The $\alpha_v\beta_3$ integrin, NF-κB, osteoprotegerin endothelial cell survival pathway: potential role in angiogenesis. *Trends Cardiovasc. Med.* 12: 83–88.

Schmieder, A.H., S.D. Caruthers, H. Zhang, T.A. Williams, J.D. Robertson, S.A. Wickline and G.M. Lanza. 2008. Three-dimensional MR mapping of angiogenesis with $\alpha_5\beta_1(\alpha_v\beta_3)$-

targeted theranostic nanoparticles in the MDA-MB-435 xenograft mouse model. *FASEB J.* 22: 4179–4189.

Scotlandi, K., S. Perdichizzi, G. Bernard, G. Nicoletti, P. Nanni, P.L. Lollini, A. Curti, M.C. Manara, S. Benini, A. Bernard and P. Picci. 2006. Targeting CD99 in association with doxorubicin: an effective combined treatment for Ewing's sarcoma. *Eur. J. Cancer* 42: 91–96.

Scott, R.C., J.M. Rosano, Z. Ivanov, B. Wang, P.L.-G. Chong, A.C. Issekutz, D.L. Crabbe and M.F. Kiani. 2009. Targeting VEGF-encapsulated immunoliposomes to MI heart improves vascularity and cardiac function. *FASEB J.* 23: 3361–3367.

Serrador, J.M., J.L. Alonso-Lebrero, M.A. Del Pozo, H. Furthmayr, R. Schwartz-Albiez, J. Calvo, F. Lozano and F. Sanchez-Madrid. 1997. Moesin interacts with the cytoplasmic region of intercellular adhesion molecule-3 and is redistributed to the uropod of T lymphocytes during cell polarization. *J. Cell Biol.* 138: 1409–1423.

Serrano, D., T. Bhowmick, R. Chadha, C. Garnacho and S. Muro. 2012. Intercellular adhesion molecule 1 engagement modulates sphingomyelinase and ceramide, supporting uptake of drug carriers by the vascular endothelium. *Arterioscler. Thromb. Vasc. Biol.* 32: 1178–1185.

Serres, S., S. Mardiguian, S.J. Campbell, M.A. Mcateer, A. Akhtar, A. Krapitchev, R.P. Choudhury, D.C. Anthony and N.R. Sibson. 2011. VCAM-1-targeted magnetic resonance imaging reveals subclinical disease in a mouse model of multiple sclerosis. *FASEB J.* 25: 4415–4422.

Setiadi, H. and R.P. Mcever. 2008. Clustering endothelial E-selectin in clathrin-coated pits and lipid rafts enhances leukocyte adhesion under flow. *Blood* 111: 1989–1998.

Shamay, Y., D. Paulin, G. Ashkenasy and A. David. 2009a. E-selectin binding peptide-polymer-drug conjugates and their selective cytotoxicity against vascular endothelial cells. *Biomaterials* 30: 6460–6468.

Shamay, Y., D. Paulin, G. Ashkenasy and A. David. 2009b. Multivalent display of quinic acid based ligands for targeting E-selectin expressing cells. *J. Med. Chem.* 52: 5906–5915.

Shao, X., H. Zhang, J.R. Rajian, D.L. Chamberland, P.S. Sherman, C.A. Quesada, A.E. Koch, N.A. Kotov and X. Wang. 2011. ^{125}I-labeled gold nanorods for targeted imaging of inflammation. *ACS Nano* 5: 8967–8973.

Shuvaev, V.V., J.Y. Han, K.J. Yu, S.H. Huang, B.J. Hawkins, M. Madesh, M. Nakada and V.R. Muzykantov. 2011. PECAM-targeted delivery of SOD inhibits endothelial inflammatory response. *FASEB J.* 25: 348–357.

Shyjan, A.M., M. Bertagnolli, C.J. Kenney and M.J. Briskin. 1996. Human mucosal addressin cell adhesion molecule-1 (MAdCAM-1) demonstrates structural and functional similarities to the $\alpha_4\beta_7$-integrin binding domains of murine MAdCAM-1, but extreme divergence of mucin-like sequences. *J. Immunol.* 156: 2851–2857.

Sikorski, E.E., R. Hallmann, E.L. Berg and E.C. Butcher. 1993. The Peyer's patch high endothelial receptor for lymphocytes, the mucosal vascular addressin, is induced on a murine endothelial cell line by tumor necrosis factor-alpha and IL-1. *J. Immunol.* 151: 5239–5250.

Springer, T.A. 1994. Traffic signals for lymphocyte recirculation and leukocyte emigration: the multistep paradigm. *Cell* 76: 301–314.

Stahn, R., C. Grittner, R. Zeisig, U. Karsten, S.B. Felix and K. Wenzel. 2001. Sialyl Lewis[x]-liposomes as vehicles for site-directed, E-selectin-mediated drug transfer into activated endothelial cells. *Cell Mol. Life Sci.* 58: 141–147.

Stayton, P.S., M.E. El-Sayed, N. Murthy, V. Bulmus, C. Lackey, C. Cheung and A.S. Hoffman. 2005. 'Smart' delivery systems for biomolecular therapeutics. *Orthod. Craniofac. Res.* 8: 219–225.

Subramaniam, M., J.A. Koedam and D.D. Wagner. 1993. Divergent fates of P-selectins and E-selectins after their expression on the plasma-membrane. *Mol. Biol. Cell* 4: 791–801.

Theoharis, S., U. Krueger, P.H. Tan, D.O. Haskard, M. Weber and A.J.T. George. 2009. Targeting gene delivery to activated vascular endothelium using anti E/P-Selectin antibody linked to PAMAM dendrimers. *J. Immunol. Methods* 343: 79–90.

Todderud, G., X. Nair, D. Lee, J. Alford, L. Davern, P. Stanley, C. Bachand, P. Lapointe, A. Marinier, A. Martel, M. Menard, J.J. Wright, J. Bajorath, D. Hollenbaugh, A. Aruffo and K.M. Tramposch. 1997. BMS-190394, a selectin inhibitor, prevents rat cutaneous inflammatory reactions. *J. Pharmacol. Exp. Ther.* 282: 1298–1304.

Torchilin, V.P. 2006. Multifunctional nanocarriers. *Adv. Drug Deliv. Rev.* 58: 1532–1555.

Tsourkas, A., V.R. Shinde-Patil, K.A. Kelly, P. Patel, A. Wolley, J.R. Allport and R. Weissleder. 2005. *In vivo* imaging of activated endothelium using an anti-VCAM-1 magnetooptical probe. *Bioconjug. Chem.* 16: 576–581.

Ujula, T., S. Salomaki, P. Virsu, P. Lankinen, T.J. Makinen, A. Autio, G.G. Yegutkin, J. Knuuti, S. Jalkanen and A. Roivainen. 2009. Synthesis, ^{68}Ga labeling and preliminary evaluation of DOTA peptide binding vascular adhesion protein-1: a potential PET imaging agent for diagnosing osteomyelitis. *Nucl. Med. Biol.* 36: 631–641.

Van Buul, J.D., J. Van Rijssel, F.P.J. Van Alphen, M. Hoogenboezem, S. Tol, K.A. Hoeben, J. Van Marle, E.P.J. Mul and P.L. Hordijk. 2010. Inside-out regulation of ICAM-1 dynamics in TNF-α-activated endothelium. *PLoS ONE* 5: e11336.

Van Mourik, J.A., O.C. Leeksma, J.H. Reinders, P.G. De Groot and J. Zandbergen-Spaargaren. 1985. Vascular endothelial cells synthesize a plasma membrane protein indistinguishable from the platelet membrane glycoprotein IIa. *J. Biol. Chem.* 260: 11300–11306.

Vernon-Wilson, E.F., F. Aurade and S.B. Brown. 2006. CD31 promotes $β_1$ integrin-dependent engulfment of apoptotic Jurkat T lymphocytes opsonized for phagocytosis by fibronectin. *J. Leukoc. Biol.* 79: 1260–1267.

Vestweber, D. and J.E. Blanks. 1999. Mechanisms that regulate the function of the selectins and their ligands. *Physiol. Rev.* 79: 181–213.

Voinea, M., I. Manduteanu, E. Dragomir, M. Capraru and M. Simionescu. 2005. Immunoliposomes directed toward VCAM-1 interact specifically with activated endothelial cells—a potential tool for specific drug delivery. *Pharmaceut. Res.* 22: 1906–1917.

Wang, Q. and C.M. Doerschuk. 2001. The p38 mitogen-activated protein kinase mediates cytoskeletal remodeling in pulmonary microvascular endothelial cells upon intracellular adhesion molecule-1 ligation. *J. Immunol.* 166: 6877–6884.

Weller, G.E., F.S. Villanueva, E.M. Tom and W.R. Wagner. 2005. Targeted ultrasound contrast agents: *in vitro* assessment of endothelial dysfunction and multi-targeting to ICAM-1 and sialyl Lewisx. *Biotechnol. Bioeng.* 92: 780–788.

Wild, M.K., M.C. Huang, U. Schulze-Horsel, P.A. Van Der Merwe and D. Vestweber. 2001. Affinity, kinetics, and thermodynamics of E-selectin binding to E-selectin ligand-1. *J. Biol. Chem.* 276: 31602–31612.

Wilder, R.L. 2002. Integrin alpha V beta 3 as a target for treatment of rheumatoid arthritis and related rheumatic diseases. *Ann. Rheum. Dis.* 61 Suppl 2: ii96–99.

Winter, P.M., A.M. Neubauer, S.D. Caruthers, T.D. Harris, J.D. Robertson, T.A. Williams, A.H. Schmieder, G. Hu, J.S. Allen, E.K. Lacy, H. Zhang, S.A. Wickline and G.M. Lanza. 2006. Endothelial $α_vβ_3$ integrin-targeted fumagillin nanoparticles inhibit angiogenesis in atherosclerosis. *Arterioscler. Thromb. Vasc. Biol.* 26: 2103–2109.

Woodfin, A., M.B. Voisin and S. Nourshargh. 2007. PECAM-1: a multi-functional molecule in inflammation and vascular biology. *Arterioscler. Thromb. Vasc. Biol.* 27: 2514–2523.

Xiao, L., Y. Zhang, Z. Yang, Y. Xu, B. Kundu, M.D. Chordia and D. Pan. 2012. Synthesis of PECAM-1-specific ^{64}Cu PET imaging agent: evaluation of myocardial infarction caused by ischemia-reperfusion injury in mouse. *Bioorg. Med. Chem. Lett.* 22: 4144–4147.

Yang, L., R.M. Froio, T.E. Sciuto, A.M. Dvorak, R. Alon and F.W. Luscinskas. 2005. ICAM-1 regulates neutrophil adhesion and transcellular migration of TNF-α-activated vascular endothelium under flow. *Blood* 106: 584–592.

Yang, L., J.R. Kowalski, P. Yacono, M. Bajmoczi, S.K. Shaw, R.M. Froio, D.E. Golan, S.M. Thomas and F.W. Luscinskas. 2006. Endothelial cell cortactin coordinates intercellular adhesion molecule-1 clustering and actin cytoskeleton remodeling during polymorphonuclear leukocyte adhesion and transmigration. *J. Immunol.* 177: 6440–6449.

Zhang, N., C. Chittasupho, C. Duangrat, T.J. Siahaan and C. Berkland. 2008. PLGA nanoparticle-peptide conjugate effectively targets intercellular cell-adhesion molecule-1. *Bioconjug. Chem.* 19: 145–152.

Zhou, Z., M. Christofidou-Solomidou, C. Garlanda and H.M. Delisser. 1999. Antibody against murine PECAM-1 inhibits tumor angiogenesis in mice. *Angiogenesis* 3: 181–188.

10

Leukocyte Transendothelial Migration: A Biophysical Event

Kimberly M. Stroka

Introduction

The innate immune response involves a sequence of highly coordinated events that allow leukocytes to exit the bloodstream and travel to the site of infection in a nearby tissue. Polymorphonuclear neutrophils are the first responders in the army of leukocytes, arriving to the infected tissue within minutes of acute trauma. Once there, they carry out their innate function of combating infectious material, through either phagocytosis or release of granular material. Neutrophils are therefore key components in successful clearance of foreign organisms from the body, but the path to reach their final destination is not without obstacles.

Transmigration through the vascular endothelium is a key step in the immune response and is the means by which neutrophils exit the luminal space and cross the endothelial cells (ECs) lining a blood vessel. The endothelium generally acts as a selectively permeable barrier between the blood and the nearby tissue, allowing soluble factors and circulating immune cells to leak out when physiologically necessary. While neutrophil

G70 Croft Hall, Institute for NanoBioTechnology, Physical Sciences Oncology Center, Department of Chemical and Biomolecular Engineering, Johns Hopkins University, Baltimore, MD 21218.
Email: kstroka@jhu.edu

transmigration engages many biological signaling pathways within both the neutrophils and the ECs, this process is also beginning to be recognized and understood as a biophysical event (Stroka and Aranda-Espinoza 2010). Although signaling is certainly a critical regulator of transmigration, an array of physical forces acts on both the ECs and neutrophils during transmigration. For example, the ECs experience shear stress due to blood flow at their apical surface, forces due to contact with neighboring ECs or even transmigrating leukocytes, and substrate effects at the basal surface. The neutrophils feel similar forces during their transmigratory journey. At the same time, both the ECs and neutrophils exert traction forces on their underlying substrate. Further, it is becoming increasingly obvious that changes in subendothelial matrix stiffness occur during cardiovascular disease and aging, which highlights the necessity to explore and understand the role of the subendothelial matrix in EC and neutrophil signaling during the normal immune response and in pathological conditions or aging.

To more fully understand the transmigration step of the leukocyte adhesion cascade, how physical forces relate to biological signaling pathways occurring in the ECs and neutrophils must be considered. Recent improvements to *in vitro* assays have utilized a combined physical sciences and biological approach to identify how subendothelial matrix stiffness affects transmigration (Stroka and Aranda-Espinoza 2011b, Stroka et al. 2012). In particular, these studies have identified myosin light chain kinase-mediated contractility in ECs as a key regulator in not only transmigration, but also in the health of the endothelium. In this chapter, we focus on quantitative aspects of EC-leukocyte interactions and force generation during transmigration.

The Leukocyte Adhesion Cascade

Before we consider transmigration as a biophysical event, it is useful to first understand the basic biology of the immune response. A neutrophil's journey from the bloodstream to a nearby tissue is widely regarded as an extremely complicated biological event, involving signaling between many different cell types. An infection induces stromal cells to release chemokines, which bind to and activate ECs lining the blood vessels. Activation of the endothelium leads to upregulation of transmembrane glycoproteins called selectins, as well as intercellular adhesion molecule-1 (ICAM-1) and vascular cell adhesion molecule-1 (VCAM-1). Meanwhile, neutrophils are normally present in the bloodstream in a quiescent state. However, in a location where the endothelium is activated, the neutrophils receive the signal to stop and transmigrate; this process is known as the leukocyte adhesion cascade (LAC).

In this cascade, neutrophils first **roll** along the endothelium, an event that is governed by binding kinetics between selectins on the endothelium and selectin receptors on neutrophils. Specifically, Sialyl Lewis X protein (SLeX) and P-selectin glycoprotein ligand-1 (PSGL-1) on neutrophils bind E-selectin and P-selectin, respectively, on the endothelium. At some point during the rolling process, the neutrophils may transition to the second step of the LAC, where they **firmly adhere** to the endothelium through binding of lymphocyte function-associated antigen-1 (LFA-1) on neutrophils to ICAM-1 on the endothelium. In the third step of the LAC, neutrophils **migrate along** the endothelium at speeds around 10 μm/min. Finally, the neutrophils complete the last step of the LAC, which is transmigration (also known as diapedesis or extravasation) through the endothelium. This may occur via one of two methods, the paracellular route (between the EC junctions) or the transcellular route (through the middle of an EC), both which have been observed *in vivo* and *in vitro*. They may migrate for some time just below the ECs, before undergoing chemotaxis and continuing through the basement membrane, remaining layers of the blood vessel, and interstitial tissue, toward the site of infection. Of course, there are important biological and biophysical considerations at all stages of the LAC; however, this chapter focuses on the least-understood stage, which is transmigration through the vascular endothelium.

In vitro Models for Transmigration: Digging into a Physical Scientist's Toolbox

Leukocyte transmigration is a highly dynamic event that can be observed in real time at high temporal and spatial resolution using advanced timelapse microscopy techniques combined with the appropriate *in vitro* assay. Recent developments in these assays have drastically improved our ability to quantify principal aspects of transmigration, such as how many cells transmigrate, how long it takes them to cross the endothelium, what forces they exert during this process, and how the underlying subendothelial matrix stiffness affects all of these parameters. In this section we describe transwell assays as the previous "standard" for transmigration assays, and then discuss the latest developments using engineered extracellular matrices, which have allowed for more quantitative dynamic information to be ascertained from the assays.

Transwell assay

A transwell assay (also known as a Boyden or modified Boyden chamber assay) consists of an activated EC monolayer, which is cultured on a cell-permeable membrane, with pore sizes ranging from hundreds of nanometers to a few microns. The EC monolayer-containing membrane is separated by two volumes of medium, the lower of which may or may not contain a chemoattractant. A known number of leukocytes is added to the upper compartment and given time to transmigrate through the endothelium, and through the pores in the membrane, to the bottom compartment. The media from the bottom compartment is then analyzed for total cells, and the total fraction of transmigrated cells can be calculated by dividing the number of counted cells in the bottom chamber at the end of the experiment, by the total number of cells plated in the top chamber at the beginning of the experiment. Alternatively, the transmigrating cells are labeled with a fluorescent intracellular dye, which can be detected in the lower compartment media upon permeabilization of the cells, and compared with an appropriate control. This assay, though useful for measuring fractions of transmigrated cells for given conditions or trans-endothelial electrical resistance (TEER), does not provide any dynamic information about the process. In addition, transwell inserts are made of plastics such as polycarbonate and polyester, which about 10^6 times stiffer than the physiological subendothelial matrix *in vivo*. As described in the following sections, engineered extracellular matrices offer many advantages over transwell assays in studying leukocyte transmigration.

Polyacrylamide gel assay

In addition to plastic transwell inserts, glass coverslips have also been used as the subendothelial surface in transmigration assays. However, glass has an elastic modulus of approximately 50 GPa, which is even further from the *in vivo* condition, being 10^7 times stiffer than the physiological subendothelial matrix, which is typically ~5 kPa (Engler et al. 2004b, Huynh et al. 2011). To address this issue, recent experiments have utilized flexible polyacrylamide gels as substrates for ECs (Stroka and Aranda-Espinoza 2011b, Stroka et al. 2012). These gels can be prepared in the kilopascal range of stiffness, which is relevant to the cardiovascular system. Importantly, the stiffness of the polyacrylamide gels can be varied systematically by adjusting the concentration of cross-linker in the pre-gel solution. For example, very soft polyacrylamide gels (0.42 kPa) can be prepared using 3% acrylamide +0.05% bis acrylamide, while very stiff (280 kPa) polyacrylamide gels can be prepared using 15% acrylamide + 1.2% bis acrylamide (Stroka and Aranda-Espinoza 2011b). The mechanical properties of the gels are typically measured using techniques such as atomic force microscopy (for local Young's modulus measurements), dynamic mechanical analysis (for bulk Young's modulus measurements), or rheology (for shear modulus measurements).

However, uncoated polyacrylamide gels are not suitable for long-term cell attachment. Thus, ECM proteins such as fibronectin, collagen, or laminin can be covalently bonded to the gels using a photoactivation procedure (Stroka and Aranda-Espinoza 2009) to improve cell attachment. In this condition, polyacrylamide gels are able to support growth of ECs into a confluent and healthy endothelium. Importantly, these gels allow leukocyte transmigration to be dynamically imaged at high resolution using phase contrast or fluorescence microscopy (Fig. 1). After transmigration occurs, leukocytes can migrate between the endothelium and the polyacrylamide gel; however, they cannot invade the gel, due to their inability to degrade the synthetic polymer, combined with the small pore size of the gel.

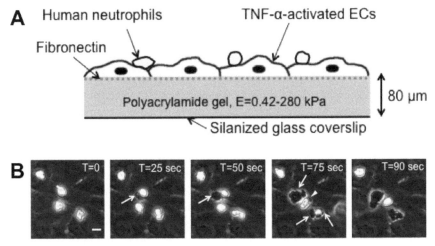

Figure 1. (A) Schematic representation of the polyacrylamide gel assay. (B) Phase contrast sequence of a human neutrophil transmigrating through a human umbilical vein endothelial cell monolayer. Neutrophils above the endothelium appear bright white, while portions of neutrophils below the endothelium appear darkened (white arrows) in phase contrast microscopy. White arrowhead points to a neutrophil that initially sends a protrusion beneath the monolayer, but then does not transmigrate. Time after initiation of transmigration is indicated in the upper right corner of each image. Scale bar is 10 μm and applies to all images. This research was originally published in *Blood* with the following reference: Kimberly M. Stroka and Helim Aranda-Espinoza. Endothelial cell substrate stiffness influences neutrophil transmigration via myosin light chain kinase-dependent cell contraction. *Blood*. 2011; 118: 1632–1640. © the American Society of Hematology.

Collagen gel assay

In other transmigration assays, thick collagen (or other ECM) gels are used as the subendothelial matrix. Like polyacrylamide gels, collagen gels are similar to the *in vivo* matrix stiffness, ranging from hundreds of Pascals to several kilopascals, depending on the concentration of collagen.

One advantage of using collagen gels over synthetic polyacrylamide gels is the ability to monitor leukocyte migration through the layers of ECM beneath the endothelium, as the leukocytes release proteases to degrade the matrix. However, the only way to manipulate collagen gel stiffness is to change the density of collagen in the gel. Further, the degradation of the collagen gel by leukocytes likely also changes the stiffness of the matrix. Thus, it is impossible to manipulate the mechanical properties of the matrix systematically without inherently changing the protein concentration and without subjecting cells to unaccounted for changes in stiffness due to degradation. As a result, it is difficult to interpret results and separate the effects of mechanical versus chemical properties on transmigration. In contrast, the stiffness of polyacrylamide gels can be tuned in a broad stiffness range, while maintaining constant ECM protein presentation on all gels, thus making them an ideal choice for mechanotransduction studies.

Micropillar assay

Recently, researchers have microfabricated matrices containing micropillars at high density and utilized them to measure forces exerted by cells in various contexts. Engineered matrices of this type have also recently been used to measure forces during neutrophil (Rabodzey et al. 2008) and monocyte (Liu et al. 2010) transmigration. In these assays, ECs are plated onto a bed of fluorescent ECM-coated micropillars, grown to confluency, and activated to induce an inflammatory response. VE-cadherin-GFP can be transfected into the ECs to facilitate the observation of transmigration with respect to EC borders; this is also possible in the polyacrylamide and collagen gel assays. Using combined timelapse fluorescent and differential contrast inference (DIC) or phase contrast microscopy, the transmigrating leukocyte, EC borders, and micropillar positions can all be observed in real time. Unique to this assay, the deflection of micropillars during transmigration can be monitored and subsequently quantified as ECs and leukocytes interact and produce contractile forces on the pillars. Knowing the micropillar height, radius, and PDMS stiffness, it is then possible to convert micropillar deflections into the corresponding forces. However, while this assay is useful to estimate the magnitude of contractile forces produced by ECs and leukocytes during transmigration, it also constrains ECs to form focal adhesions only in predetermined (and possibly non-physiological) places (i.e., the tops of micropillars), and of predetermined sizes, which could alter the course of transmigration. Indeed, recently it was discovered that the focal adhesion proteins paxillin and focal adhesion kinase are lost at sites of transmigration (Parsons et al. 2012). Thus, before choosing an appropriate assay, it is important to first identify what parameters need to be quantified during transmigration, and also to take into account how the assay itself will affect the results.

Quantitative aspects of transmigration dynamics

Using the assays described above, in combination with timelapse microscopy (Fig. 1B), it is possible to measure quantitative parameters associated with transmigration at high temporal and spatial resolution. After neutrophils adhere to the endothelium, they migrate along the apical surface of the endothelium at speeds around 15 µm/min, which is faster than migration on endothelium-free fibronectin-coated polyacrylamide gels (Stroka and Aranda-Espinoza 2009). This period of migration lasts for several minutes on average. *In vitro*, the number of neutrophils that have transmigrated increases rapidly from the time of plating until about 20 minutes after plating, and saturates by 30 minutes. Meanwhile, monocytes interact with the apical surface of the endothelium for more than an hour prior to initiating transmigration. The difference in response time may be indicative of the physiological role of neutrophils as the first responders to a site of infection. Once a neutrophil or monocyte initiates transmigration by sending a protrusion under the endothelium, the transmigration event lasts approximately one minute (Stroka and Aranda-Espinoza 2011b). In comparison, lymphocyte transmigration is much slower, taking 5–6 minutes (Carman et al. 2007).

Biophysical Role of the Endothelium in Transmigration

During transmigration, both leukocytes and the endothelium play an active biophysical role. Thus, neither cell type is simply passive as the other does the work. Instead, transmigration involves biological signaling, as well as biomechanical force production, in both cells. In this section, we first focus on the biophysical role of the endothelium in transmigration.

Endothelial cells as mechano-transducers

As neutrophils adhere to the surface of the endothelium via LFA-1/ICAM-1 interactions, a cytosolic calcium-dependent signaling cascade is initiated within the ECs (Garcia et al. 1998, Huang et al. 1993, Saito et al. 1998, Sans et al. 2001). In this pathway, Ca^{2+} binds to calmodulin, leading to activation of myosin light chain kinase (MLCK). MLCK phosphorylates myosin light chains to activate myosin motors, which generate force along the actin stress fibers to promote EC contractility. Importantly, the ECs' production of traction forces induces EC gap formation, which leads to regulation of neutrophil transmigration (Garcia et al. 1998, Saito et al. 1998). In addition to MLCK, Rho-associated protein kinase (ROCK) can also phosphorylate myosin light chains, providing another pathway to induce EC contractility (Saito et al. 2002). Thus, even before transmigration initiates, adherence of neutrophils to the surface of the endothelium commences a signaling cascade that leads to enhanced EC contractile forces and opening of intercellular gaps. The role of specific cell-cell adhesion molecules in this process will be further discussed in a later section.

However, global increases in contractility may not explain what happens locally at the site of transmigration. Attempts to measure changes in EC biomechanics locally during transmigration have led to mixed results. For example, using a "nanosurgery" technique (Riethmuller et al. 2008) via atomic force microscopy (AFM) combined with fluorescent staining, F-actin depolymerization, as well as localized softening of the endothelium, have been observed at the site of neutrophil transmigration (Isac et al. 2010). In contrast, the first group to use nanosurgery showed that leukocytes leave behind a "footprint" 8–12 µm wide and 1 µm deep in the endothelium, but that these footprints are not associated with EC softening or F-actin depolymerization (Riethmuller et al. 2008). Meanwhile, stiffening of both ECs and neutrophils has been measured during transmigration (Wang et al. 2001, Wang and Doerschuk 2000). In addition, using a fabricated micropillar assay, a three-fold increase in EC traction forces was measured locally at the region of penetration during neutrophil transmigration, in comparison with only adhesion to the endothelium (Rabodzey et al. 2008). Obviously, more work is required to identify what cytoskeletal and biomechanical changes occur locally at the site of transmigration, and how these changes relate to global alterations in EC contractility.

Endothelial cell-cell adhesion

An array of homophilic proteins specifically localize at the sites of endothelial cell-cell junctions, including vascular endothelial-cadherin (VE-cadherin), junction adhesion molecules (JAMs), and platelet endothelial cell adhesion molecule-1 (PECAM-1) (for a comprehensive review see (Vestweber 2007)). VE-cadherin plays an important role in transmigration, as it dislocates away from the junction as a leukocyte initiates penetration through the endothelium (Shaw et al. 2001). The short-lived gap in the endothelial barrier created during this process is necessary for successful transmigration (Alcaide et al. 2008) and depends on LFA-1/ICAM-1 binding (Wee et al. 2009). Using single molecule AFM, the unbinding force of two VE-cadherin molecules has been measured as 35–55 pN (Baumgartner et al. 2000); though, whether this force is produced by the endothelium or the leukocyte (or both) during transmigration remains unknown. Importantly, leukocytes are capable of exerting force on the underlying substrate (Oakes et al. 2009, Smith et al. 2007), so it is plausible that they can "push" their way through the VE-cadherin bonds holding the ECs together. However, VE-cadherin is also linked to the EC actin cytoskeleton through a mechanism that is under debate (Vestweber et al. 2009). Because F-actin, along with myosin II, is involved in producing contractile forces within the ECs, the connection between actin and VE-cadherin could serve to physically pull the VE-cadherin away from the junctions. In the next section, we will focus on how regulation of endothelial cell-cell adhesion is also modulated by adhesions with the substrate.

Endothelial cell-substrate adhesion and subendothelial matrix stiffness

It is known that a balance between cell-cell and cell-substrate adhesions determine EC biomechanical properties (Stroka and Aranda-Espinoza 2011a). Therefore, in addition to cell-cell adhesions, cell-substrate adhesions would also be expected to regulate leukocyte transmigration. Indeed, the focal adhesion proteins paxillin and focal adhesion kinase (FAK) play a role in neutrophil transmigration. Down-regulation of endothelial paxillin blocks neutrophil transmigration while having no effect on rolling or adhesion, and down-regulation of total FAK protein or FAK signaling reduces transmigration (Parsons et al. 2012). In addition, both paxillin and FAK are lost from focal adhesions at the site proximal to transmigration. Meanwhile, other focal adhesion proteins such as vinculin and β1 integrin are unaffected during transmigration. Thus, endothelial cell-substrate adhesions certainly play an important role in leukocyte transmigration.

As previously discussed, careful selection of an appropriate *in vivo* assay can lead to more physiologically-relevant experimental results. For example, an important parameter governing the immune response is the stiffness of the subendothelial matrix, as recently discovered using the polyacrylamide gel assay (Huynh et al. 2011, Stroka and Aranda-Espinoza 2011b, Stroka et al. 2012). ECs directly lie on a basement membrane composed of ECM proteins such as laminin and collagen; other ECM proteins and also stromal cells are found in the intima and media layers beneath the basement membrane. Importantly, matrix stiffening is associated with progression of cancer, atherosclerosis, and aging, even in the microvasculature (Erler and Weaver 2009, Huynh et al. 2011, Majno and Joris 1996, Paszek et al. 2005). Many different cell types, including smooth muscle cells (Engler et al. 2004a, Harley et al. 2008, Peyton and Putnam 2005), fibroblasts (Lo et al. 2000, Pelham and Wang 1997, Yeung et al. 2005), neurons (Flanagan et al. 2002, Gunn et al. 2005, Norman and Aranda-Espinoza 2010), stem cells (Engler et al. 2006), glioma cells (Ulrich et al. 2009), macrophages (Fereol et al. 2006), and neutrophils (Jannat et al. 2010, Oakes et al. 2009, Stroka and Aranda-Espinoza 2009) are capable of "mechanosensing" changes in matrix stiffness. In addition, ECs are also mechanosensitive (Califano and Reinhart-King 2008, Reinhart-King et al. 2008, Sieminski et al. 2004, Stroka and Aranda-Espinoza 2011b, Yeung et al. 2005), and therefore any disease- or aging-related changes in subendothelial matrix stiffness will affect the health and homeostasis of the endothelium. Next, we focus on how subendothelial matrix stiffness affects both a healthy and activated endothelium, as well as leukocyte transmigration through an activated endothelium.

Prior to monolayer formation, single EC morphology, cytoskeletal arrangement, focal adhesion structure, and traction forces depend on the stiffness of their underlying substrate. Interestingly, however, once ECs form a healthy, confluent monolayer, their morphology (spreading area and aspect ratio) and adhesions with the substrate (focal adhesion size and density) no longer depend on matrix stiffness (Stroka and Aranda-Espinoza 2011b). It is possible that at this point, cell-cell adhesions become more important in dictating cellular organization than cell-matrix adhesions. Meanwhile, similar to single ECs, the F-actin cytoskeleton in the confluent endothelium is more mature and arranges more densely on stiffer substrates. In addition, the endothelium is, on average, stiffer as matrix stiffness increases (Stroka and Aranda-Espinoza 2011b). Therefore, even before onset of an immune response, a diseased or aging blood vessel will likely house an endothelium with different biomechanical properties than a healthy vessel.

Upon activation by the cytokine TNF-α, which is released by stromal cells during an immune response, the endothelium undergoes significant biological changes, including upregulation of ICAM-1 and redistribution of junctional proteins. However, the endothelium also experiences morphological, cytoskeletal, and biomechanical changes. The individual cells in the endothelium enlarge, elongate, and align with local neighbors. The F-actin network transitions from circumferential orientation in individual cells to alignment along the length of the major cell axis, as the cells soften (Stroka and Aranda-Espinoza 2011b). Notably, matrix stiffness-dependent variance in cytoskeletal organization and cell stiffness is nearly abrogated after the ECs are exposed to the cytokine. In addition, the amount of ICAM-1 expressed on the surface of the endothelium does not depend on the subendothelial matrix stiffness. These properties are key to understand in relation to transmigration, especially considering neutrophils are mechanosensitive (Stroka and Aranda-Espinoza 2009). Because the activated endothelium "looks" and "feels" the same to a neutrophil, regardless of matrix stiffness, it might be reasonable to hypothesize that transmigration does not depend on subendothelial matrix stiffness. However, recent *in vitro* and *in vivo* work has demonstrated that transmigration does indeed depend on the stiffness of the surface below the endothelium. Specifically, the fraction of transmigrating neutrophils increases linearly with increasing subendothelial matrix stiffness (from 0.42 to 5 kPa), reaching a maximum of ~90% around 5 kPa (Huynh et al. 2011, Stroka and Aranda-Espinoza 2011b) (Fig. 2). Further stiffening of the subendothelial matrix (up to 280 kPa) has no additional effect on transmigration. An understanding of this phenomenon is necessary, since cardiovascular disease and aging are both associated with blood vessel stiffening (Erler and Weaver 2009, Huynh et al. 2011, Majno and Joris 1996). Neutrophil mechanosensing of this matrix does not occur until the neutrophils are beneath the endothelium, at which point they have already penetrated the endothelium. Instead, the explanation for this phenomenon lies at the heart of the intersection between the biological and biophysical properties of the endothelium and its individual cell-cell adhesions during transmigration.

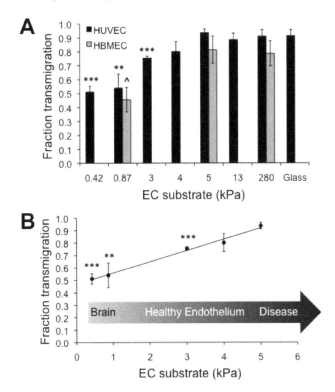

Figure 2. Effects of subendothelial matrix stiffness on neutrophil transmigration. Data is shown for transmigration through both human umbilical vein endothelial cells (HUVECS) and human brain microvascular endothelial cells (HBMECs). This research was originally published in *Blood* with the following reference: Kimberly M. Stroka and Helim Aranda-Espinoza. Endothelial cell substrate stiffness influences neutrophil transmigration via myosin light chain kinase-dependent cell contraction. *Blood*. 2011; 118: 1632–1640. © the American Society of Hematology.

Leukocyte binding to the surface of the endothelium through LFA-1/ICAM-1 interactions induces a signaling cascade that depends on MLCK- and ROCK-mediated activation of actomyosin contractility. Ultimately, this pathway leads to production of intercellular gaps (or "loosening" of junctions) and rearrangement of VE-cadherin away from the endothelial cell-cell junctions; leukocytes can take advantage of these gaps and use them to cross the endothelium. A key piece of data in connecting the relationship between cell-cell adhesion and cell-substrate adhesion in leukocyte transmigration is that ECs are capable of exerting larger traction forces on stiffer substrates (Krishnan et al. 2011, Lo et al. 2000), possibly due to enhanced integrin-ECM connections. The physical exertion of traction forces on the substrate opens intercellular gaps; larger forces on the substrates (i.e., on stiff substrates) translate to larger (or more) gaps as the ECs pull apart from each other (Huynh et al. 2011, Krishnan et al. 2011, Stroka and Aranda-Espinoza 2011b, Stroka et al. 2012). This explains why more neutrophils are able to transmigrate through an endothelium on a stiffer matrix and is depicted in Fig. 3. Further, when the endothelium is treated

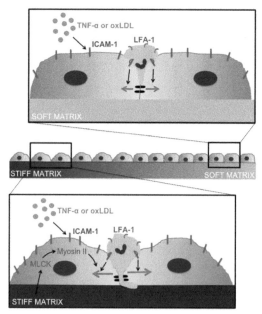

Figure 3. Effects of soft versus stiff subendothelial matrices on endothelial cell signaling and force production are depicted in this schematic. TNF-α or oxLDL upregulates ICAM-1 on the surface of the endothelium. Leukocyte LFA-1 binding to endothelial ICAM-1 initiates a signaling cascade inside the endothelial cells which ultimately leads to increased contractility (red arrows). Stiff matrices further promote the endothelial cell contractile forces, while soft substrates suppress the contractile forces.

Color image of this figure appears in the color plate section at the end of the book.

with pharmacological drugs to inhibit MLCK or ROCK, transmigration is reduced only on stiffer (≥5 kPa) substrates and no longer depends on subendothelial matrix stiffness (Huynh et al. 2011, Stroka and Aranda-Espinoza 2011b, Stroka et al. 2012). Thus, blood vessels that are stiffened due to cardiovascular disease or aging are predisposed to increased leukocyte transmigration, a condition that could lead to further breakdown of endothelial homeostasis.

Paracellular vs. Transcellular transmigration

Thus far we have largely focused on the role of the endothelium in mediating transmigration through endothelial cell-cell junctions, which is known as the "paracellular" route. However, leukocytes may also take a pathway straight through the body of the EC, in a less-understood process known as "transcellular" transmigration (Carman and Springer 2004). Both the paracellular and transcellular modes of transmigration have been observed *in vitro* and *in vivo* (Alcaide et al. 2008, Carman et al. 2007, Carman and Springer 2004, Shaw et al. 2001, Wee et al. 2009, Wojcikiewicz et al. 2008) but are believed to occur by different biological mechanisms. The mechanical role of the endothelium likely also differs between the routes, since the EC's mechanical machinery (e.g., the cytoskeleton and myosin II) must adapt to support the leukocyte as it transmigrates through different parts of the EC.

The endothelium actively contributes to transcellular transmigration by forming a "docking structure" or "transmigratory cup" comprised of microvilli-like apical membrane protrusions that surround adherent leukocytes (Barreiro et al. 2002, Carman et al. 2003, Carman and Springer 2004). These docking structures are enriched in ICAM-1, F-actin, and actin binding proteins (Carman et al. 2003, Shaw et al. 2004), which form rings at the adhesion site even before the apical membrane protrusions occur (van Rijssel et al. 2012). ICAM-1 clustering activates the guanine nucleotide factor Trio (GEF Trio), which also localizes at the sites of ICAM-1 clustering and helps to regulate local F-actin polymerization (van Rijssel et al. 2012). It has been suggested that the docking structures may protect leukocytes from hemodynamic shear forces, thus promoting transmigration (Carman and Springer 2004). Thus, signaling within the ECs plays an important role during transcellular transmigration. However, it is not known how (or even whether) these docking structures vary in composition or dynamics, depending on the endothelium bed, type of leukocyte, or subendothelial matrix stiffness.

The factors involved in a leukocyte's decision to exploit the paracellular versus transcellular route are quite unclear at this point. Most of the current literature indicates that the paracellular route is preferable to leukocytes.

However, it is also known that microvascular ECs (rather than larger vessel ECs), as well as higher ICAM-1 expression on ECs (induced by longer TNF-α treatment or shear stress), increases the rate of transcellular transmigration (Carman et al. 2007, Yang et al. 2005). For example, human umbilical vein ECs (HUVECs) support ~10% transcellular transmigration, while human dermal or human lung microvascular ECs support ~30% transcellular transmigration. Whether this change in the rate of transcellular transmigration is due to differences in large vessel versus microvascular EC contractility, ICAM-1 expression, or other factors such as cell-cell adhesion molecule distribution, remains to be investigated. Interestingly, though, selectively blocking the paracellular pathway *in vivo* by stabilizing the VE-cadherin-catenin complex completely inhibits all neutrophil or lymphocyte transmigration and vascular permeability in the cremaster, lung, and skin; these results indicate that, in these tissues, leukocytes do not switch to the transcellular pathway when the paracellular pathway is blocked (Schulte et al. 2011).

The Leukocyte's Biophysical Role in Transmigration

While the endothelium is a key regulator of transmigration, the leukocyte also plays an important active, biophysical role. Leukocytes sense the mechanical properties of their environment, exert measurable traction forces on their substrate, and signal to the endothelium through adhesion molecules. Thus, they are not simply viscous blobs squeezing through the endothelium. Here we focus on the ways in which leukocytes actively contribute to the transmigration event.

Mechanosensitivity of leukocytes

Like many other cell types, neutrophils are mechanosensitive (Jannat et al. 2010, Oakes et al. 2009, Stroka and Aranda-Espinoza 2009); their adhesion, spreading area, traction forces, and migration all depend on the stiffness of their substrate. When plated on soft (<3 kPa) fibronectin-coated polyacrylamide gels, neutrophils activated by the chemoattractant N-Formyl-Met-Leu-Phe are unable to migrate, unable to adhere strongly, and unable to exert significant traction forces; rather, they undergo random fluctuations in a localized position (Stroka and Aranda-Espinoza 2009). On stiff substrates (13 kPa) neutrophils are thinly spread, form tight adhesions to the substrate, exert large traction forces, and migrate slowly but most persistently. Likely, migration on stiff substrates is slow due to the fact that the neutrophils must disassemble the strong adhesions formed with the substrate in order to move forward; increased adhesion probably decreases the rate at which this process can occur. On intermediate substrates (4–7 kPa), neutrophils find a "happy-medium": they develop enough traction and adhesion to migrate, but not so much that it impedes their random migration (Oakes et al. 2009, Stroka and Aranda-Espinoza 2009). Thus, neutrophil migration speed and random motility (i.e., diffusion) coefficient are both biphasic with substrate stiffness, reaching an optimum around a substrate stiffness of 4–7 kPa. Meanwhile, the amount of ECM protein functionalized to the surface of the gel also is important. Specifically, the optimum stiffness for migration is a function of ECM surface density and shifts to a lower stiffness when more protein is present (Stroka and Aranda-Espinoza 2009). A similar trend has been observed in smooth muscle cells (Peyton and Putnam 2005). Therefore, neutrophil migration, like smooth muscle cells, depends on a balance between the mechanical and chemical properties of the surface on which they are migrating. We expect that this trend also applies to the case where neutrophils migrate along and through the vascular endothelium. However, an appropriate assay has yet to be developed to study this hypothesis.

Leukocyte contractile forces

In addition to sensing the mechanical properties of their substrate, leukocytes are also able to transmit force to the underlying substrate. These forces have been measured using E-selectin- and ICAM-1-coated polyacrylamide gels embedded with fluorescent marker beads that displace as cells exert forces on the gel. Importantly, the polyacrylamide gels are engineered to be of similar stiffness to the vascular endothelium (several kilopascals). The bead displacements are converted to forces using one of several algorithms that have been recently developed, such as Fourier transform traction cytometry or traction force microscopy (Butler et al. 2002, Dembo and Wang 1999). The magnitude of the force produced by activated neutrophils depends on the mode of migration; they exert an average force of 28 nN during chemokinesis (random migration), and 67 nN during chemotaxis (directed migration) (Smith et al. 2007). These forces are concentrated in the uropod, located at the rear of the cell, and the neutrophils are able to rapidly reorient the traction forces as they make a turn (Smith et al. 2007).

Thus, the soldiers of the immune system are capable of both sensing and exerting forces as they migrate along physiologically relevant engineered substrates. Because the chemical and mechanical properties of the polyacrylamide gels are specifically chosen to closely mimic the native endothelium in these assays, it is likely that the traction forces measured on the gels are similar to those exerted on an actual endothelium. Though, engineered substrates cannot recapitulate all aspects of EC signaling that occur during transmigration. To more fully understand how leukocytes play an active biophysical role in transmigration, we must turn to other *in vitro* assays that actually incorporate the endothelium.

We have already discussed how MLCK-mediated contractile forces in the *endothelium* regulate leukocyte transmigration. Interestingly, inhibition of MLCK by the drug ML-7 in *neutrophils*, while leaving the endothelium untreated, reduces the fraction of neutrophils that transmigrate through the endothelium by approximately 20–50%, depending on subendothelial matrix stiffness (Stroka and Aranda-Espinoza 2011b). This clearly indicates that MLCK-mediated contractile forces in neutrophils at least partially contribute to the ability of neutrophils to transmigrate. Of note, for the case where neutrophils are treated with ML-7, transmigration is reduced to the same extent as when the endothelium is treated with ML-7 on stiff substrates (~30% on 280 kPa), suggesting that EC and neutrophil contractile forces are equally important for transmigration on stiff subendothelial matrices. Meanwhile, on softer subendothelial matrices (0.87 kPa, where ECs are less contractile) transmigration is reduced by ~44% when neutrophils are treated with ML-7, but transmigration is unaffected when the endothelium is treated with ML-7 (Stroka and Aranda-Espinoza 2011b). These results suggest that when the endothelium is less contractile, as on a soft subendothelial matrix, with less (or smaller) intercellular gaps, the neutrophils need to use their contractile machinery more in order to squeeze through tighter endothelial cell-cell junctions.

In addition to MLCK, the small GTPase RhoA in leukocytes is also important in the context of transmigration. In monocytes, decreasing activity of RhoA (without affecting the GTPases Rac1 or Cdc42) by loading with the C3 exoenzyme prevents complete retraction of the tail end of the cell under the endothelium (Worthylake et al. 2001). Specifically, inhibition of RhoA does not block invasion of monocytes into the endothelium, but causes monocytes to leave a tail above the endothelium, thus preventing the cells from completing transmigration. These results are especially interesting in light of the fact that neutrophil traction forces are concentrated at the rear end of the cell on polyacrylamide gel substrates (Smith et al. 2007), assuming that monocytes produce similar traction patterns as neutrophils. In inflamed vessels *in vivo*, elongation of the uropod is the final step during neutrophil, monocyte, and T-cell transmigration across the endothelium (Hyun et al. 2012), further highlighting the importance of uropod dynamics during transmigration. RhoA activates ROCK (RhoA kinase), which, like MLCK, is responsible for phosphorylating myosin light chains to promote contractility. Thus, by decreasing RhoA activity, the leukocyte experiences a reduced ability to generate force at the rear end during transmigration. In other work, RhoA depletion results in a loss of T-cell polarity during migration, such that the cells no longer have a leading edge or uropod; rather, delocalized protrusions are produced (Heasman et al. 2010).

Further, the spatial and temporal components of RhoA activity have been measured in T-cells during transmigration using a fluorescence resonance energy transfer (FRET) RhoA biosensor, combined with live cell multiphoton lifetime fluorescence imaging microscopy (FLIM) (Heasman et al. 2010). In these experiments, RhoA is activated at both the leading and trailing edges of the transmigrating T-cell. At the leading edge, RhoA activity initiates before both extension and retraction occur, while at the uropod, RhoA activity is associated with ROCK-mediated cell contractile forces (Heasman et al. 2010). Meanwhile, the Rho guanine nucleotide exchange factor GEF-H1 is involved in uropod contraction but not in extension of the leading edge. Future research will further elucidate what pathway(s) regulates leukocyte tail retraction, in the context of transmigration through the vascular endothelium.

Leukocytes Play "Follow the Leader" through the Endothelium

The question of how and where leukocytes decide to transmigrate is one that has received little attention. Interestingly, the answer to this question may be as much biophysical as it is biological. Recently, it has been shown that monocyte transmigration enhances subsequent transmigration events (Hashimoto et al. 2011). In other words, after one monocyte transmigrates at a particular location in the endothelium, other monocytes are likely to transmigrate in the same spot. For example, in an assay where two waves of monocytes were plated 30 minutes apart, there was more transmigration when the second wave was introduced (Shaw et al. 2001). Biologically speaking, this could be explained by observation of local accumulations of adhesion molecules such as ICAM-1 (Shaw et al. 2004) and PECAM-1 (Hashimoto et al. 2011) at the site of transmigration. The first monocyte to cross the endothelium in a particular location induces increased ICAM-1 and PECAM-1 at that site, and as subsequent leukocytes migrate along the endothelium, they could be attracted to transmigrate in the presence of these adhesion molecule accumulations, creating a positive feedback loop. In this case, where a leukocyte decides to transmigrate is dictated by biological signaling between the leukocyte and endothelium.

However, in addition to signaling by adhesion molecules, transmigration is known to induce gaps in endothelial VE-cadherin. The first leukocyte to transmigrate through the endothelium either crosses through a pre-existing gap in VE-cadherin at the endothelial cell-cell junctions, or it forms a new gap (Shaw et al. 2001). The transient gap in VE-cadherin formed by a leukocyte as it transmigrates is 4–6 μm wide and seals within five minutes after the leukocyte passes (Shaw et al. 2001). The first leukocyte appears to push the VE-cadherin aside, creating a point of less resistance, through which subsequent leukocytes can transmigrate. In other words, the first leukocyte makes a hole and "opens the door" through which other cells can easily follow. In this case, where a leukocyte decides to transmigrate could be dictated by the physical properties of the endothelium. The size and distribution of these holes across the endothelium are likely to be important factors in directing transmigration.

Transmigration in Cardiovascular Disease

Because transmigration is a critical step in the innate immune response, the endothelium is able to dynamically adapt to this event. For example, during a normal immune response, the endothelium upregulates adhesion molecules to cue the leukocyte, the leukocyte signals to the endothelium to open a gap, the leukocyte squeezes through, and the endothelium seals the hole. However, in diseased conditions, the endothelium may remain permanently damaged as a result of this process. In this section we focus on the biophysics of transmigration in cardiovascular disease.

Like TNF-α, oxidized low-density lipoprotein (oxLDL) upregulates endothelial ICAM-1, leading to increased neutrophil (Stroka et al. 2012) and monocyte (Hashimoto et al. 2007) transmigration. Physiologically, the cardiovascular disease atherosclerosis (specifically, coronary artery disease and plaque formation) is associated with oxidation of LDLs in the body (Berliner et al. 2001, Diaz et al. 1997, Holvoet et al. 2001, Holvoet et al. 1998, Levitan et al. 2010, Toshima et al. 2000), and the elevation of leukocyte transmigration levels could be partially responsible for this correlation. During atherosclerosis, blood vessels stiffen, leading to increased subendothelial matrix stiffness. At the same time, EC exposure to oxLDL results in biomechanical changes such as increased EC contractility (Byfield et al. 2006) and stiffness (Byfield et al. 2006, Chouinard et al. 2008).

As previously mentioned, the subendothelial matrix is an important regulator of EC biomechanics and neutrophil transmigration during a normal immune response (i.e., when the endothelium is activated by TNF-α) (Stroka and Aranda-Espinoza 2011b). Subendothelial matrix stiffness also plays an important role when the endothelium is activated via oxLDL rather than TNF-α. Specifically, oxLDL and subendothelial matrix stiffness together promote neutrophil transmigration through the vascular endothelium; that is, more neutrophil transmigration is observed *in vitro* through oxLDL-treated ECs on stiffer subendothelial matrices (Stroka et al. 2012). Inhibition of MLCK in the endothelium abrogates effects of subendothelial matrix stiffness on neutrophil transmigration. Thus, as in the case of TNF-α-treated endothelium (Stroka and Aranda-Espinoza 2011b), the increase in transmigration with substrate stiffness is not due to differences in EC morphology or ICAM-1 expression on the apical surface;

rather, the effect is due to MLCK-dependent contractile forces in ECs on stiffer matrices (Stroka et al. 2012).

An additional effect of enhanced EC contractility on stiffer matrices after oxLDL treatment is an increased risk of endothelial injury following leukocyte transmigration *in vitro*. Interestingly, this relationship also holds for TNF-α-treated endothelium. Large holes with an area of 1000–5000 mm^2 and a density of 10–30 holes/mm^2 (increasing as subendothelial matrix stiffness varies from 0.87 to 280 kPa) begin forming in the endothelium around the same time when neutrophil transmigration has reached a maximum (about 25 minutes after plating neutrophils onto the endothelium) (Stroka and Aranda-Espinoza 2011b, Stroka et al. 2012). However, a major difference between the exposure to oxLDL versus TNF-α is that the oxLDL-treated endothelium does not heal itself following injury (Stroka et al. 2012), while the holes in the TNF-α-treated endothelium seal within 2 hours (Stroka and Aranda-Espinoza 2011b) after plating neutrophils. Inhibition of MLCK in the ECs completely eliminates hole formation (Stroka et al. 2012), indicating that the injury is related to the contractile properties of the ECs. Likely, increased contractility on stiffer substrates, in conjunction with severance of cell-cell and cell-substrate adhesion as the leukocytes migrate between and under the ECs, pulls the ECs apart from each other, forming a large hole and baring the underlying matrix. The holes remain even 2 hours after plating neutrophils onto the oxLDL-treated endothelium, and 1.5 hours after a maximum of neutrophils transmigrate. These results suggest that the presence of oxLDL and stiffened vasculature, both which are present in cardiovascular disease, have the potential to induce permanent and significant damage to the barrier function of the endothelium.

Conclusions and Future Directions

In this chapter we have explored transmigration as a biophysical event, involving not only biological signaling between the endothelium and leukocytes, but also force production by both types of cells. Individually, both leukocytes and ECs are able to sense mechanical changes in their microenvironment and also to produce traction forces. As a result, transmigration requires a coordinated effort between both cells in order to exert the necessary forces to achieve successful transmigration. This process is mediated by MLCK in both the ECs and the leukocytes and therefore is heavily dependent on subendothelial matrix stiffness. Future work should focus on understanding the mechanistic differences between paracellular and transcellular transmigration and the specific biophysical signaling mechanisms activated during each of the pathways. In addition, the field will benefit from development of *in vitro* models that incorporate other aspects of the physiological microenvironment within the blood vessel, such as shear stress due to blood flow, multiple ECM components, and, in the case of large vessels, smooth muscle cell activity and signaling. It is likely that a combination of all these factors contribute to the biophysics of leukocyte transmigration. The work should also be translated to the *in vivo* situation, where the biophysical roles of the neutrophils and the endothelium could depend on the location of transmigration (e.g., brain, liver, muscle, etc.).

The role of mechanical forces in transmigration is extremely important in the context of not only the normal immune response, but also in pathological conditions. In this chapter, we emphasized aspects of cardiovascular disease, which is associated with vascular stiffening and increased EC contractility. However, in addition to immune cells, metastatic cancer cells must also cross the endothelium during their journey from the primary tumor to a distal metastatic site. In fact, they cross the endothelium twice during this process, as they undergo both intravasation (migration into the bloodstream) and extravasation (exit from the bloodstream). However, the biological and/or biophysical mechanisms by which cancer cells complete these processes are barely understood in comparison with leukocytes. It is quite possible that the mechanisms are very different between cancer cells and leukocytes. Future work will likely identify the specific roles of cancer cells and the endothelium during cancer cell intravasation and extravasation.

References

Alcaide, P., G. Newton, S. Auerbach, S. Sehrawat, T.N. Mayadas, D.E. Golan, P. Yacono, P. Vincent, A. Kowalczyk and F.W. Luscinskas. 2008. p120-Catenin regulates leukocyte transmigration through an effect on VE-cadherin phosphorylation. *Blood* 112: 2770–2779.

Barreiro, O., M. Yanez-Mo, J.M. Serrador, M.C. Montoya, M. Vicente-Manzanares, R. Tejedor, H. Furthmayr and F. Sanchez-Madrid. 2002. Dynamic interaction of VCAM-1 and ICAM-1 with moesin and ezrin in a novel endothelial docking structure for adherent leukocytes. *J. Cell Biol.* 157: 1233–1245.

Baumgartner, W., P. Hinterdorfer, W. Ness, A. Raab, D. Vestweber, H. Schindler and D. Drenckhahn. 2000. Cadherin interaction probed by atomic force microscopy. *Proc. Natl. Acad. Sci. USA* 97: 4005–4010.

Berliner, J.A., G. Subbanagounder, N. Leitinger, A.D. Watson and D. Vora. 2001. Evidence for a role of phospholipid oxidation products in atherogenesis. *Trends Cardiovasc. Med.* 11: 142–147.

Butler, J.P., I.M. Tolic-Norrelykke, B. Fabry and J.J. Fredberg. 2002. Traction fields, moments, and strain energy that cells exert on their surroundings. *Am. J. Physiol. Cell Physiol.* 282: C595–605.

Byfield, F.J., S. Tikku, G.H. Rothblat, K.J. Gooch and I. Levitan. 2006. OxLDL increases endothelial stiffness, force generation, and network formation. *J. Lipid Res.* 47: 715–723.

Califano, J.P. and C.A. Reinhart-King. 2008. A balance of substrate mechanics and matrix chemistry regulates endothelial cell network assembly. *Cell. Mol. Bioeng.* 1: 122–132.

Carman, C.V., C.D. Jun, A. Salas and T.A. Springer. 2003. Endothelial cells proactively form microvilli-like membrane projections upon intercellular adhesion molecule 1 engagement of leukocyte LFA-1. *J. Immunol.* 171: 6135–6144.

Carman, C.V., P.T. Sage, T.E. Sciuto, M.A. de la Fuente, R.S. Geha, H.D. Ochs, H.F. Dvorak, A.M. Dvorak and T.A. Springer. 2007. Transcellular diapedesis is initiated by invasive podosomes. *Immunity* 26: 784–797.

Carman, C.V. and T.A. Springer. 2004. A transmigratory cup in leukocyte diapedesis both through individual vascular endothelial cells and between them. *J. Cell Biol.* 167: 377–388.

Chouinard, J.A., G. Grenier, A. Khalil and P. Vermette. 2008. Oxidized-LDL induce morphological changes and increase stiffness of endothelial cells. *Exp. Cell Res.* 314: 3007–3016.

Dembo, M. and Y.L. Wang. 1999. Stresses at the cell-to-substrate interface during locomotion of fibroblasts. *Biophys. J.* 76: 2307–2316.

Diaz, M.N., B. Frei, J.A. Vita and J.F. Keaney, Jr. 1997. Antioxidants and atherosclerotic heart disease. *N Engl. J. Med.* 337: 408–416.

Engler, A., L. Bacakova, C. Newman, A. Hategan, M. Griffin and D. Discher. 2004a. Substrate compliance versus ligand density in cell on gel responses. *Biophys. J.* 86: 617–628.

Engler, A.J., L. Richert, J.Y. Wong, C. Picart and D. Discher. 2004b. Surface probe measurements of the elasticity of sectioned tissue, thin gels and polyelectrolyte multilayer films: Correlations between substrate stiffness and cell adhesion. *Surf. Sci.* 570: 142–154.

Engler, A.J., S. Sen, H.L. Sweeney and D.E. Discher. 2006. Matrix elasticity directs stem cell lineage specification. *Cell* 126: 677–689.

Erler, J.T. and V.M. Weaver. 2009. Three-dimensional context regulation of metastasis. *Clin. Exp. Metastasis* 26: 35–49.

Fereol, S., R. Fodil, B. Labat, S. Galiacy, V.M. Laurent, B. Louis, D. Isabey and E. Planus. 2006. Sensitivity of alveolar macrophages to substrate mechanical and adhesive properties. *Cell Motil. Cytoskeleton* 63: 321–340.

Flanagan, L.A., Y.E. Ju, B. Marg, M. Osterfield and P.A. Janmey. 2002. Neurite branching on deformable substrates. *Neuroreport* 13: 2411–2415.

Garcia, J.G.N., A.D. Verin, M. Herenyiova and D. English. 1998. Adherent neutrophils activate endothelial myosin light chain kinase: role in transendothelial migration. *J. Appl. Physiol.* 84: 1817–1821.

Gunn, J.W., S.D. Turner and B.K. Mann. 2005. Adhesive and mechanical properties of hydrogels influence neurite extension. *J. Biomed. Mater. Res. A* 72: 91–97.

Harley, B.A., H.D. Kim, M.H. Zaman, I.V. Yannas, D.A. Lauffenburger and L.J. Gibson. 2008. Microarchitecture of three-dimensional scaffolds influences cell migration behavior via junction interactions. *Biophys. J.* 95: 4013–4024.

Hashimoto, K., N. Kataoka, E. Nakamura, K. Hagihara, M. Hatano, T. Okamoto, H. Kanouchi, Y. Minatogawa, S. Mohri, K. Tsujioka and F. Kajiya. 2011. Monocyte transendothelial migration augments subsequent transmigratory activity with increased PECAM-1 and decreased VE-cadherin at endothelial junctions. *Int. J. Cardiol.* 149: 232–239.

Hashimoto, K., N. Kataoka, E. Nakamura, K. Tsujioka and F. Kajiya. 2007. Oxidized LDL specifically promotes the initiation of monocyte invasion during transendothelial migration with upregulated PECAM-1 and downregulated VE-cadherin on endothelial junctions. *Atherosclerosis* 194: e9–17.

Heasman, S.J., L.M. Carlin, S. Cox, T. Ng and A.J. Ridley. 2010. Coordinated RhoA signaling at the leading edge and uropod is required for T cell transendothelial migration. *J. Cell Biol.* 190: 553–563.

Holvoet, P., A. Mertens, P. Verhamme, K. Bogaerts, G. Beyens, R. Verhaeghe, D. Collen, E. Muls and F. Van de Werf. 2001. Circulating oxidized LDL is a useful marker for identifying patients with coronary artery disease. *Arterioscler. Thromb. Vasc. Biol.* 21: 844–848.

Holvoet, P., J. Vanhaecke, S. Janssens, F. Van de Werf and D. Collen. 1998. Oxidized LDL and malondialdehyde-modified LDL in patients with acute coronary syndromes and stable coronary artery disease. *Circulation* 98: 1487–1494.

Huang, A.J., J.E. Manning, T.M. Bandak, M.C. Ratau, K.R. Hanser and S.C. Silverstein. 1993. Endothelial-Cell Cytosolic Free Calcium Regulates Neutrophil Migration across Monolayers of Endothelial-Cells. *J. Cell Biol.* 120: 1371–1380.

Huynh, J., N. Nishimura, K. Rana, J.M. Peloquin, J.P. Califano, C.R. Montague, M.R. King, C.B. Schaffer and C.A. Reinhart-King. 2011. Age-Related Intimal Stiffening Enhances Endothelial Permeability and Leukocyte Transmigration. *Sci. Transl. Med.* 3.

Hyun, Y.M., R. Sumagin, P.P. Sarangi, E. Lomakina, M.G. Overstreet, C.M. Baker, D.J. Fowell, R.E. Waugh, I.H. Sarelius and M. Kim. 2012. Uropod elongation is a common final step in leukocyte extravasation through inflamed vessels. *J. Exp. Med.* 209: 1349–1362.

Isac, L., G. Thoelking, A. Schwab, H. Oberleithner and C. Riethmuller. 2010. Endothelial f-actin depolymerization enables leukocyte transmigration. *Anal. Bioanal. Chem.* 399: 2351–2358.

Jannat, R.A., G.P. Robbins, B.G. Ricart, M. Dembo and D.A. Hammer. 2010. Neutrophil adhesion and chemotaxis depend on substrate mechanics. *J. Phys-Condens. Mat.* 22: -.

Krishnan, R., D.D. Klumpers, C.Y. Park, K. Rajendran, X. Trepat, J. van Bezu, V.W. van Hinsbergh, C.V. Carman, J.D. Brain, J.J. Fredberg et al. 2011. Substrate stiffening promotes endothelial monolayer disruption through enhanced physical forces. *Am. J. Physiol. Cell Physiol.* 300: C146–154.

Levitan, I., S. Volkov and P.V. Subbaiah. 2010. Oxidized LDL: diversity, patterns of recognition, and pathophysiology. *Antioxid. Redox Signal.* 13: 39–75.

Liu, Z.J., N.J. Sniadecki and C.S. Chen. 2010. Mechanical Forces in Endothelial Cells during Firm Adhesion and Early Transmigration of Human Monocytes. *Cell. Mol. Bioeng.* 3: 50–59.

Lo, C.-M., H.-B. Wang, M. Dembo and Y.-L. Wang. 2000. Cell movement is guided by the rigidity of the substrate. *Biophys. J.* 79: 144–152.

Majno, G. and I. Joris. 1996. Cells, Tissues, and Disease: Principles of General Pathology (Worcester, Massachusetts, Blackwell Science).

Norman, L.L. and H. Aranda-Espinoza. 2010. Cortical Neuron Outgrowth is Insensitive to Substrate Stiffness. *Cell. Mol. Bioeng.* 3: 398–414.

Oakes, P.W., D.C. Patel, N.A. Morin, D.P. Zitterbart, B. Fabry, J.S. Reichner and J.X. Tang. 2009. Neutrophil morphology and migration are affected by substrate elasticity. *Blood* 114: 1387–1395.

Parsons, S.A., R. Sharma, L. Roccamatisi, H. Zhang, B. Petri, P. Kubes, P. Colarusso and K.D. Patel. 2012. Endothelial paxillin and focal adhesion kinase (FAK) play a critical role in neutrophil transmigration. *Eur. J. Immunol.* 42: 436–446.

Paszek, M.J., N. Zahir, K.R. Johnson, J.N. Lakins, G.I. Rozenberg, A. Gefen, C.A. Reinhart-King, S.S. Margulies, M. Dembo, D. Boettiger et al. 2005. Tensional homeostasis and the malignant phenotype. *Cancer Cell* 8: 241–254.

Pelham, R.J., Jr. and Y. Wang. 1997. Cell locomotion and focal adhesions are regulated by substrate flexibility. *Proc. Natl. Acad. Sci. USA* 94: 13661–13665.

Peyton, S.R. and A.J. Putnam. 2005. Extracellular matrix rigidity governs smooth muscle cell motility in a biphasic fashion. *J. Cell Physiol.* 204: 198–209.

Rabodzey, A., P. Alcaide, F.W. Luscinskas and B. Ladoux. 2008. Mechanical forces induced by the transendothelial migration of human neutrophils. *Biophys. J.* 95: 1428–1438.

Reinhart-King, C.A., M. Dembo and D.A. Hammer. 2008. Cell-cell mechanical communication through compliant substrates. *Biophys. J.* 95: 6044–6051.

Riethmuller, C., I. Nasdala and D. Vestweber. 2008. Nano-surgery at the leukocyte-endothelial docking site. *Pflugers Arch.* 456: 71–81.

Saito, H., Y. Minamiya, M. Kitamura, S. Saito, K. Enomoto, K. Terada and J. Ogawa. 1998. Endothelial myosin light chain kinase regulates neutrophil migration across human umbilical vein endothelial cell monolayer. *J. Immunol.* 161: 1533–1540.

Saito, H., Y. Minamiya, S. Saito and J. Ogawa. 2002. Endothelial Rho and Rho kinase regulate neutrophil migration via endothelial myosin light chain phosphorylation. *J. Leukocyte. Biol.* 72: 829–836.

Sans, E., E. Delachanal and A. Duperray. 2001. Analysis of the roles of ICAM-1 in neutrophil transmigration using a reconstituted mammalian cell expression model: Implication of ICAM-1 cytoplasmic domain and Rho-dependent signaling pathway. *J. Immunol.* 166: 544–551.

Schulte, D., V. Kuppers, N. Dartsch, A. Broermann, H. Li, A. Zarbock, O. Kamenyeva, F. Kiefer, A. Khandoga, S. Massberg et al. 2011. Stabilizing the VE-cadherin-catenin complex blocks leukocyte extravasation and vascular permeability. *Embo J.* 30: 4157–4170.

Shaw, S.K., P.S. Bamba, B.N. Perkins and F.W. Luscinskas. 2001. Real-time imaging of vascular endothelial-cadherin during leukocyte transmigration across endothelium. *J. Immunol.* 167: 2323–2330.

Shaw, S.K., S. Ma, M.B. Kim, R.M. Rao, C.U. Hartman, R.M. Froio, L. Yang, T. Jones, Y. Liu, A. Nusrat et al. 2004. Coordinated redistribution of leukocyte LFA-1 and endothelial cell ICAM-1 accompany neutrophil transmigration. *J. Exp. Med.* 200: 1571–1580.

Sieminski, A.L., R.P. Hebbel and K.J. Gooch. 2004. The relative magnitudes of endothelial force generation and matrix stiffness modulate capillary morphogenesis *in vitro*. *Exp. Cell Res.* 297: 574–584.

Smith, L.A., H. Aranda-Espinoza, J.B. Haun, M. Dembo and D.A. Hammer. 2007. Neutrophil traction stresses are concentrated in the uropod during migration. *Biophys. J.* 92: L58–60.

Stroka, K.M. and H. Aranda-Espinoza. 2009. Neutrophils display biphasic relationship between migration and substrate stiffness. *Cell Motil. Cytoskeleton* 66: 328–341.

Stroka, K.M. and H. Aranda-Espinoza. 2010. A biophysical view of the interplay between mechanical forces and signaling pathways during transendothelial cell migration. *FEBS J.* 277: 1145–1158.

Stroka, K.M. and H. Aranda-Espinoza. 2011a. Effects of Morphology vs. Cell-Cell Interactions on Endothelial Cell Stiffness. *Cell. Mol. Bioeng.* 4: 9–27.

Stroka, K.M. and H. Aranda-Espinoza. 2011b. Endothelial cell substrate stiffness influences neutrophil transmigration via myosin light chain kinase-dependent cell contraction. *Blood* 118: 1632–1640.

Stroka, K.M., I. Levitan and H. Aranda-Espinoza. 2012. OxLDL and substrate stiffness promote neutrophil transmigration by enhanced endothelial cell contractility and ICAM-1. *J. Biomech.* 45: 1828–1834.

Toshima, S., A. Hasegawa, M. Kurabayashi, H. Itabe, T. Takano, J. Sugano, K. Shimamura, J. Kimura, I. Michishita, T. Suzuki et al. 2000. Circulating oxidized low density lipoprotein levels. A biochemical risk marker for coronary heart disease. *Arterioscler. Thromb. Vasc. Biol.* 20: 2243–2247.

Ulrich, T.A., E.M.D. Pardo and S. Kumar. 2009. The Mechanical Rigidity of the Extracellular Matrix Regulates the Structure, Motility, and Proliferation of Glioma Cells. *Cancer Research* 69: 4167–4174.

van Rijssel, J., J. Kroon, M. Hoogenboezem, F.P.J. van Alphen, R.J. de Jong, E. Kostadinova, D. Geerts, P.L. Hordijk and J.D. van Buul. 2012. The Rho-guanine nucleotide exchange factor Trio controls leukocyte transendothelial migration by promoting docking structure formation. *Mol. Biol. Cell* 23: 2831–2844.

Vestweber, D. 2007. Adhesion and signaling molecules controlling the transmigration of leukocytes through endothelium. *Immunol. Rev.* 218: 178–196.

Vestweber, D., M. Winderlich, G. Cagna and A.F. Nottebaum. 2009. Cell adhesion dynamics at endothelial junctions: VE-cadherin as a major player. *Trends Cell Biol.* 19: 8–15.

Wang, Q., E.T. Chiang, M. Lim, J. Lai, R. Rogers, P.A. Janmey, D. Shepro and C.M. Doerschuk. 2001. Changes in the biomechanical properties of neutrophils and endothelial cells during adhesion. *Blood* 97: 660–668.

Wang, Q. and C.M. Doerschuk. 2000. Neutrophil-induced changes in the biomechanical properties of endothelial cells: roles of ICAM-1 and reactive oxygen species. *J. Immunol.* 164: 6487–6494.

Wee, H., H.M. Oh, J.H. Jo and C.D. Jun. 2009. ICAM-1/LFA-1 interaction contributes to the induction of endothelial cell-cell separation: implication for enhanced leukocyte diapedesis. *Exp. Mol. Med.* 41: 341–348.

Wojcikiewicz, E.P., R.R. Koenen, L. Fraemohs, J. Minkiewicz, H. Azad, C. Weber and V.T. Moy. 2008. LFA-1 binding destabilizes the JAM-A homophilic interaction during leukocyte transmigration. *Biophys. J.*

Worthylake, R.A., S. Lemoine, J.M. Watson and K. Burridge. 2001. RhoA is required for monocyte tail retraction during transendothelial migration. *J. Cell Biol.* 154: 147–160.

Yang, L., R.M. Froio, T.E. Sciuto, A.M. Dvorak, R. Alon and F.W. Luscinskas. 2005. ICAM-1 regulates neutrophil adhesion and transcellular migration of TNF-alpha-activated vascular endothelium under flow. *Blood* 106: 584–592.

Yeung, T., P.C. Georges, L.A. Flanagan, B. Marg, M. Ortiz, M. Funaki, N. Zahir, W. Ming, V. Weaver and P.A. Janmey. 2005. Effects of substrate stiffness on cell morphology, cytoskeletal structure, and adhesion. *Cell Motil. Cytoskeleton* 60: 24–34.

Index

Color Plate Section

Chapter 2

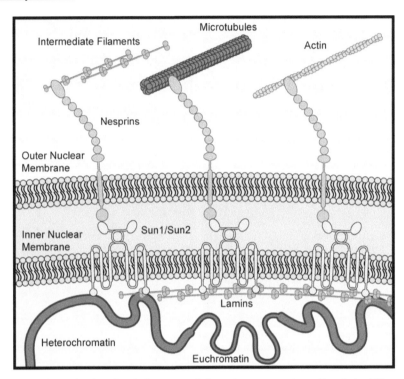

Figure 1. Schematic of nucleoskeleton-cytoskeleton interconnections. Cytoskeletal filament systems connect to the outer uclear membrane via nesprin proteins. The direct connection is then maintained through SUN 1/2 protein complexes, which bind to inner nuclear membrane nesprins, transmembrane proteins and lamins. Lamins then bind directly and indirectly to chromatin in the nuclear interior.

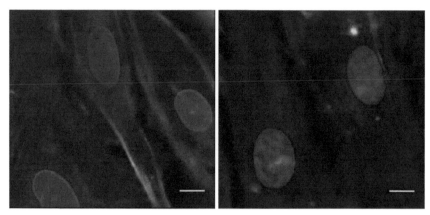

Figure 2. Response of nucleus to shear stress stimulation. Control HUVECs (left) show actin stress fibers (green; phalloidin) with a centralized nucleus with DNA (blue; DAPI) and lamin A (red; immunolabeling for lamin A/C). HUVECs treated with 10 dyn/cm^2 for 12 hours showed nuclear reorganization in the direction of flow along with actin filament structures.

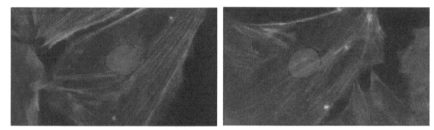

Figure 3. Response of nucleus to VEGF stimulation. Control HUVECs (left) show actin stress fibers (green; phalloidin) with a centralized nucleus with DNA (blue; DAPI) and low levels of lamin A (red; immunolabeling for lamin A/C). HUVECs treated with 50 ng/mL of VEGF for 24 hours showed increase lamin labeling suggesting reorganization of nuclear structure.

Chapter 3

Metastasis formation

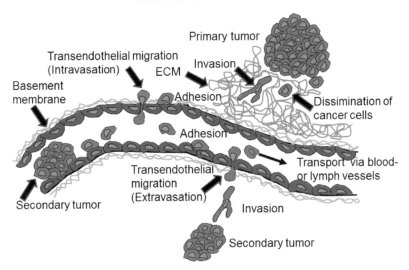

Figure 1. Process of metastasis formation.

How do cancer cells transmigrate?

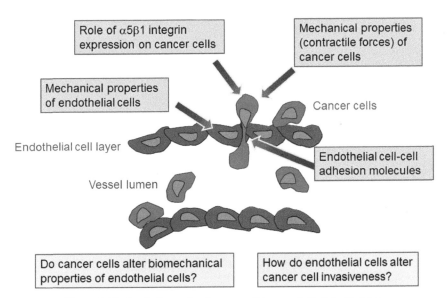

Figure 2. Mechanical aspects of cancer cell transendothelial migration.

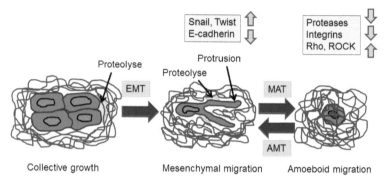

Figure 3. EMT and AMT or MAT transition.

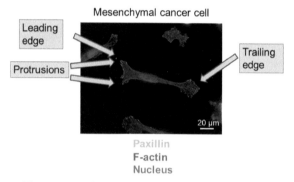

Figure 4. Morphology of a mesenchymal cancer cell.

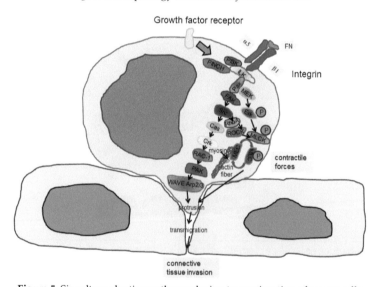

Figure 5. Signaltransduction pathway during transmigration of cancer cells.

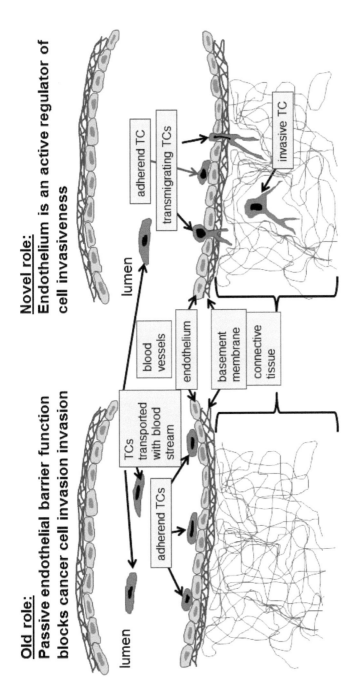

Figure 6. Old and novel role of the endothelium.

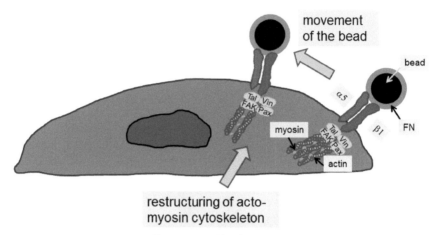

Figure 7. Bead binding to cells.

Chapter 4

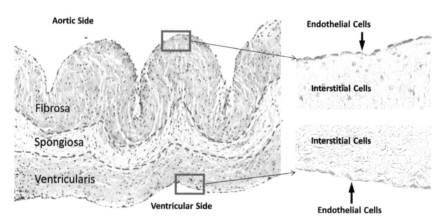

Figure 1. Cross-section of a porcine aortic valve leaflet demonstrating the striated three layer structure, which contains valvular interstitial cells and is lined by the valvular endothelial cells. Adapted from Simmons et al. (2005).

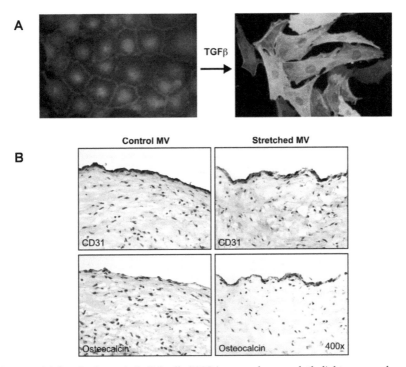

Figure 2. Adult valvular endothelial cells (VECs) can undergo endothelial-to-mesenchymal transformation (EMT). (A) Ovine aortic VEC clone treated without (left) or with (right) transforming growth factor-β1 for 5 days. The loss of endothelial marker expression (CD31 in red), the loss of cell-cell contacts, and the induction of α-smooth muscle actin expression (green) are hallmarks of EMT. Adapted from Bischoff and Aikawa (2011) (B) In ovine mitral valves subjected to elevated mechanical stress *in vivo*, VECs co-express CD31 and osteocalcin, suggesting that they have osteogenic differentiation potential that can be induced mechanically. Adapted from Wylie-Sears et al. (2011).

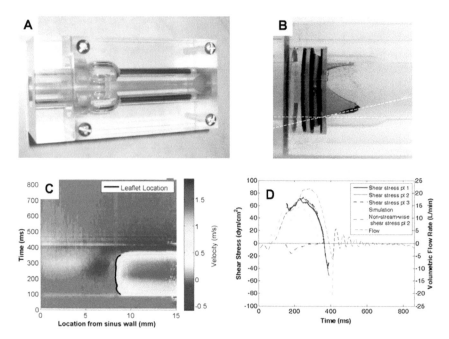

Figure 4. Experimental system to estimate shear stresses on the surface of the aortic valve. (A) Acrylic chamber used to measure shear stresses on polymeric aortic valve leaflets. (B) Image of polymeric valve during opening. Dotted lines depict how opening angles were quantified. (C) Velocity map obtained using laser Doppler velocimetry in stream wise direction. (D) Plot of fluid shear stress measured on the ventricular surface of the valve leaflets at three separate points. Adapted from Yap et al. (2011).

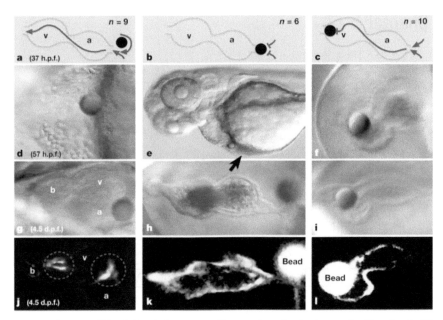

Figure 5. Impaired blood flow influences cardiogenesis and valvulogenesis. Glass beads (50 μmdiameter) were inserted into zebrafish embryos in one of three positions: (a) close to the entrance of the primitive heart tube, but not blocking flow (sham); (b) in front of the heart tube, blocking inflow; or (c) into the outflow tract to block blood efflux. (d-f) Blockage was successful as evidenced by the lack of blood in the atrium when the inflow was blocked (e) and the accumulation of blood when the outflow was blocked (f). The heart (g) and valves (j) developed normally in the sham operated embryos, but cardiogenesis (h, i) and valve formation (k, l) were severely disrupted independent of blockage location. From Hove et al. (2003).

Chapter 5

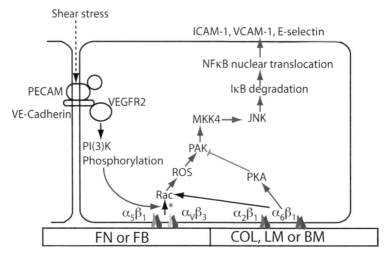

Figure 3. Schematic of signaling pathways by which shear stress activates integrins leading either to promotion or inhibition of NFκB nuclear transmigration and expression of leukocyte adhesion molecules ICAM-1, VCAM-1 and E-selectin in ECs. Adapted from (Hahn et al. 2009).

Chapter 6

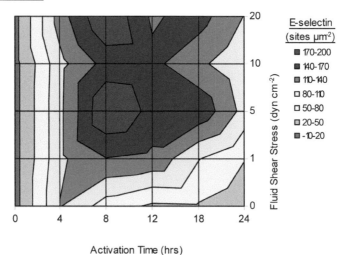

Figure 2. E-selectin expression (sites μm^{-2}) on 0–20 dyn cm^{-2} shear-cytokine activated monolayers. HUVEC monolayers activated with 0.1 ng ml^{-1} IL-1ß either under static conditions (no fluid shear) or simultaneously exposed to 1–20 dyn cm^{-2} of laminar fluid shear at activation periods ranging from 0 to 24 hr.

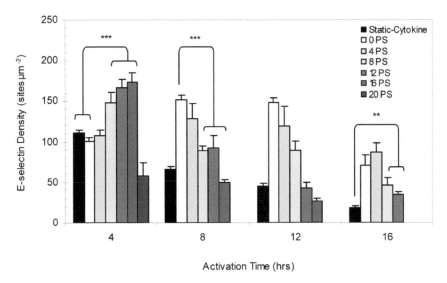

Figure 3. E-selectin expression (sites μm⁻²) on 10 dyn cm⁻² shear-cytokine activated preconditioned HUVEC monolayers. Monolayers were subjected to 4, 8, 12, 16, or 20 hrs of 10 dyn cm⁻² magnitude shear preconditioning (4 PS, 8 PS, 12 PS, 16 PS, and 20 PS, respectively) followed by up to 16 hrs of shear-cytokine activation with 0.1 ng mL⁻¹ of IL-1β. E-selectin density values were measured after each 4 hr time point. Non-preconditioned monolayers (0 PS) and static-activated monolayers (Static-Cytokine) were used as controls.

Chapter 7

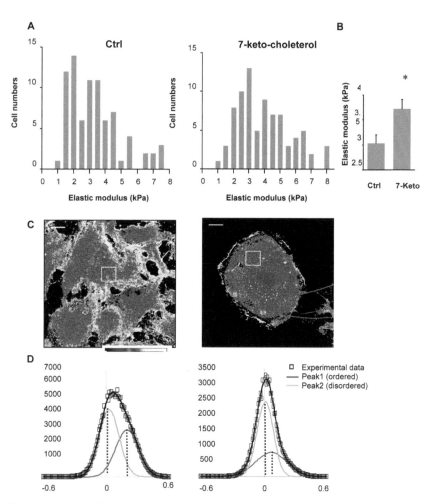

Figure 2. Impact of 7-ketocholesterol on endothelial stiffness and lipid packing. A: Histograms of elastic modulus measured in control (Ctrl) and 7-keto-cholesterol-treated cells. **B:** Average elastic modulus for control (Ctrl) and 7-keto-cholesterol–treated cells. (means± SEM, n=80 cells for each experimental condition). **C:** Typical GP images of control cells (Ctrl), 7-keto-cholesterol-treated cells. Scale bar is 11.2μm. **D:** GP histograms for the corresponding image fitted by a double-Gaussian distribution with the curve shifted to the right representing ordered domains (red) and the curve shifted to the left representing fluid domains (green). The sum of the Gaussians is shown in black (From (Shentu et al. 2010)

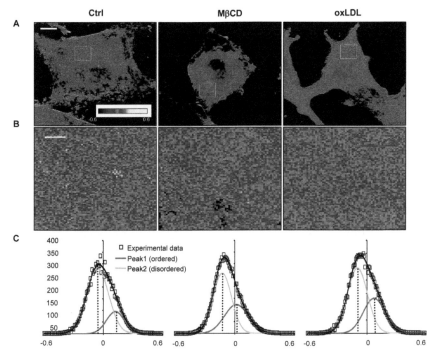

Figure 5. Impact of oxLDL on lipid packing of membrane domains in BAECs. A: Typical GP images of control cells (Ctrl), MβCD-treated cells (MβCD), or oxLDL-treated cells (oxLDL). Scale bar is 5.6µm. **B**: The zoom-in representative regions of the GP images shown above (the zoomed regions are 5.6µm×5.6µm) Scale bar is 1µm. **C**: GP histograms for the three experimental cell populations (dots) fitted by a double-Gaussian distribution with the curve shifted to the right representing ordered domains (red) and the curve shifted to the left representing fluid domains (green). The sum of the Gaussians is shown in black. GP distribution is obtained from the region –0.6 to +0.6 as shown in the x-axis and the number of counts is normalized (sum=10000) as shown in y-axis. (n=23-25 images per experimental condition, 4 independent experiments). (From Shentu et al. 2010).

Chapter 8

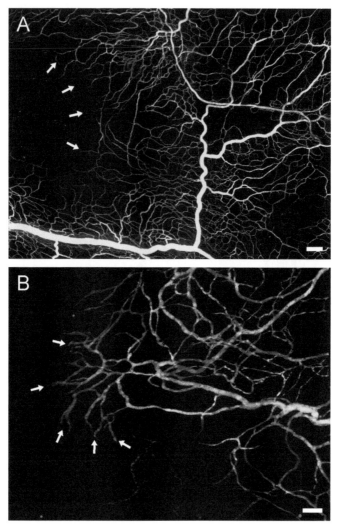

Figure 3. The presence of fluid flow in microvessels across the hierarchy of angiogenic adult rat mesenteric microvascular networks. Antibody labeling against PECAM (red) identifies all endothelial cells across the hierarchy of microvascular networks. Prior to tissue harvesting, vessel lumens were identified via intra-femoral vein injection of a FITC-conjugated fixable 40 kDa dextran (green). A, B) Representative images of growing networks 10 days post mast cell degranulation stimulation. Network growth is apparent by the increase in capillary density. Arrows indicate capillary sprouts. Dextran is present along capillaries in high vessel density regions and capillary sprouts indicating that new capillaries and sprouts have lumens. Scale bars = 200 µm (A), 100 µm (B).

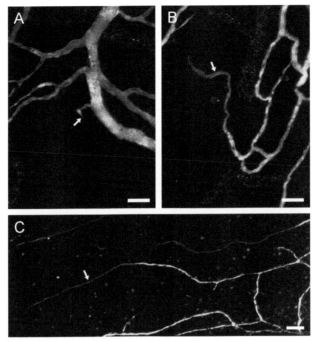

Figure 5. Representative images that support the presence of lumens along capillary sprouts in adult rat mesenteric microvascular networks. Antibody labeling against PECAM (red) identified all endothelial cells across the hierarchy of adult rat mesenteric microvascular networks. Prior to tissue harvesting, vessel lumens were identified via intra-femoral vein injection of a FITC-conjugated fixable 40 kDa dextran (green). Networks were stimulated by mast cell degranulation. Arrows indicate capillary sprouts. A) Example of a short capillary sprout off a venule. B) Example of a capillary sprout off an existing capillary. C) Example of a longer capillary sprout off a capillary. Faint PECAM labeling in (C) identifies blunt ended lymphatic vessels. In some cases the presence of the injected dextran extended the length of the sprout and in some cases stopped before the end of the sprout. Scale bars = 50 μm (A), 50 μm (B), 100 μm (C).

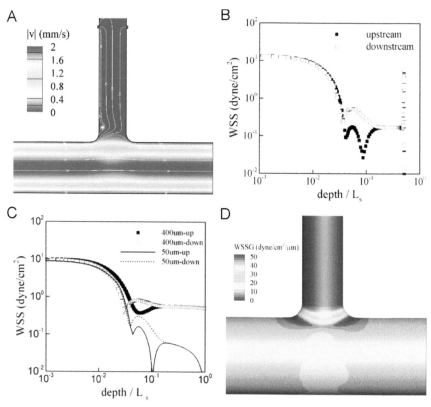

Figure 6. Computational predictions of hemodynamics within a capillary sprout. A) Plasma velocity profile for a 50 μm long, 6 μm diameter capillary sprout permeable at endothelial cell cleft regions. B) Predicted wall shear stress magnitudes along the length (depth) of the sprout (A). The shear stress distribution was normalized by the length of the sprout, "L_s", and shown for the upstream and downstream walls of the sprout. C) Wall shear stress distribution along the walls of 50 μm and 400 μm long capillary sprouts with a 6 μm diameter uniformly permeable walls. Notice the effect of sprout length on the elevated shear stress magnitudes along the length of the sprout. D) Wall shear stress gradient distribution for a 50 μm long, 6 μm diameter capillary sprout with uniform wall permeability. High shear stress gradients appear at the sprout entrance. Images were adapted from Stapor et al. 2011.

Chapter 10

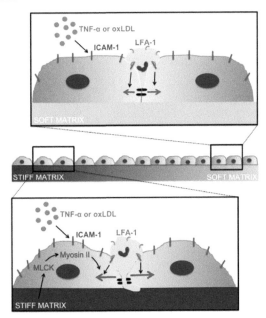

Figure 3. Effects of soft versus stiff subendothelial matrices on endothelial cell signaling and force production are depicted in this schematic. TNF-α or oxLDL upregulates ICAM-1 on the surface of the endothelium. Leukocyte LFA-1 binding to endothelial ICAM-1 initiates a signaling cascade inside the endothelial cells which ultimately leads to increased contractility (red arrows). Stiff matrices further promote the endothelial cell contractile forces, while soft substrates suppress the contractile forces.

T - #0403 - 071024 - C17 - 234/156/12 - PB - 9780367377786 - Gloss Lamination